Plumbing Services Series

Basic Skills

3rd Edition

John Dnistriansky, Bruce Paulsen

R.J. Puffett, L.J. Hossack, R. Kavanagh
Project Coordinators

National Library of Australia Cataloguing-in-Publication entry

Title: Basic skills: Mcgraw-hill plumbing services series / John Dnistriansky [et al.]
Edition: 3rd ed.
ISBN: 9780071015042 (pbk.)
Subjects: Plumbing–Textbooks.
Other Authors/Contributors:
Kavanagh, Robert James.
Dewey Number: 696.1

Published in Australia by
McGraw-Hill Australia Pty Limited
Level 2, 82 Waterloo Road, North Ryde, NSW 2113, Australia
Publisher: Norma Angeloni-Tomaras
Editorial coordinator: Carolina Pomilio
Production editor: Claire Linsdell
Permissions editor: Haidi Bernhardt
Copyeditor: Peter Grant
Proofreader: Nicole McKenzie
Indexer: Olive Grove Indexing
Cover design: Luke Causby, Blue Cork
Internal design: Natalie Bowra, David Rosemeyer and Dominic Giustarini
Typeset in Scala 9.75 pt by diacriTech, Chennai, India
Printed in China on 90 gsm matt art by CTPS

Contents

Preface

The McGraw-Hill *Plumbing Services Series* has been designed to complement the education and training of apprentices in Australia and New Zealand, and to provide a valuable reference for practising plumbers. When it was created in 1982, the local plumbing industry welcomed this excellent plumbing textbook series; the first to be written by qualified and practising plumbers; the first to be so comprehensively illustrated; and the first to be truly national and which provided information about the various rules and regulations.

This series of plumbing texts has served the plumbing industry well now for 30 years. With this latest edition it has been kept relevant and brought up-to-date. The text deals with the principles of plumbing while observing the integration of local rules and the latest national standards.

Basic skills and knowledge are the cornerstones of all trades. To build a foundation of excellence in plumbing it is first necessary to understand and appreciate the questions of 'how' and 'why'. *Basic Skills* sets out in its easy-to-understand format the methods and explanations used in this vital trade.

Our vision some 30 years ago was to contribute to the plumbing industry and its learners with a truly Australian text to complement the teaching of this trade. Our dream has now been realised and this edition becomes another platform for the continuing development of skills excellence in plumbing.

Len Hossack and Bob Puffett

Acknowledgements

The authors would like to thank the following people for their input into the creation of this text:

- Andrew Craine, Head Teacher, Plumbing, Gymea TAFE for reviewing the manuscript and providing invaluable feedback and advice
- Bryan Robinson at Cody Corporation, Phil Sutton at Iplex, and Gerry Smith at PPI Corporation for their helpful technical suggestions
- Carolina Pomilio for her research, interviews and writing of the Plumber profiles.

In addition, McGraw-Hill would like to thank the following for permission to reproduce their images:

- Andrew Richardson, State Water Corporation (Fig. 1.24)
- Blackwoods (Fig. 3.13)
- Paramount Brown (Fig. 4.11(a))
- Hilti (Aust) (Figs 5.1, 5.2, 5.3)
- BOC Limited (Figs 5.14, 5.15, 5.16, 5.18). *Reproduced with the kind permission of BOC Limited, a member of the Linde Group. Copyright.*
- Georg Fischer Piping Systems (Figs 5.50, 5.52).
- Vinidex (Figs 5.51, 5.53)
- Flexible Learning Toolboxes (Fig. 6.5)
- General Tools & Instruments Pty Ltd (Fig. 6.6b)
- Cody Corporation, South Australia (Figs 6.7, 6.9, 6.11, 6.18)
- www.irwin.com.au (Fig. 6.8)
- www.signblitz.com.au (Fig. 6.12)
- Istock (Fig 7.1)
- The Maine Watersheds Project: Teaching for Sustainability, University of Southern Maine (Fig. 7.2)
- Daryl Andrall, Missouri River Energy Services (Fig 7.4)
- M. Turner, Clear the Air, www.cleartheair.com.au/tag/newcastle-property/ (Fig. 7.6)
- Temple University (Fig. 7.8)
- The New Yorker (Fig. 8.1)
- *Community Eye Health Journal*, International Centre for Eye Health, London School of Hygiene and Tropical Medicine (Fig. 8.2)
- Rossignol, K. L. (1999). *Communication Skills in the Workplace*. Croydon: Eastern House, p.53 (Fig. 8.3)
- http://theteemingbrain.wordpress.com (Fig 8.4)
- Shutterstock (Figs 8.5, 8.6, 8.7)
- ABS (2006) Work-Related Injuries, Australia (cat no. 6324.0) ABS:Canberra (Fig. 9.1)
- Cartoon Stock (Fig 9.2)
- Emo Direct (Fig. 9.4)
- Department of the Prime Minister and Cabinet. (Fig. 9.5) © *Copyright Commonwealth of Australia 2010*
- www.lifesavingwa.com.au (Fig. 9.7)

About the team

John Dnistriansky

John Dnistriansky contributed chapters 1 to 6 of *Plumbing Basic Skills*. He was the inaugural principal co-author of the first edition and the updated second edition texts.

John has been a TAFE lecturer in plumbing programmes ranging from basic trade to advanced certificate levels for 30 years of his working life. John was also engaged over a period of 23 years as a consulting part-time lecturer in the architectural faculty of the University of Adelaide, teaching sanitary science and building services (plumbing). He was also involved with teaching programmes in environmental health and water industry operations. John is well qualified with previous experience in urban and rural water supply as well as broad acreage irrigation design and farm hydraulics.

John is the author of the *Pump Series of Handbooks* including *Pump Basics for Plumbers, Pump Basics Workbook, A Guide to Installation and Selection, Art of Teaching Pump Basics* and *A Guide to Pump Systems in Plumbing Services.*

He has contributed articles and his work has been reprinted in national magazines on pump-related topics.

John's tertiary qualifications include: Bachelor of Arts, Bachelor of Education, Graduate Diploma in Educational Technology, Diploma in Teaching Technical and the Royal Society of Health Diploma in Public Health. He also holds a Certificate IV in Training and Assessment and the Advanced Certificate in Plumbing.

Len Hossack

Len Hossack is a former Head of School of Plumbing at the South Australian Department of Technical and Further Education and has been actively involved in his local community and with the plumbing industry in South Australia for many years.

Robert Kavanagh

Robert Kavanagh contributed chapters 1 to 6 of *Basic Skills* and currently holds the position of Training Co-ordinator at the Plumbing Industry Association of South Australia. Robert has been a VET plumbing teacher since 1987 and is a qualified Master Plumber with a lifetime of plumbing experience, having entered the industry at the age of 17.

His qualifications include the Advanced Certificate in Plumbing, a Bachelor of Teaching in Adult Education, Certificate IV in Workplace Education and Certificate IV in Training and Assessment. He was responsible for introducing a Certificate I in Plumbing to high schools throughout Adelaide in partnership with the Department of Education, Training and Employment (DETE).

Bruce Paulsen

Bruce Paulsen contributed chapters 7 to 9 of *Basic Skills.* Bruce has been in the plumbing industry since 1973—starting out as an apprentice in the family plumbing business, H R Paulsen and Sons P/L, where he gained extensive experience in construction work and maintenance in the domestic, commercial and industrial fields. He then moved into other areas of the building industry, including domestic design and building renovation.

Over the course of his career Bruce has worked in most areas of the plumbing industry, including greywater and recycling systems, rainwater and water efficiency areas. He is currently a full-time Plumbing and Sustainability teacher at North Sydney TAFE, and was involved in writing the course material for the Certificate IV in Business Sustainability Assessment course offered there.

Bruce's qualifications include Certificate IV in Plumbing Technology, a Bachelor of Education (Honours) in Adult Education, Certificate IV in Training and Assessment and Certificate IV in Business Sustainability Assessment.

Bob Puffett

After serving as the Head of School of Plumbing and Sheetmetal in NSW, Bob Puffett went on to be Director of Staff, Principal and Assistant Director General, TAFE. Bob was made a Member of the Order of Australia (AM) for his contribution to Technical Education as Director of the Sydney Institute of Technology. Following his 'retirement' Bob became National Chairman of Worldskills Australia. He now serves on local community organisations and is a Board member of a NSW plumbing training organisation.

E-student/E-instructor

www.mhhe.com/au/plumbing

The Online Learning Centre (OLC) that accompanies this text is an integrated online product to assist you in getting the most from your course, providing a powerful learning experience beyond the printed page.

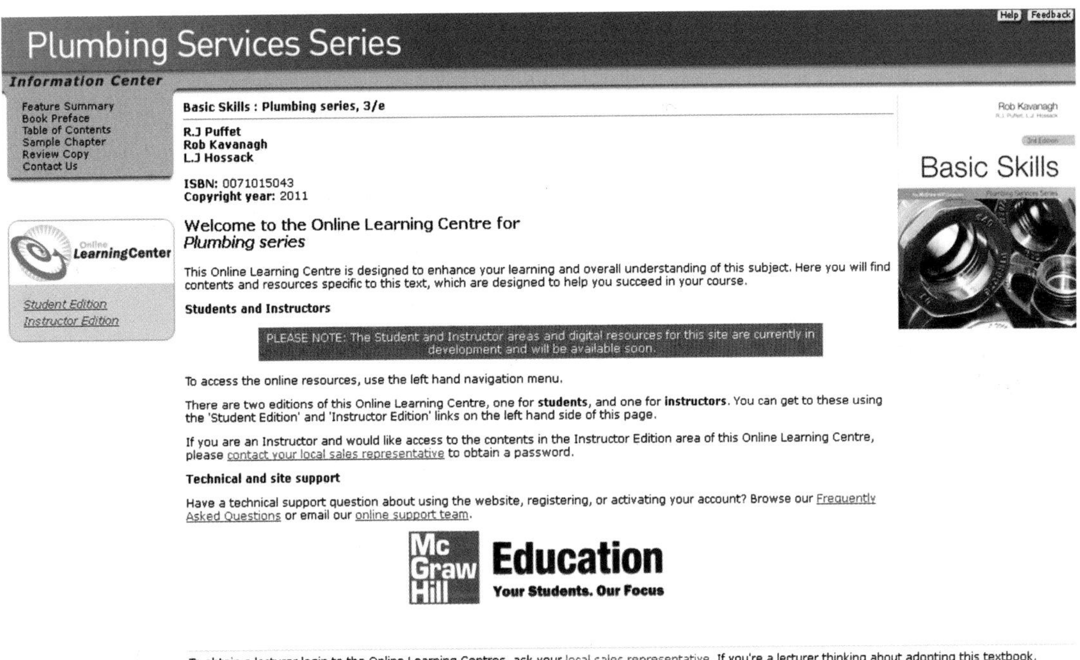

PowerPoint® slides

A set of PowerPoint slides accompanies each chapter and features items that provide a lecture outline, plus key figures and tables from the text.

Art Library

All the illustrations from the text are provided in a convenient ready-to-use format.

Solutions Manual

Worked solutions to the end-of-chapter exercises are included for instructors.

Plumbing profiles

Profiles of Australian plumbers sharing their experiences inspire students and help to inform them about the career pathways that are available after study.

Measurement

- The metric (or SI) system may be viewed as a 'base 10' system of measurement; each unit is derived from another unit

TABLE 1.3

Prefix	Fraction
deci	one-tenth (0.1)
centi	one-hundredth (0.01)
milli	one-thousandth (0.001)
Prefix	**Multiple**
deca	10 times
hecto	100 times
kilo	1000 times

Copyright © 2012 McGraw-Hill Australia Pty Ltd
PowerPoint slides to accompany Kavanagh, Puffett & Hossack, *Basic Skills 3e* 1-2

LEARNING OBJECTIVES

PART 1 TRADE MATHS ESSENTIALS

Chapter 1		Chapter 2	
1.1	Understand basic mathematical symbols	2.1	Understand drawing equipment and layout
1.2	Understand mathematical processes	2.2	Understand geometric concepts
1.3	Make accurate calculations of measurements	2.3	Understand orthogonal projection
1.4	Calculate fluid capacity	2.4	Understand isometric drawing
1.5	Understand plumbing applied calculations	2.5	To be able to read and interpret plans
		2.6	To apply pattern development

PART 2 TRADE SKILL ESSENTIALS

Chapter 3		Chapter 4	
3.1	Understand cutting with hand tools	4.1	Recognise types of joints in sheet metal
3.2	Understand sharpening of drills	4.2	Identify types of sheet metal machinery
3.3	Recognise types of drilling machines	4.3	Understand pipe and tube bending
3.4	To apply tapping and threading	4.4	Understand forming of copper tube
		4.5	Understand junction forming of copper tube

Chapter 5		Chapter 6	
5.1	Recognise types of anchor fixings	6.1	Understand the levelling process
5.2	Understand powder-powered fastening	6.2	Understand levelling terms
5.3	Understand soft soldering	6.3	Identify levelling tools
5.4	Understand oxyacetylene welding	6.4	Recognise types of optical levels
5.5	Identify welding techniques	6.5	Apply general levelling procedures
5.6	Understand oxyacetylene cutting	6.6	Identify the rise and fall method
5.7	Understand hard soldering techniques		
5.8	Understand arc welding		
5.9	Identify welding safety		
5.10	Recognise screwed pipe joints and fittings		
5.11	Understand UPVC pressure pipe applications		
5.12	Understand polyethylene pipe applications		
5.13	Understand application of cast-iron pipe and fittings		
5.14	Identify the use of sealants		

PART 3 THE PLUMBER AT WORK

Chapter 7		Chapter 8	
7.1	To define sustainability	8.1	To define communication
7.2	Understand how sustainability affects the plumbing industry	8.2	Understand nonverbal communication
7.3	Learn what plumbers can do to operate in a sustainable manner	8.3	Recognise various forms of communication

Chapter 9	
9.1	Identify the OH&S Act 2000, and OH&S Regulation 2001
9.2	Understand how OH&S operates in the workplace
9.3	Identify duty of care responsibilities: yours and your employer's
9.4	Identify hazards and the risk assessment process
9.5	Learn how to deal with common hazards in the plumbing industry

PLUMBING COMPETENCIES

CHAPTER	CERTIFICATE I TO III	MODULE UNIT CODE
PART 1 TRADE MATHS ESSENTIALS		
1	**Trade calculations** • Carry out measurements and calculations • Read plans and calculate plumbing quantities	CPCCCM1005A CPCPCM2004A
2	**Basic graphics** • Mark out materials • Read and interpret plans and specification – Read plans and calculate plumbing quantities	CPCCCM2001A CPCPCM2004A
PART 2 TRADE SKILL ESSENTIALS		
3	**Cutting and drilling** • Use construction tools and equipment • Handle construction materials – Use plumbing hand and power tools	CPCCCM2005A CPCCCM2004A CPCPCM2006A
4	**Forming and bending** • Cut and join Sheet metal • Use construction tools and equipment – Use plumbing hand and power tools	CPCPCM2008A CPCCCM2005A CPCPCM2006A
5	**Fastening, joining and sealing** • Fabricate and install non-ferrous pressure piping • Use construction tools and equipment • Weld using oxy-LPG acetylene equipment • Weld using arc welding equipment	CPCPCM3003A CPCCCM2005A CPCPCM2012A CPCPCM2013A
6	**Levelling** • Apply basic levelling procedures • Carry out levelling	CPCCCM2006A CPCPCM2007A
PART 3 THE PLUMBER AT WORK		
7	**The green plumber** • Work effectively and sustainably in the construction Industry – Work effectively in the plumbing and services sector	CPCCCM1002A CPCPCM2001A
8	**Effective communication** • Conduct workplace communication	CPCCCM1004A CPCPCM2002A
9	**A safe and healthy workplace** • Work safely in the construction industry • Carry out OHS requirements	CPCCOHS1001A CPCPCM2003A

AUSTRALIAN STANDARDS

AS 1100	Technical Drawing
1100.101	Part 101: General principles (1992)
1100.201	Part 201: Mechanical engineering drawing (1992)
1100.301	Part 301: Architectural drawing (1985)
1100.401	Part 401: Engineering survey and engineering survey design drawing (1984)
1100.501	Part 501: Structural engineering drawing (2002)
AS/NZS 1337:1992	Eye protectors for industrial applications
AS/NZS 1554.4:2010	Structural steel welding
AS 1579:2001	Arc-welded steel pipes and fittings for water and waste-water
AS 1674.1:1997	Safety in welding and allied processes Part 1: Fire precautions
AS 1722.2-1992	Pipe threads of Whitworth form Part 2. Fastening pipe threads
AS 1834.1:1991	Material for soldering – soldering alloys
AS/NZS 1873:2003	Powder-actuated (PA) hand-held fastening tools
AS 2280:2004	Ductile iron pressure pipes and fittings
AS 2397:1993	Safe use of lasers in the building and construction industry
AS 2528:1982	Bolts, studbolts and nuts for flanges and other high and low temperature applications
AS/NZS 3500:2003	Plumbing and Drainage
3500.0	Part 0: Glossary of terms
3500.1	Part 1: Water services
3500.2	Part 2: Sanitary plumbing and drainage
3500.3	Part 3: Stormwater drainage
3500.4	Part 4: Heated water systems
3500.5	Part 5: Domestic installations
AS 4087:2004	Metallic flanges for waterworks purposes
AS/NZS 4130:1997	Polyethylene (PE) pipes for pressure applications
AS 4603:1999	Flashback arrestors. Safety devices for use with fuel gasses and oxygen or compressed air
AS/NZS 4680:2006	Hot-dip galvanized (zinc) coatings on fabricated ferrous articles
AS 4839:2001	The safe use of portable and mobile oxy, fuel gas systems for welding, cutting, heating and allied processes
AS/NZS 6400:2005	Water-efficient products-Rating and labelling

PART 1 TRADE MATHS ESSENTIALS

Plumbing Services Series

CHAPTER **ONE**

Trade calculations

LEARNING OBJECTIVES

In this chapter you will learn about:

1.1 basic mathematical symbols

1.2 mathematical processes

1.3 calculation of measurements

1.4 calculation of fluid capacity

1.5 plumbing applied calculations.

INTRODUCTION

The main aim of this chapter is to explain some of the basic mathematical rules, methods and procedures needed by students. The topic includes:

1 mathematical symbols
2 the four basic skills of:
- addition
- subtraction
- multiplication
- division.

SOME BASIC SYMBOLS

- two dots, as in 1:50, indicate a ratio (in this case, one in 50)
- three dots as shown by ∴ mean 'therefore' but when inverted, as in ∵, mean 'because'
- 'proportional to' is expressed by the symbol $\propto$
- 'less than' is represented by the symbol $<$
- 'greater than' is represented by the symbol $>$
- the 'square' of a number is shown by a small raised '2'; for example, 10^2 means 10×10
- the 'square' root of a number is expressed by the symbol $\sqrt{\ }$
- the Greek letter π (pi) represents a constant approximately equal to 3.142. It represents the number of times the diameter of a circle goes around that circle's circumference. It is used, for example, in calculating the area of a circle.
- angles are represented by the symbol °, as in 30°, which indicates an angle of inclination between two lines. The same symbol is used to express temperature, but a letter is added as in 30 °C (Celsius) or 30 °F (Fahrenheit).

The body of the chapter concentrates on laws of mathematics that are of particular use to plumbers and may be applied to practical situations. Essential information is presented using examples in a simple and sequential order.

THE FOUR MATHEMATICAL PROCESSES

Every calculation made to obtain an answer to a problem requires the use of one or more of the basic processes: addition, subtraction, multiplication and division.

Addition (+)

Addition means the sum of two or more numbers and is the most basic of the mathematical processes. Accurate addition can be mastered only by practice.

Accuracy in the use of numbers is essential. All calculations should be checked after each problem is attempted.

Subtraction (−)

Subtraction is the opposite operation to addition. 'Adding to' and 'taking away from' can be thought of as opposite processes. The two most commonly practised rules used in subtraction exercises are 'borrowing the ten' and 'complementary addition'.

Borrowing the ten

Example

	Hundreds	*Tens*	*Units*
	6 ~~7~~	11 ~~2~~	16
−	5	6	8
answer =	1	5	8

Commencing with the units column:

8 from 6 is not possible.
∴ 8 from 16 is 8.

Using the tens column:

6 from 1 is not possible.
∴ 6 from 11 is 5.

Using the hundreds column:

5 from 6 is 1.

Note: It is not necessary to cross out the digits 2 and 7 when borrowing. This is done in this example simply to emphasise the process.

Complementary addition

Example

	Hundreds	Tens	Units
	7	2	6
−	5	6	8
answer =	1	5	8

Multiplication (×)

Multiplication of whole numbers

The process of multiplication is a short form of addition, and is an easy method of adding groups of the same numbers. Multiplication gives the product of two or more numbers. To be efficient at multiplication, you must have an automatic knowledge of the multiplication table shown in Table 1.1.

Commencing with the units column in Example 2, above:

8 and 8 is 16 (carry 1).
7 and 5 is 12 (carry 1).
6 and 1 is 7.

This is how to use the table, for example, to find 9 × 7:

Find 9 in column A (down the side).
Find 7 in column B (along the top).

Run your finger down this column to find where it intersects the '9' row. This is at 63.

Then 9 × 7 = 63.
Check with 7 × 9.

Multiplication by numbers of more than one digit should always commence with the figure on the right.

Examples

1 Multiply 6347 by 278.

```
           6 347
           ×278
          50 776
          444 29
         12694
answer = 1764 466
```

2 Multiply 623 by 204.

```
           623
          ×204
          2492
          000
        1246
answer = 127092
```

Multiplication of decimals

Decimal fractions will be used here in preference to vulgar fractions, for example, 3.142 instead of 3 1/7

Consider the following:

0.3 × 7
0.3 × 0.7
0.03 × 0.07

Now the digits in the answers are derived by the ordinary process of multiplication. However, care must be

TABLE 1.1 Multiplication of whole numbers

	A											
B	1	2	3	4	5	6	7	8	9	10	11	12
	2	**4**	6	8	10	12	14	16	18	20	22	24
	3	6	**9**	12	15	18	21	24	27	30	33	36
	4	8	12	**16**	20	24	28	32	36	40	44	48
	5	10	15	20	**25**	30	35	40	45	50	55	60
	6	12	18	24	30	**36**	42	48	54	60	66	72
	7	14	21	28	35	42	**49**	56	63	70	77	84
	8	16	24	32	40	48	56	**64**	72	80	88	96
	9	18	27	36	45	54	63	72	**81**	90	99	108
	10	20	30	40	50	60	70	80	90	**100**	110	120
	11	22	33	44	55	66	77	88	99	110	**121**	132
	12	24	36	48	60	72	84	96	108	120	132	**144**

Note: The numbers in **bold** typeface are perfect squares.

exercised in placing the decimal point to give digits the correct place value:

$\therefore\ 0.3 \times 7 = 2.1$
$0.3 \times 0.7 = 0.21$
$0.03 \times 0.07 = 0.0021$

In the answer, the number of digits after the decimal point is equal to the total number of digits after the decimal points in the question.

Example

$$\begin{array}{r} 6.2 \\ \times 5.9 \\ \hline 558 \\ 310 \\ \hline \end{array}$$

answer $= 36.58$

Check by approximation, the answer is roughly $6 \times 6 = 36$. The exact answer is therefore 36.58.

Division ($\div$)

Division is the opposite process to multiplication and can be thought of as 'undoing' a multiplication operation. The number to be divided is called the numerator, whereas the number by which it is divided is the denominator (or divisor). The number of times that the divisor is contained in the numerator is referred to as the quotient.

In the example $10 \div 2 = 5$, the numerator is 10, the denominator is 2 and the quotient is 5. The fact that multiplication and division are inverse operations provides a method for checking division.

Example

$10 \div 2 = 5$
$5 \times 2 = 10$

Division by multiples of ten

The value of any number depends upon its position (see Table 1.2). Consequently, dividing by 10, 100 or 1000 simply means shifting the decimal point one space to the left for every power of ten in the denominator.

Examples

$1235 \div 10 = 123.5$
$1235 \div 100 = 12.35$
$1235 \div 1000 = 1.235$

TABLE 1.2

Thousands	1000
Hundreds	100
Tens	10
Units	1
Tenths	0.1
Hundredths	0.01
Thousandths	0.001

Division by other whole numbers

Example

$950 \div 6$

Step 1 Divide 9 by 6.
(a) Write 1 above 9 of numerator.
(b) Subtract: $9 - 6 = 3$.
(c) Bring down 5.

$$\begin{array}{r|l} & 1 \\ \cline{2-2} 6 & 950 \\ & -6\downarrow \\ \cline{2-2} & 35 \end{array}$$

Step 2 Divide 35 by 6.
(a) Write 5 above 5 of numerator.
(b) Subtract: 30 (6×5) from 35.
(c) Bring down 0.

$$\begin{array}{r|l} & 15 \\ \cline{2-2} 6 & 950 \\ & -6\ \downarrow \\ \cline{2-2} & 35 \\ & -30 \\ \cline{2-2} & 50 \end{array}$$

Step 3 Divide 50 by 6.
(a) Write 8 above 0 of numerator.
(b) Subtract 48 (6×8) from 50.
(c) Write remainder 2.

$$\begin{array}{r|l} & 158 \\ \cline{2-2} 6 & 950 \\ & -6 \\ \cline{2-2} & 35 \\ & -30 \\ \cline{2-2} & 50 \\ & -48 \\ \cline{2-2} & 2 \end{array}$$

answer = 158 remainder 2

Check that:
whole number $\times$ denominator + remainder = numerator
$158 \times 6 + 2 = 950$

Division of decimals

A decimal divisor is first made into a whole number by shifting the decimal point to the right-hand side. The decimal point in the numerator is shifted to the right the same number of places. This fixes the decimal point for the answer at the same time. The calculation can now proceed by long division.

Example

$457.1 \div 0.35$

Step 1 Shift the decimal point in the divisor two places to the right. The number thus becomes 35.

Step 2 Shift the decimal point in the numerator two places to the right. The number now becomes 45 710.

Step 3 Complete long division.

```
       1306
 35 ) 45710
     -35↓
      107
     -105↓↓
        210
      - 210
          0
```

answer = 1306

ADDITIONAL MATHEMATICS

Use of indices or powers of 10

The important fact, when we group 10 units together to make one ten, is that we can proceed to group 10 tens to make one hundred, 10 hundreds to make one thousand, and so on. The number 100 can be factorised and written as 10×10. If indices are used, 10×10 is written as 10^2. Similarly, 1000 can be factorised as $10 \times 10 \times 10$, or 10^3.

Using powers of 10 allows us to remove most of the noughts before or after the decimal point in multiplication and division calculations. This makes arithmetical problems easier and reduces the likelihood of mistakes.

Some powers of 10 are:

$10 = 10^1$	10 to the first power
$10 \times 10 = 100 = 10^2$	10 to the second
$10 \times 10 \times 10 = 1000 = 10^3$	10 to the third

Division using indices

Example

Consider $\frac{10^5}{10^2}$

By expansion, this can be written:

$$\frac{10 \times 10 \times 10 \times \cancel{10} \times \cancel{10}}{\cancel{10} \times \cancel{10}}$$

By cancellation of two tens (top and bottom), the answer becomes $10 \times 10 \times 10 = 10^3$.

An alternative answer is 1000.

PERCENTAGES

The importance and relevance of understanding percentages in the plumbing industry is being able to calculate pay rises, allowance rises, interest rates and discounts on sale items.

What is a percentage?

Percentage means 'out of 100', the meaning of 'out of' in maths terms means 'divide by'. We write the symbol (%) as an easy way to write a fraction that has 100 as the denominator.

For example, the easy way of saying 85 out of the 100 students were present, would be to say there is 85% attendance.

So per cent means divided by 100, which makes the decimal point move two places to the left.

Example

85% = 85/100 = 0.85

You can even change a number to the percentage value by multiplying by 100. This means you move the decimal point two places to the right.

Example

$0.85 \times 100 = 85\%$

Formula for calculating percentages

The formula for calculating percentage is multiply by 100.

Example

$1/5 \times 100 = 100 \div 5 = 20$

To convert a percentage to a fraction, divide by 100 and reduce the fraction (where possible).

Example

60% = 60/100 = 3/5

THE SQUARE ROOT (√)

The square root of a given number is that number which, when multiplied by itself produces the given number, for example, the square root of 16 is 4 because 4×4 or $4^2 = 16$. The squares and square roots of certain numbers are used in finding the first number in the answer to any square root problem. This is the only step in the entire process in which the principles of square root are used.

The following squares are used in finding the first number in a square root problem.

$1^2 = 1$	$4^2 = 16$	$7^2 = 49$
$2^2 = 4$	$5^2 = 25$	$8^2 = 64$
$3^2 = 9$	$6^2 = 36$	$9^2 = 81$

The corresponding square roots are:

$\sqrt{1} = 1$	$\sqrt{16} = 4$	$\sqrt{49} = 7$
$\sqrt{4} = 2$	$\sqrt{25} = 5$	$\sqrt{64} = 8$
$\sqrt{9} = 3$	$\sqrt{36} = 6$	$\sqrt{81} = 9$

A good check procedure with square root problems is to square the quotient. The arithmetical procedure for more difficult examples is set out below.

Example

Find the square root of 441.

```
        21.0
  2 ) 441.00
     -4 ↓↓
 41    041
      - 41
        00
```

answer = 21.0

Step 1 (a) Mark off in pairs the digits to the right and left of the decimal point.

(b) Place the decimal point in position above the numerator.

Step 2 (a) Consider the first digit of the numerator. Determine the largest perfect square that is not greater than this number (in this case, 4 is the appropriate perfect square). Write the square root of the perfect square (i.e. $\sqrt{4} = 2$) above the digit, to give the start of the quotient.
(b) Write the square root (2) to the left of the numerator to give a divisor.
(c) Multiply the divisor and the quotient together ($2 \times 2 = 4$) and subtract the result ($4 - 4 = 0$).
(d) Bring down the next pair of digits (41).

Step 3 (a) Double the quotient ($2 + 2 = 4$) and place the result in the column as the new divisor, leaving a space for one more digit.
(b) Place a digit in the space provided in the divisor column and in the quotient so that the digit, when multiplied by the new divisor, gives the largest possible number that can be subtracted (i.e. $1 \times 41 = 41$).
(c) Subtract to give no remainder.

The answer is 21.0 (check $21^2 = 441$)

Finding the square root of a number by factors

The process of extracting the square root of a number can sometimes be simplified by splitting up the number into suitable factors. These factors should be numbers whose square roots are easily recognised.

Example

Using the process of factorisation, determine the square root of 1296.

$$\begin{aligned}\sqrt{1296} &= \sqrt{4 \times 324}\\ &= \sqrt{4 \times 9 \times 36}\\ &= \sqrt{4} \times \sqrt{9} \times \sqrt{36}\\ &= 2 \times 3 \times 6\end{aligned}$$

answer = 36

SQUARE ROOT RULES

1. If a whole number has one or two digits (1 to 99), the square root of that number will contain one digit.
2. If a whole number is a three-digit or four-digit number, the square root of that number contains two digits.
3. The square root of a decimal fraction is greater than the decimal fraction under consideration.
4. The square root of a number is the product of the square roots of its factors.

MEASUREMENT

The metric, or System International (SI), system may be viewed as a 'base 10' system of measurement; each unit is derived from another unit. Units are broken down by factors of 10, and it is this feature that makes calculation easy.

Fractions and multiples of the fundamental units in the metric system are identified by using prefixes. This principle is illustrated in Table 1.3.

TABLE 1.3 Prefixes

Prefix	Fraction
deci	one-tenth (0.1)
centi	one-hundredth (0.01)
milli	one-thousandth (0.001)
Prefix	**Multiple**
deca	10 times
hecto	100 times
kilo	1000 times

Length

If we use the metre (m) as a standard unit for length, by applying prefixes we can obtain the following:

1 millimetre (mm) = 0.001 m
1 centimetre (cm) = 0.01 m
1 kilometre (km) = 1000 m

This method is used because the units differ from each other by factors of ten; conversion from one unit to another is quite simple. The only requirement is to choose the correct power of ten and either multiply or divide to achieve the correct answer. A conversion is set out below.

Example

Assume the distance of a pipeline from a reservoir to a header tank is 10 000 m. How many kilometres is this?

$$\begin{aligned}1\,\text{km} &= 1000\,\text{m}\\ \therefore 10\,000\,\text{m} &= \frac{10^4}{10^3}\\ &= \frac{10 \times 10 \times 10 \times 10}{10 \times 10 \times 10} \quad \text{(by expansion)}\\ \therefore \text{answer} &= 10\,\text{km}\end{aligned}$$

Note: The number of metres was divided by 1000.

Weight and volume

The preferred units of weight in the SI system are:

kilogram (kg)	1 kg = 1000 g
gram (g)	1 g = 1000 mg
milligram (mg)	1 mg = 0.001 g

The preferred units of volume are:

megalitre (mL)	1 mL = 1 000 000 L
kilolitre (kL)	1 kL = 1000 L
litre (L)	1 L = 1000 mL
millilitre (mL)	1 mL = 0.001 L

Note: 1000 L of water will fill a cube 1 m × 1 m × 1 m (1 m^3) and weighs 1000 kg.

The next two sections of this chapter involve the areas and volumes of the various shapes (for example, the measurement of perimeters) to be encountered by the student plumber.

Plumbing installations that contain pipes, fittings and equipment, including the structures in which these components are installed, have familiar enclosed geometric shapes such as squares, rectangles, triangles, circles, cylinders and cubes.

Because the measurement of these shapes fall within the scope of plumbing work, the student must be capable of determining such factors as length, height, area and volume of the geometric shapes in order to lay out a job properly.

Perimeter

The word 'perimeter' means the total distance bounding a figure or shape. It is derived from the Greek words 'peri', meaning 'around' and 'meter', meaning 'to measure'. In practice, it simply involves measuring the length of each side of the figure and totalling them.

The square

The square is a figure having four equal sides that meet at right angles. Hence, a square has four equal sides and four equal angles (right angles). If the length of one side of a square is l unit, the perimeter (the sum of all the sides) becomes $4l$, that is, $4 \times$ the length of one side (Figure 1.1). We can use the letter P to represent 'perimeter'.

Example

Suppose the length of one side of a square is 4 mm. What is the perimeter of the square?

$$P = 4l$$
$$= 4 \times 4$$
$$\text{answer} = 16 \text{ mm}$$

The length of one side of a square is one-quarter of the perimeter.

FIG 1.1 The perimeter of a square

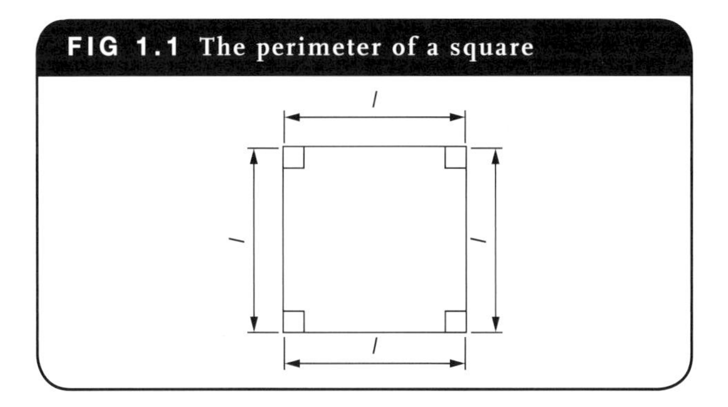

The rectangle

The rectangle is a four-sided figure with sides at right angles to each other but with only opposite sides equal (Figure 1.2). There are four equal angles (right angles) and two pairs of equal sides.

The perimeter rule for a rectangle is: 2(length + width) that is, $P = 2(l + w)$.

FIG 1.2 The perimeter of a rectangle

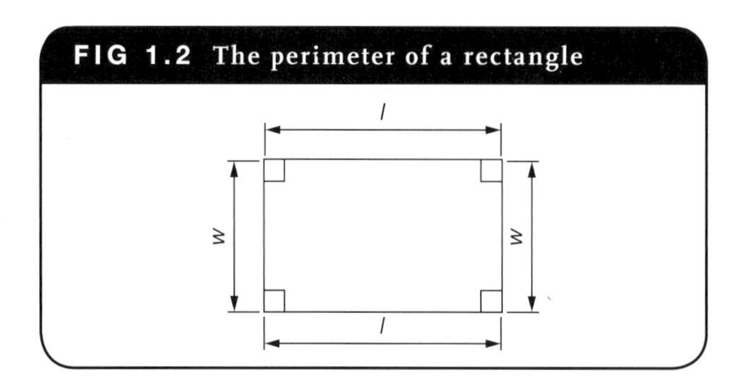

Example

Suppose the length of a rectangle is 6 mm and the width is 4 mm. What is the perimeter?

$$P = 2(l + w)$$
$$= 2(6 + 4)$$
$$= 2 \times 10$$
$$\text{answer} = 20 \text{ mm}$$

Other plane figures

Other plane figures include:

- the triangle (a three-sided figure)
- the hexagon (a six-sided figure)
- the octagon (an eight-sided figure).

The circle

A circle is a figure with an outline which at any point is the same distance from a centre point. The line which forms the circle is called the circumference or perimeter of the circle. Practically, the circumference of a circle may be measured by wrapping a tape around it and reading off the length.

Alternatively, the circumference of a circle may be derived mathematically by multiplying the diameter by a constant represented by the Greek symbol π (pi). This constant is approximately equal to 3.142.

The three important parts of a circle are illustrated in Figure 1.3.

FIG 1.3 The parts of a circle

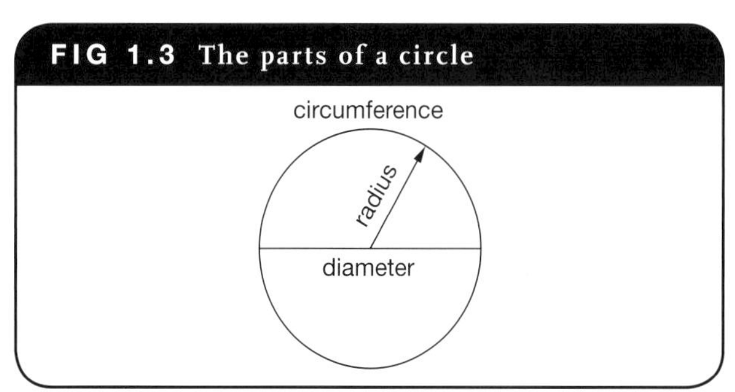

The circumference (C) is the distance around the circle.

$$C = \pi \times d \text{ or } 2 \times \pi \times r$$

The diameter (d) is the distance of a line that goes across the circle and passes through the centre.

$$d = 2 \times r$$

The radius (r) is the distance from the centre of a circle to the circumference.

It is equal to half the diameter.

$$r = d \div 2$$

Example

The diameter of a circle is 25 mm. What is its circumference?

$$C = \pi \times d$$
$$= 3.142 \times 25$$
$$\text{answer} = 78.5 \text{ mm}$$

Note: Since 3.142 is only an approximation for π, all calculations involving π give approximate answers. However, the answers obtained using 3.142 are usually accurate enough for plumbing work.

Area

As mentioned above, the total length of lines enclosing a space is called the perimeter. The shape formed is the figure and the total space enclosed is called the area. The area is the amount of surface, and all figures are assumed to have a flat surface. The area is a useful concept since we often need to know, for example, the floor area of a room or the area of a flat roof. We can represent area by the letter A.

The rectangle

A rectangle has two dimensions—length and width (see Figure 1.4), and if we multiply one measurement by the other we obtain the area of the rectangle.

The rule for area of rectangle is:

length (l) × width (w)

$$A = l \times w$$

FIG 1.4 The area of a rectangle

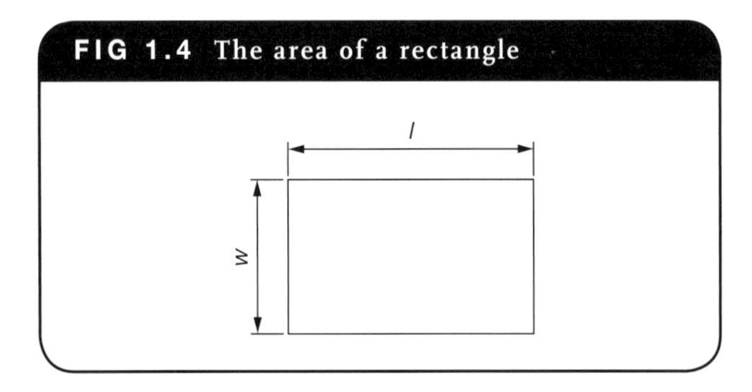

Example

Consider a rectangular courtyard measuring 15 m long and 8 m wide. Calculate the area of the courtyard.

$$A = l \times w$$
$$= 15 \times 8$$
$$\text{answer} = 120 \text{ m}^2$$

The parallelogram

A parallelogram is a quadrilateral having opposite sides equal and parallel. Squares and rectangles are special types of parallelogram. However, the term parallelogram is usually applied to a quadrilateral having its opposite sides equal and parallel, and angles that are not right angles. This is illustrated in Figure 1.5.

FIG 1.5 A parallelogram

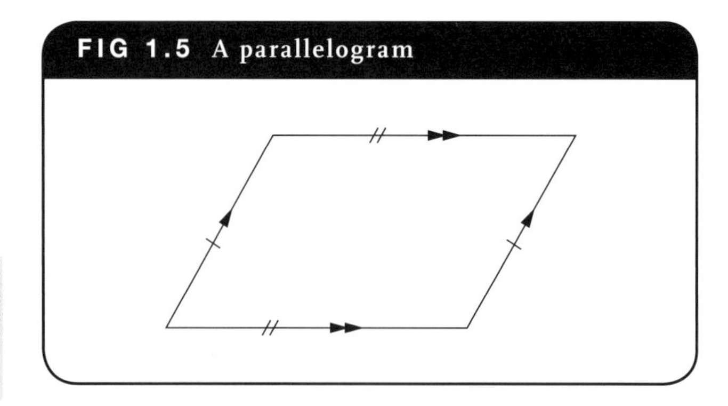

Notice that Figures 1.6(a) and 1.6(b) have a common length at the base and the same vertical width. If we cut off the shaded section in Figure 1.6(b) and move it to the opposite end of the figure, shown as a dashed outline, the result is a rectangle. By this process, we can change any parallelogram into a rectangle of the same area. This new figure has the same length and width as the original parallelogram.

Thus, the area of a parallelogram = length × vertical width.

$$A = l \times w$$

FIG 1.6 (a) Rectangle, (b) parallelogram

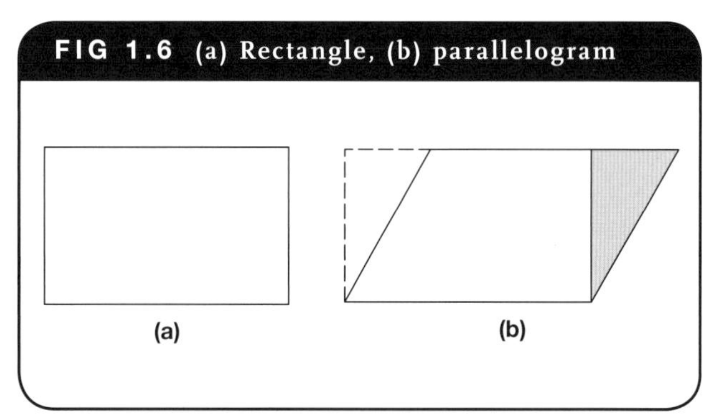

The triangle

The most useful triangle to examine is the right-angled triangle. The right-angled triangle is frequently used in plumbing calculations and it is for this reason that it is included in this chapter.

In Figure 1.7 we can see that the rectangle is made up of two right-angled triangles having the same base and same height as the rectangle. As previously mentioned, the area of a rectangle is calculated by 'length × width'. By drawing in a diagonal, as shown in Figure 1.7, we can see that the area of a triangle must be:

$$\frac{1}{2}(\text{length} \times \text{width})$$

or

$$\frac{1}{2}(\text{base} \times \text{height})$$

$$A = \frac{1}{2}(l \times w) \text{ or } A = \frac{1}{2}(b \times h)$$

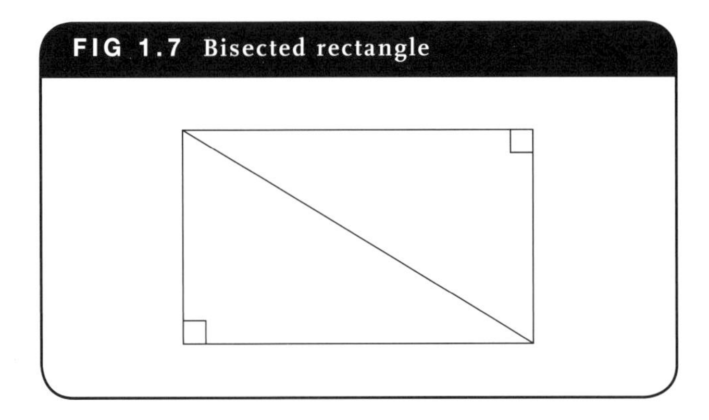

FIG 1.7 Bisected rectangle

Example

The base of a right-angled triangle is 15 cm long and the vertical height is 8 cm. What is the area of the triangle?

$$A = \frac{1}{2}(b \times h)$$
$$= \frac{15 \times 8}{2}$$
$$\text{answer} = 60 \text{ cm}^2$$

Further reference will be made throughout this chapter to calculations using the right-angled triangle rule (Pythagoras' theorem).

Irregular figures

To determine the area of any polygon, it is necessary to break the figure into regular shapes—rectangles, triangles and so on, and add up the areas of all the parts of the irregular shape.

Example

Find the area of a trapezoid having the dimensions as shown in Figure 1.8(a).

Drop lines from A and B perpendicular to the base of the trapezium to give points C and D as shown in Figure 1.8(b).

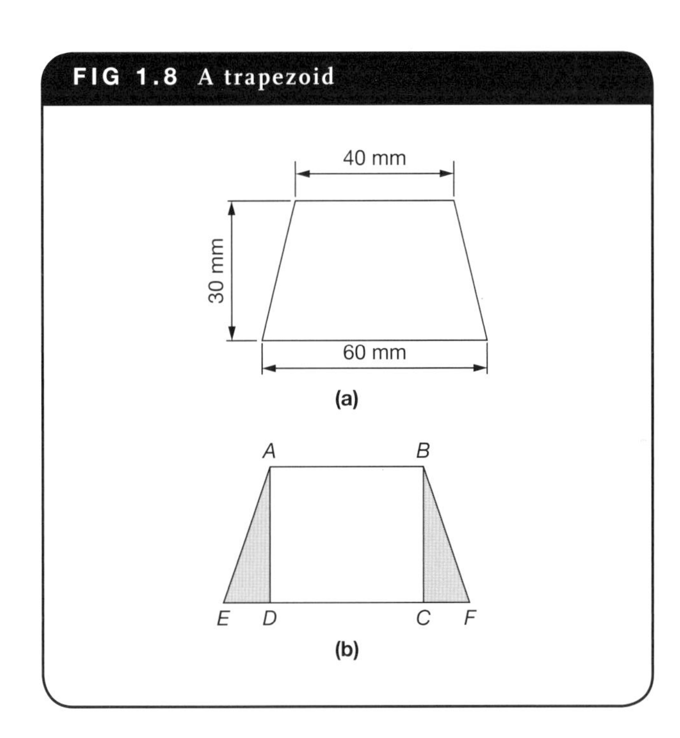

FIG 1.8 A trapezoid

$ABCD$ thus becomes a rectangle 40 mm × 30 mm. Shaded areas ADE and BCF are triangles 30 mm × 10 mm.

Now the total surface area of Figure 1.8(b) is:

$$ABCD = l \times w$$
$$\text{plus } ADE = \frac{1}{2}(b \times h)$$
$$\text{plus } BCF = \frac{1}{2}(b \times h)$$

$$ABCD = 40 \times 30 = 1200\text{mm}^2$$
$$ADE = \frac{1}{2}(10 \times 30) = 150\text{mm}^2$$
$$BCF = \frac{1}{2}(10 \times 30) = 150\text{mm}^2$$

$\therefore$ total surface area of $ABFCDE$ = 1500 mm^2

The circle

We have mentioned the relationship between the circumference and the diameter of a circle. If we divide a circle into a number of equal parts by drawing diameters, we will find that the resulting segments resemble triangles, except for their curved bases. See Figure 1.9.

If this arrangement could be opened out, it would give us a number of triangles, each of which has height equal to the radius of the circle. See Figure 1.10. The length of the base is the circumference of the circle.

$$\text{area of a circle} = \frac{1}{2}(\text{sum of base lengths} \times \text{height } r)$$
$$= \frac{1}{2}(2\pi r \times r)$$
$$\therefore A = \pi r^2$$

Note: The alternative formula for finding the area of a circle is $A = D^2 \times 0.7854$ where 0.7854 represents the factor of a circle to square of the same size.

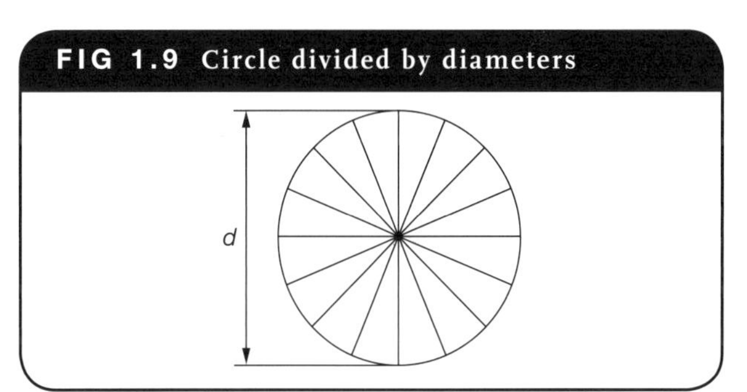

FIG 1.9 Circle divided by diameters

Examples

1 Find the area of a circle having radius 20 mm, given that $\pi = 3.142$.

$$A = \pi r^2$$
$$= 3.142 \times (20)^2$$
$$= 3.142 \times 400$$
$$\text{answer} = 1256.8 \text{ mm}^2$$

2 Calculate the cross-section area of a 100 mm diameter stack.

$$A = \pi r^2$$
$$= 3.142 \times (50)^2$$
$$= 3.142 \times 2500$$
$$\text{answer} = 7855 \text{ mm}^2$$

FIG 1.10 Base length is equal to circumference of circle

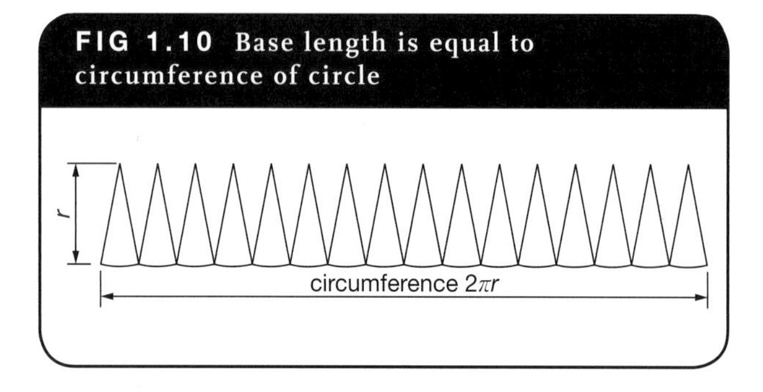

MEASUREMENTS OF SOLIDS

All geometrical solids occupy space commonly referred to as 'capacity' or 'volume'. Unlike plane figures, which enclose an area and have the basic dimensions of length and width, solids have three dimensions.

To calculate surface area

The cylinder

The total surface of a cylinder consists of a curved surface and two circular plane surfaces which form the top and bottom (Figures 1.11 and 1.12).

FIG 1.11 Drinking fountain connected to a stack

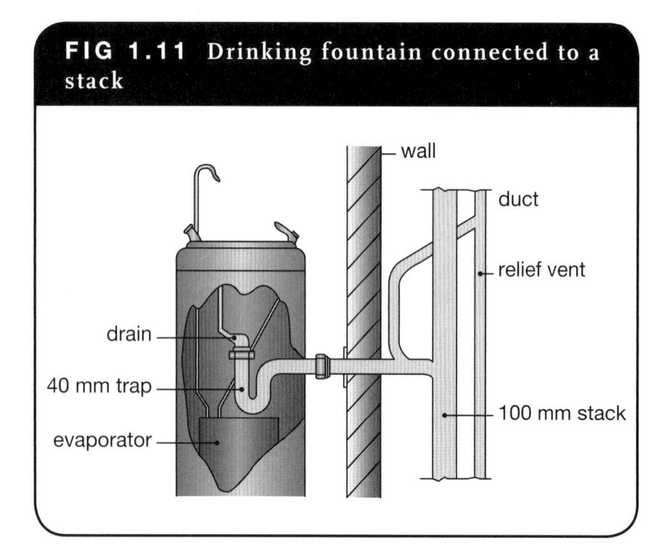

FIG 1.12 Total surface area of a cylinder

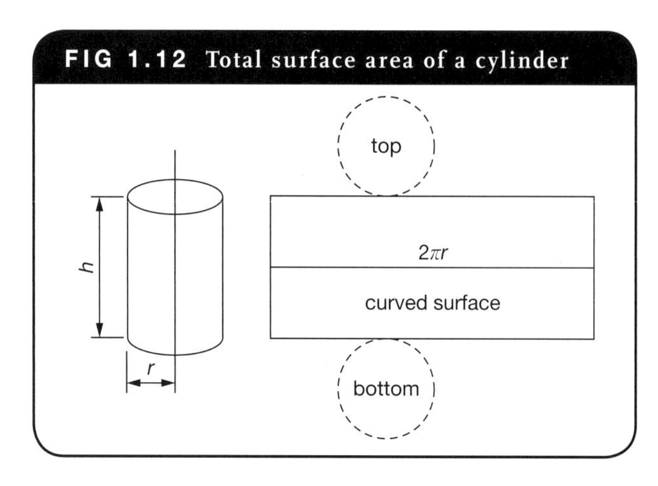

$$\text{total surface area} = \text{area (top and bottom)} + \text{area of curved surface}$$
$$= 2\pi r^2 + 2\pi rh$$
$$SA = 2\pi r(r + h)$$

The sphere

In calculating the surface area of a sphere or any part of a spherical surface, the relationship between the surface areas of a sphere and cylinder must be remembered. If a cylinder has its height equal to its diameter, the area of its curved surface is equal to the surface of a sphere of the same diameter. See Figure 1.13.

$$SA = 4\pi r^2$$

FIG 1.13 (a) Area of curved surface of cylinder = $4\pi r^2$, sphere = $4\pi r^2$, (b) area of sphere = $4\pi r^2$

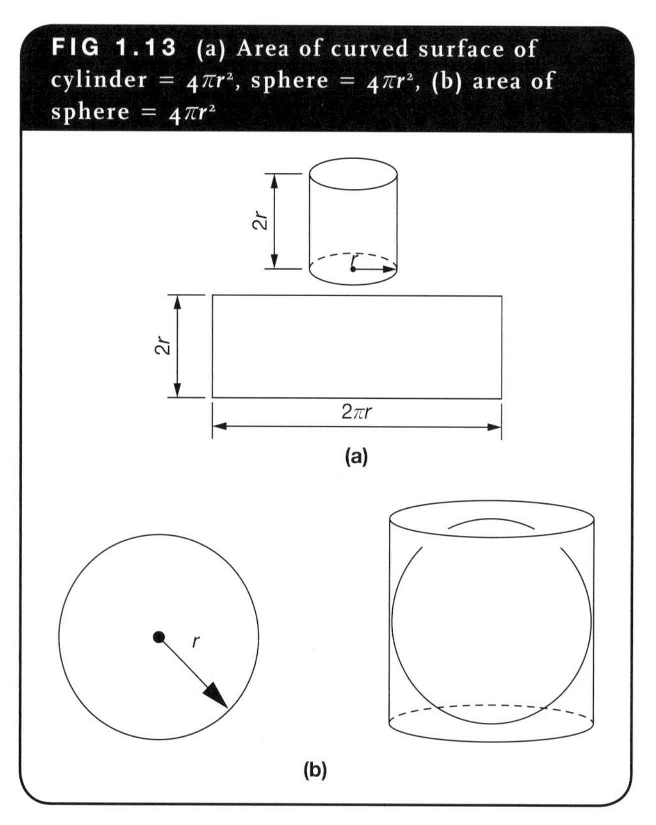

To calculate volume

The cube

The simplest of all solid shapes is the cube. It is a square block which has its length, width and height all equal to each other and is built up of six faces and twelve edges. Figure 1.14 illustrates the shape of the cube.

The volume (V) of a cube where l = length of one side is given by:

$$V = l^3$$

Example

A cube has length 100 mm. What is its volume?

$$V = l^3$$
$$= 100^3$$
$$\text{answer} = 1\,000\,000 \text{ mm}^3 \text{ or } 10^6 \text{ mm}^3$$

To calculate the dimensions of a cube given the volume, the formula $l = \sqrt[3]{V}$ is used. To find the cube root of a number, it is advisable to use cube root tables or a calculator with a $\sqrt[3]{\ }$ function.

FIG 1.14 A cube

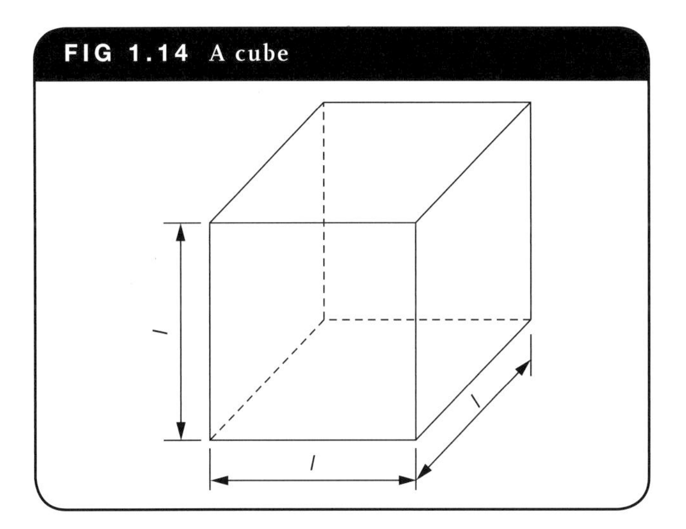

Example

A cube has volume 1000 mm^3. What is the length of the edge?

$$l = \sqrt[3]{V}$$
$$= \sqrt[3]{1000}$$
$$\text{answer} = 10 \text{ mm}$$

The cuboid

A cuboid, also referred to as a rectangular prism, is a solid having six faces, each one a rectangle (see Figure 1.15). The volume of a cuboid is given by:

$$V = l \times w \times h$$

FIG 1.15 A cuboid

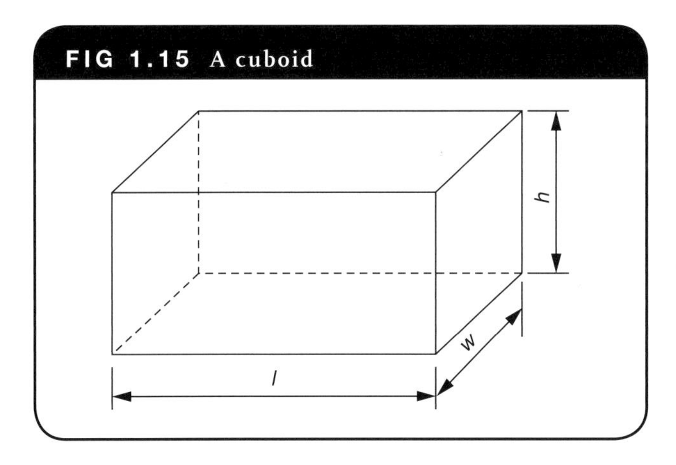

Example

A cuboid has dimensions length 10 cm, width 6 cm and height 8 cm. What is its volume?

$$V = l \times w \times h$$
$$= 10 \times 6 \times 8$$
$$\text{answer} = 480 \text{ cm}^2$$

The cylinder

A cylinder is a solid of uniform circular cross-section. The volume of a cylinder is equal to the area of the cross-section multiplied by length or height, as indicated in Figures 1.16(a) and (b).

$$V = \pi r^2 l \text{ or } V = \pi r^2 h$$

FIG 1.16 Area of cross-section and volume of a cylinder

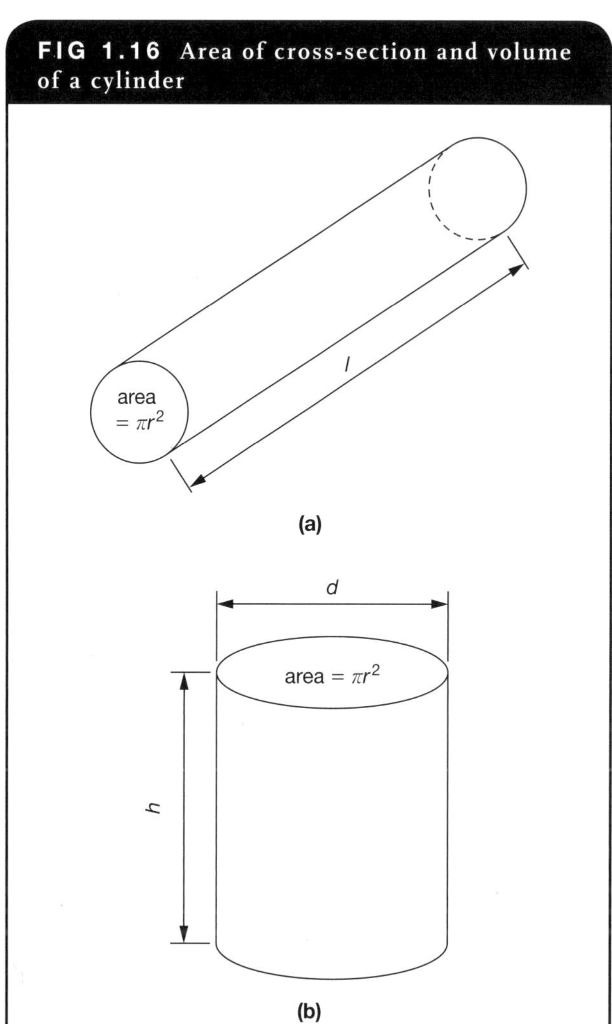

Examples

1 A copper tube is 300 mm long and has an inner diameter 18 mm. Calculate the volume of the copper tube.

$$V = \pi r^2 l$$
$$= 3.142 \times 9^2 \times 300$$
$$\text{answer} = 76351 \text{mm}^3$$

2 A semicircular trough (see Figure 1.17) is 2.0 m long and has a diameter 0.450 m. What is the volume of the trough?

$$A(\text{or circle}) = \pi r^2$$
$$A(\text{or semicircle}) = \frac{\pi r^2}{2}$$
$$\therefore V = \frac{\pi r^2}{2} \times l$$
$$= \frac{3.142 \times (0.225)^2}{2} \times 2$$
$$\text{answer} = 0.159 \text{ m}^3$$

FIG 1.17 A semicircular trough

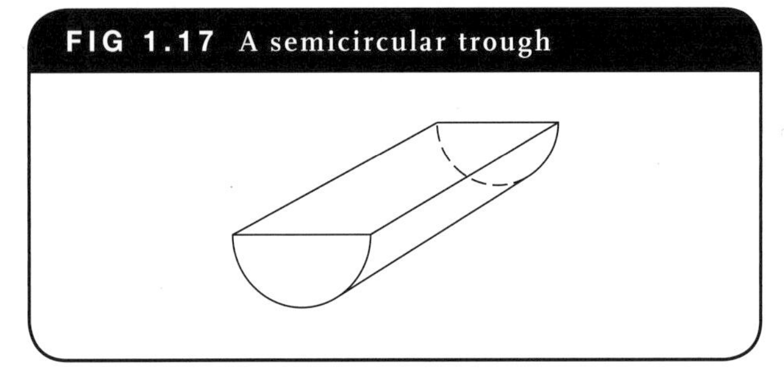

The right cone

A right cone can be defined as a surface which has a circular base and a sloping side which radiates from an apex located vertically above the centre of the base (Figure 1.18).

FIG 1.18 A right cone

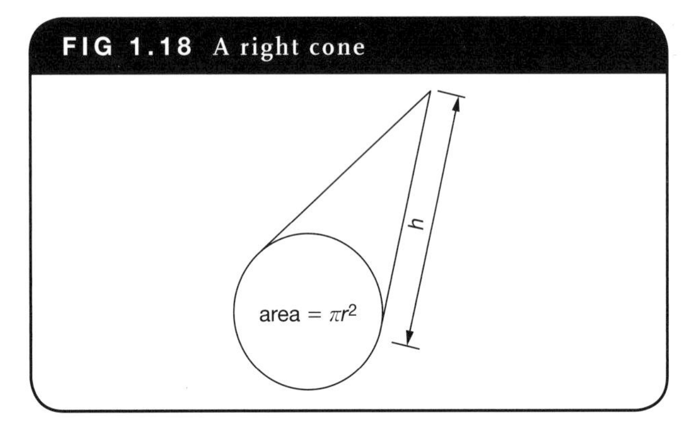

If we cut down a slanting side of the cone from the vertex, B (Figure 1.19(a)) and 'unwrap' the curved surface, we obtain a sector of a circle with centre B. The base of the cone is the curved line '*AA*' as shown in Figure 1.19(b).

The volume of a right cone is $\frac{1}{3}$ area of base × height.

$$V = \frac{1}{3}\pi r^2 h$$

By using Pythagoras' theorem, we also have:

$$l^2 = r^2 + h^2$$

Example

Find the volume of a cone having base diameter 14 cm and height 20 cm.

$$\begin{aligned} V &= \frac{1}{3}\pi r^2 h \\ &= \frac{1}{3} \times 3.142 \times 7^2 \times 20 \\ \text{answer} &= 1026 \text{ cm}^3 \end{aligned}$$

FIG 1.19 (a) A right cone, (b) sector of a circle

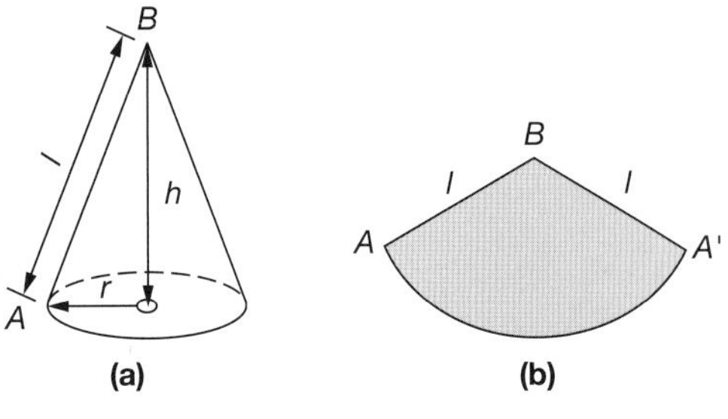

The sphere

It has been previously stated that the circumference of a circle is πd. It can also be written as $2\pi r$. Because the circumference is a length, it has one dimension. A circle has two dimensions, so we have r^2 in the equation for area of a circle. A sphere, however, occupies space and therefore must have three dimensions (r^3).

The formula used for the volume of a sphere is:

$$V = \frac{4}{3}\pi r^3$$

FIG 1.20 Geometrical solids

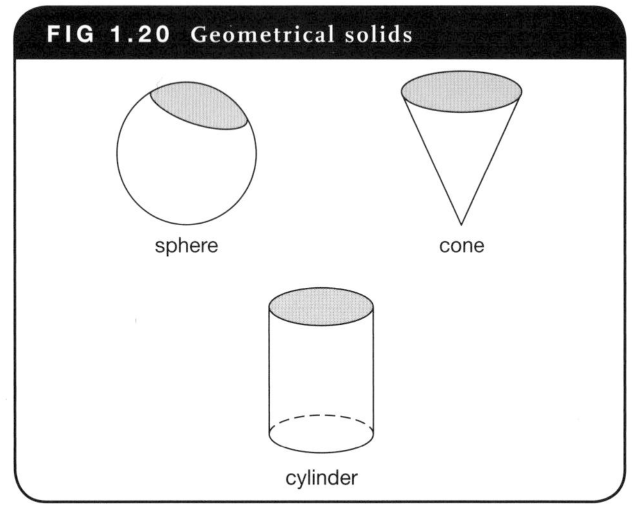

Example

Calculate the volume of a sphere having a diameter 10 cm. Substituting in the formula above, we have:

$$\begin{aligned} V &= \frac{4}{3}\pi r^3 \\ &= \frac{4}{3} \times 3.142 \times 5^3 \\ \text{answer} &= 524 \text{ cm}^2 \end{aligned}$$

If we compare the volumes of three geometrical solids—a sphere, a cone and a cylinder (Figure 1.20), all having a common diameter and vertical height, we will arrive at a remarkable result.

If we filled the cone with fine running sand and poured its contents into the sphere, we would find that two cones are required to fill it. Now, if we filled the cone with sand but this time poured the sand into a cylinder, we would find that three cones of sand are required to fill the cylinder. This relationship was discovered by Archimedes, the Greek mathematician and physicist.

To calculate fluid capacity

The storage of liquids prior to delivery is an important function of any water distribution system. Appropriate structures therefore need to be provided to store water for domestic, industrial or rural watering systems. Whether they are designed on a large scale (such as dams to regulate water levels) or a small scale (such as rainwater tanks) it is important for the designers to be able to calculate the storage capacities.

The standard unit for measuring liquids or liquid capacity in water engineering and in the metric system is the litre (L). A litre is the amount of fluid which can be contained in a cube having volume 1000 cm^3. For large capacities, such as annual water consumption by householders, the kilolitre (kL) measure is used (1 kL = 1000 L).

For underground water and storage reservoirs serving major distribution systems, the measure is the megalitre (ML) (1 ML = 10^6 L = 10^3 m^3).

Examples

1 Calculate the number of litres of water that the container (e.g. rainwater tank) in Figure 1.21 will hold when filled.

$$\begin{aligned} \text{capacity (in L)} &= \text{volume (in m}^3) \times 1000 \\ &= l \times w \times h \times 1000 \\ &= 2 \times 1.5 \times 3 \times 1000 \\ \text{answer} &= 9000\text{ L} \end{aligned}$$

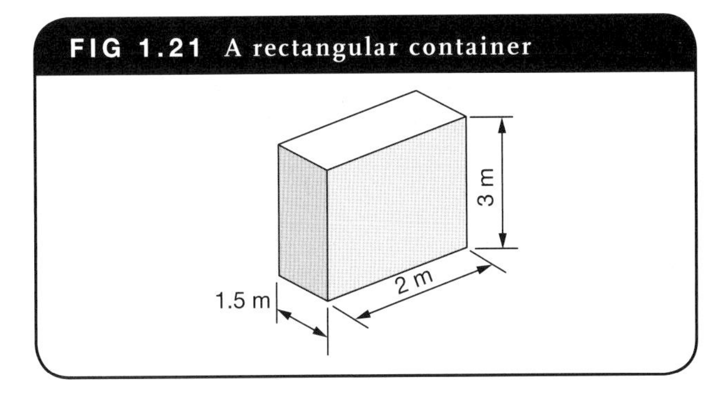

FIG 1.21 A rectangular container

2 Find the capacity in litres of the horizontal storage tank (e.g. solar hot water tank) shown in Figure 1.22.

$$\begin{aligned} \text{capacity (in L)} &= \text{volume (in m}^3) \times 1000 \\ &= \pi r^2 l \times 1000 \\ &= 3.142 \times 0.75^2 \times 4 \times 1000 \\ \text{answer} &= 7069.5\text{ L} \end{aligned}$$

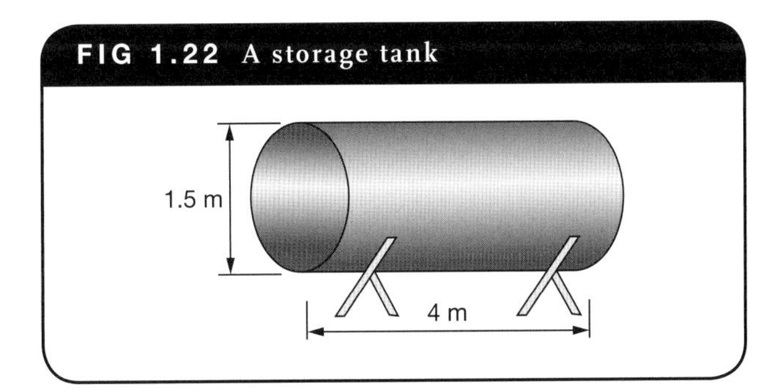

FIG 1.22 A storage tank

3 Calculate the number of litres to be added to completely fill the rectangular rain water tank in Figure 1.23.

$$\begin{aligned} \text{depth of water to be added} &= 3\text{m} \\ \text{volume (in m}^3\text{) of empty section} &= l \times w \times d \\ &= 8 \times 6 \times 3 \\ \therefore \text{capacity (in L)} &= 8 \times 6 \times 3 \times 1000 \\ \text{answer} &= 144000\text{ L} \end{aligned}$$

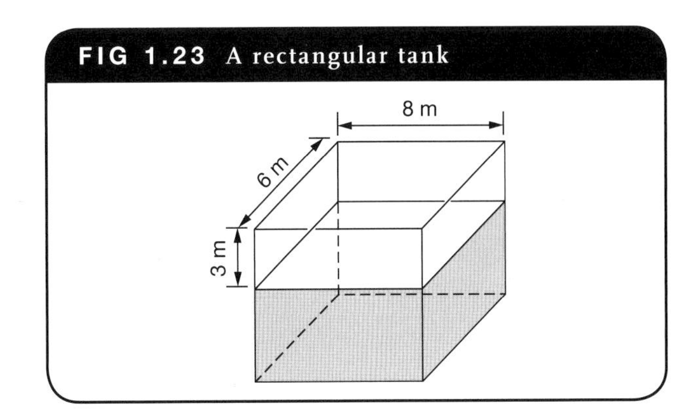

FIG 1.23 A rectangular tank

4 Consider that a reservoir has a capacity of 12 700 000 (1.27×10^7) kL. How many litres does it hold?

$$\begin{aligned} 1\text{ kL} &= 1000\text{ L} \\ \therefore 12\,700\,000\text{ kL} &= 12\,700\,000 \times 1000\text{ L} \\ \text{answer} &= 12\,700\,000\,000\text{ L or } 1.27 \times 10^{10}\text{ L} \end{aligned}$$

Usually, the capacity of a reservoir is expressed in megalitres (ML).

Example

12 700 000 ÷ 1000

answer = 12 700 ML

FIG 1.24 Carcoar Dam, a concrete arch dam

APPLIED CALCULATIONS

This section will deal with the application of mathematical rules, laws and formulae to practical plumbing situations.

Pythagoras' theorem

The most useful and frequently used law of mathematics in plumbing problem-solving is Pythagoras' theorem. It can be applied to such areas of work as:

1 roof work
2 drainage excavations
3 pipework.

Let us examine this law of mathematics. Pythagoras' theorem states that 'In a right-angled triangle the square of the hypotenuse is equal to the sum of the squares on the other two sides.' Figure 1.25 overleaf, illustrates this principle.

FIG 1.25 Pythagoras' theorem

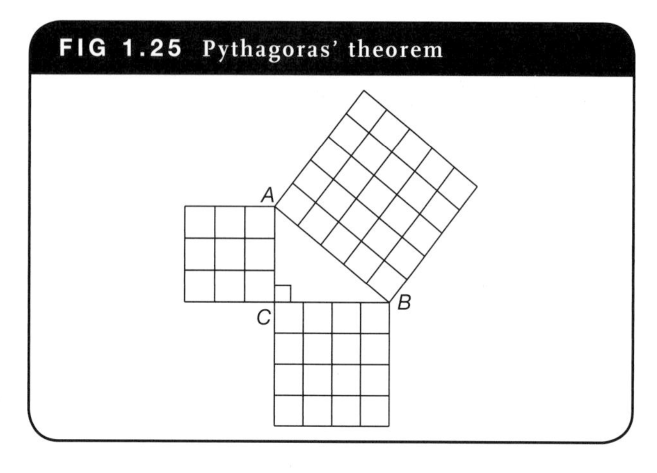

Proof

$$AB^2 = 25 \text{ square units}$$
$$BC^2 = 16 \text{ square units}$$
$$AC^2 = 9 \text{ square units}$$
$$\therefore BC^2 + AC^2 = AB^2$$
$$\because 4^2 + 3^2 = 5^2$$
$$16 + 9 = 25$$

Roof work calculations

In the building industry, areas of roof coverage are expressed in square metres (m^2). To calculate the area of a roof, as in Figure 1.26, the length of the common rafter must first be determined. The right-angled triangle principle is useful in calculating the length of the common rafter (see Figure 1.27).

To calculate the length of the common rafter, add the squares of half the span and the rise, then find the square root.

FIG 1.26 New home under construction with fitted roof sheets

The formula for the length of the common rafter (gable roof) is:

$$CR = \sqrt{\frac{1}{2}\text{span}^2 + \text{rise}^2}$$

where CR = length of the common rafter.

FIG 1.27 The common rafter is the first roof calculation to be made

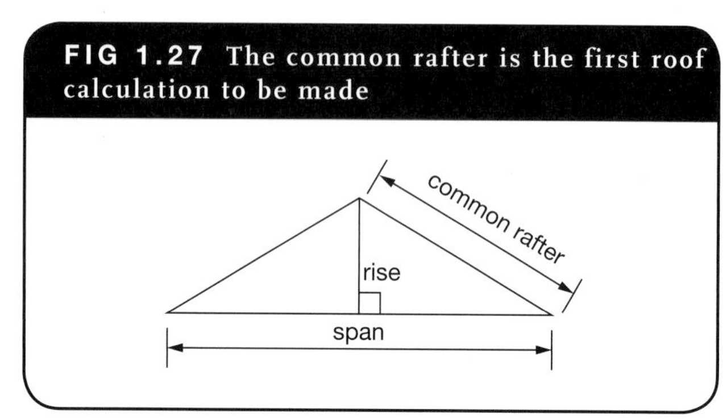

Example

The span of a roof is 10 m and the rise is 2.5 m. Find the length of the common rafter.

$$\begin{aligned} CR &= \sqrt{\frac{1}{2}\text{span}^2 + \text{rise}^2} \\ &= \sqrt{5^2 + 2.5^2} \\ &= \sqrt{25 + 6.25} \\ &= \sqrt{31.25} \\ \text{answer} &= 5.6\text{ m (approx.)} \end{aligned}$$

Note the following formulae:

$$\text{roof area} = \text{common rafter} \times \text{gutter length} \times 2$$

$$\text{roof pitch} = \frac{\text{rise}}{\text{span}}$$

Drainage calculations

In all drainage work the main objective of the plumber is the effective removal of offensive domestic and industrial waste materials. This is achieved by providing a proper and even grade, such that sewage or other waste materials in a

FIG 1.28 Typical timber or steel roof truss

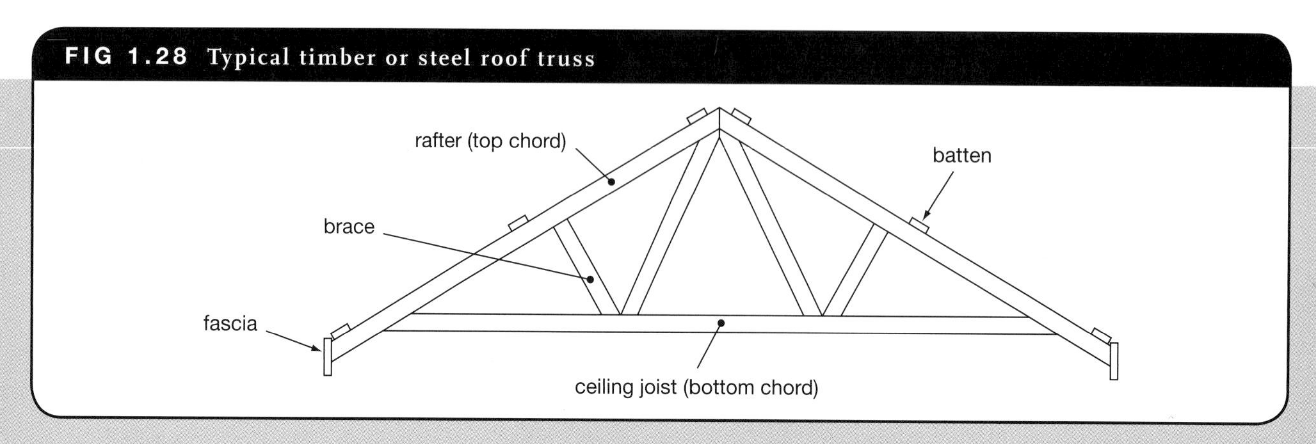

drain will flow at a 'self-cleansing velocity' by the force of gravity and discharge into a sewer main.

A knowledge of site conditions (e.g. the rise and fall of the ground surface, the depths at the sewer connection end and the head of the drain, as well as other depths along the drain) is essential if proper waste disposal is to be achieved. This section deals with drainage calculations and the formulae which may be used for various site conditions.

FIG 1.29 Drain excavation

Grade and fall formulae

The following alphabetical symbols will assume these meanings:

C = depth at deep end (connection)
D = distance (length of drain)
G = grade of drain
H = depth at head of drain
F = fall
R = rise

The grade in a drain may be expressed either as a ratio or as a percentage. The degree of inclination illustrated in Figure 1.30 could therefore be written as either 1:50 or 2%.

Grade (expressed as a ratio):

$$\text{grade} = \frac{\text{fall or rise}}{\text{distance}}$$

$$\text{i.e. } G = \frac{F}{D} \text{ or } G = \frac{R}{D}$$

By transposition:

$$F = G \times D \text{ or } R = G \times D$$

Grade (expressed as a percentage):

$$\%\text{ grade} = \frac{\text{fall or rise}}{\text{distance}} \times 100$$

$$\text{i.e. } \%\text{G} = \frac{\text{F}}{\text{D}} \times 100 \text{ or } \frac{\text{R}}{\text{D}} \times 100$$

By transposition:

$$F \text{ or } R = \frac{\%G \times D}{100}$$

FIG 1.30 Expressing grade as a ratio or a percentage

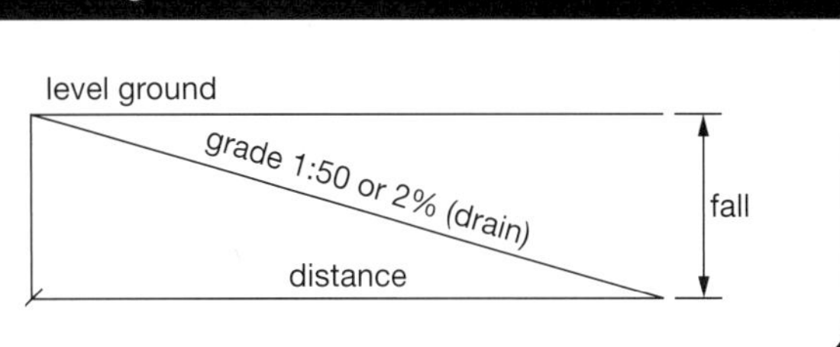

Examples

Note: Most authorities insist on a minimum soil coverage of 300 mm at the head of a drain.

1 Calculation of fall. Refer to Figure 1.31.

(a) Calculation for depth at connection using grade as a ratio:

FIG 1.31 Calculating fall to determine depth at C

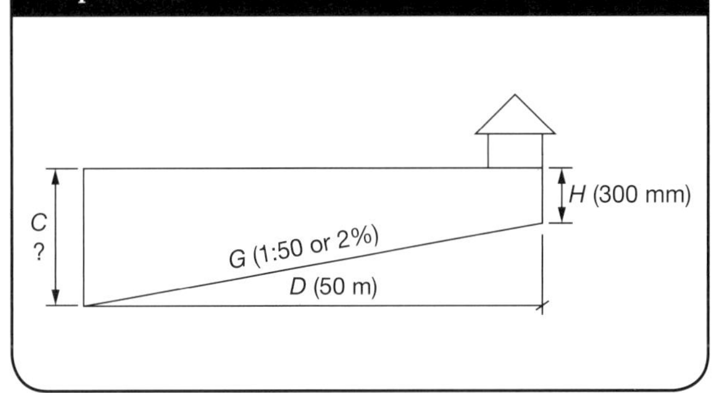

Assume $D = 50$ m
$G = 1{:}50$
$H = 300$ mm

$$\begin{aligned} F &= G \times D \\ &= \frac{1}{50} \times 50 \\ &= \frac{50}{50} \\ &= 1\text{m} \end{aligned}$$

but $H = 300$ mm
$\therefore C = 1.3$ m

(b) Calculation for depth at connection using grade as a percentage:

Assume $D = 50$ m
$G = 2\%$
$H = 300$ mm

$$\begin{aligned} F &= \frac{\%G \times D}{100} \\ &= \frac{2 \times 50}{100} \\ &= \frac{100}{100} \\ &= 1\text{m} \end{aligned}$$

but $H = 300$ mm
$\therefore C = 1.3$ m

2 Calculation for depth at head of drain.

FIG 1.32 Calculating fall to determine depth at H

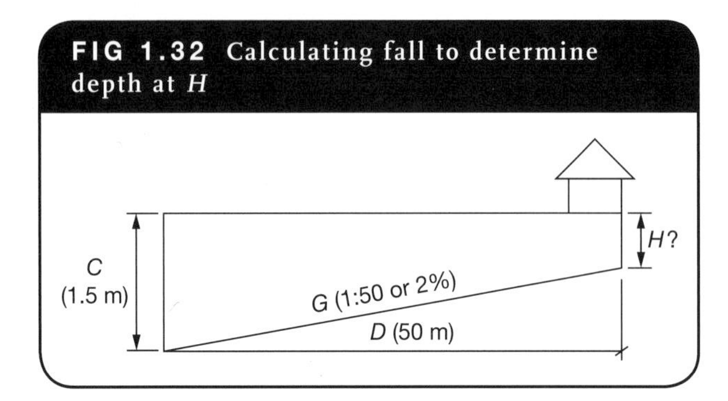

(a) Using grade as a ratio:

Assume $D = 50$ m
$G = l{:}50$
$C = 1.5$ m

$$R = (G \times D)$$
$$= \frac{1}{50} \times 50$$
$$= \frac{50}{50}$$
$$= 1 \text{ m}$$

but $C = 1.5$ m
$\therefore H = 500$ mm

(b) Using grade as a percentage:

Assume $D = 50$ m
$G = 2\%$
$C = 1.5$ m

$$R = \frac{\%G \times D}{100}$$
$$= \frac{2 \times 50}{100}$$
$$= \frac{100}{100}$$
$$= 1 \text{ m}$$

but $C = 1.5$ m
$H = 500$ mm

3 Calculation for grade.

(a) As a ratio:

FIG 1.33 Calculating gradient G

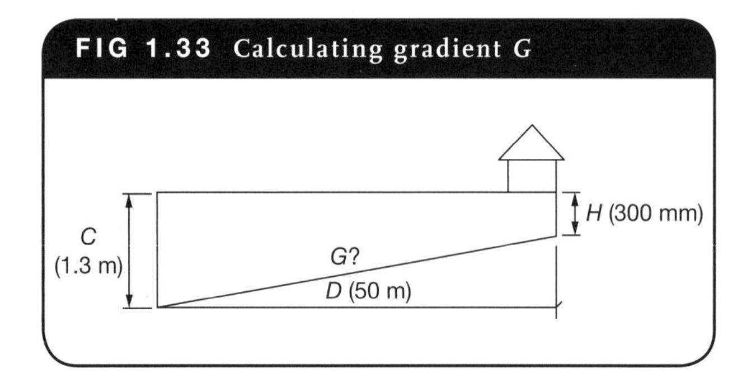

Assume $D = 50$ m
$C = 1.3$ m
$H = 300$ mm

$$G = \frac{F}{D}$$

But $F = C - H$
$= 1.3 - 0.3$
$= 1$ m

$$\therefore G = \frac{1}{50}$$
$$= 1{:}50$$

(b) As a percentage.

Assume $D = 50$ m
$C = 1.3$ m
$H = 300$ mm

$$\%G = \frac{F}{D} \times 100$$
$$= \frac{1}{50} \times 100$$
$$= \frac{100}{50}$$
$$\therefore G = 2\%$$

Composite example

Observe the drainage layout in Figure 1.34 and the longitudinal section in Figure 1.35.

A = depth of connection = 1.5 m
D = depth at head of drain = 300 mm
Section AB is 20 m long.
Section BC is 15 m long.
Section CD is 10 m long.
BC and CD are laid at grade of 1:50.

FIG 1.34 A drainage layout

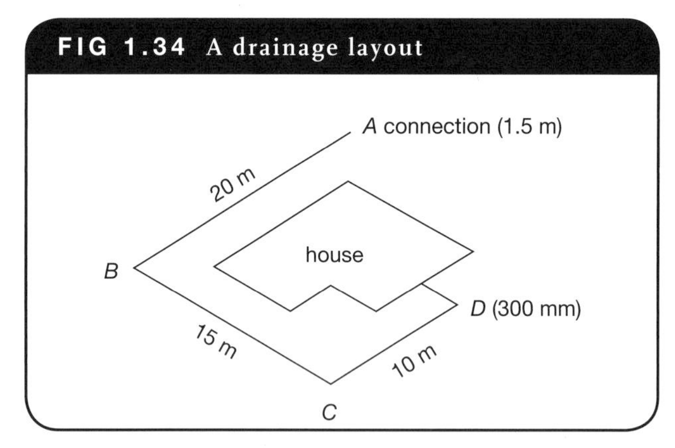

Calculate:

1 the depths at C and B;

2 the grade for section AB.

1 Depth at C:

$$F = G \times D$$
$$= \frac{1}{50} \times 100$$
$$= \frac{10}{50} \text{m}$$
$$= 200\text{mm}$$

$\therefore$ depth at C = fall between DC + depth at D
$= 200 + 300$
$= 500$ mm

FIG 1.35 Longitudinal section

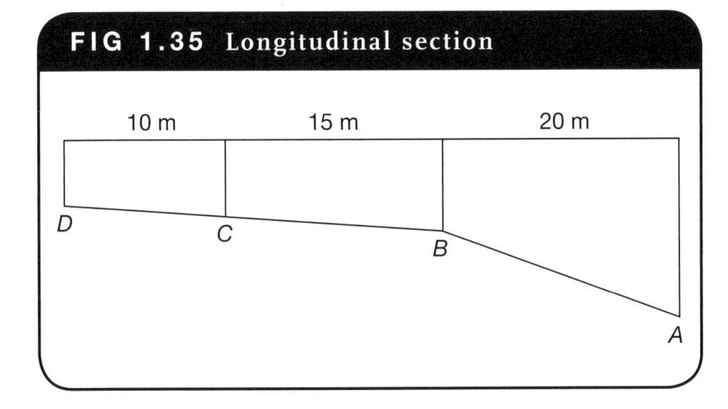

Depth at *B*:

$$\begin{aligned} F &= G \times D \\ &= \frac{1}{50} \times 15 \\ &= \frac{15}{50} \text{m} \\ &= 300\text{mm} \\ \therefore \text{depth at B} &= \text{fall between CB + depth at C} \\ &= 300 + 500 \\ &= 800 \text{ mm} \end{aligned}$$

2 Grade for section *AB*:

$$\begin{aligned} G &= \frac{F}{D} \\ F &= \text{depth at A} - \text{depth at B} \\ &= 1.5 - 0.8 \\ &= 0.7 \text{ m} \\ &= 700 \text{ mm} \\ &= 1{:}28.6 \end{aligned}$$

∴ grade for *AB*

$$\begin{aligned} AB &= \frac{700}{200000} \\ &= 1{:}28.6 \end{aligned}$$

Differences in elevation between two points, or at other intermediate points along a chain, can also be calculated from a given elevation level above or below a given datum level. The remaining calculations will be presented in this manner.

For further information pertaining to the theoretical aspects of elevation levels derived from a given datum level, refer to the section dealing with levelling (Chapter 6).

Example

Given the line of drain as shown in Figure 1.36, calculate intermediate elevation levels (*EL*) at points *B*, *C* and *D* from the given details:

A to *B* = 9 m
B to *C* = 10 m
C to *D* = 16 m
Ground level datum level is *EL* 4.000.
Invert level at connection *A* is *EL* 2.800.
Drain to be laid at 1:50 grade (2%).

EL at *B*:

$$\begin{aligned} D &= 9 \\ G &= 1{:}50 \\ F &= G \times D \\ \text{depth at } B &= \frac{1}{50} \times \frac{9}{1} \\ &= \frac{9}{50} \text{ m} \\ &= 180 \text{ mm} \\ \therefore EL \text{ at } B &= 2.800 + 0.180 \\ &= 2.980 \end{aligned}$$

EL at *C*:

$$\begin{aligned} D &= 10 \\ G &= 1{:}50 \\ \text{depth at } C &= \frac{1}{50} \times \frac{10}{1} \\ &= \frac{10}{50} \text{ m} \\ &= 200 \text{ mm} \\ \therefore EL \text{ at C} &= 2.980 + 0.200 \\ &= 3.180 \end{aligned}$$

(*continued*)

FIG 1.36 (a) Line of a drain, (b) longitudinal L section

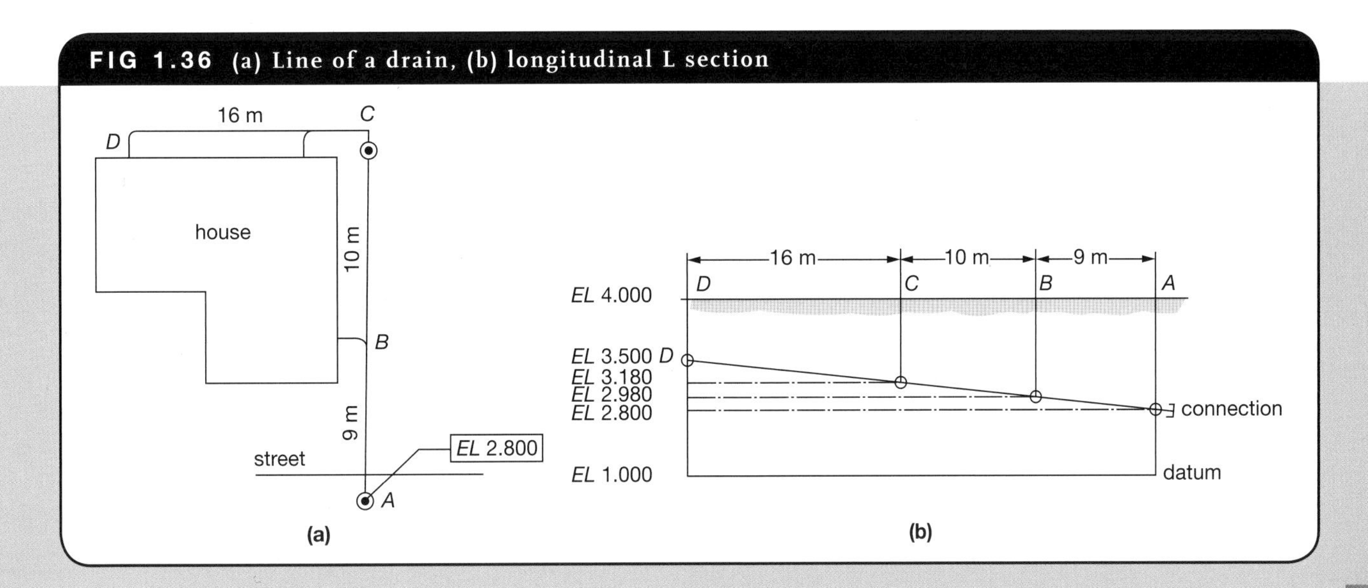

EL at *D*:

$$D = 16$$
$$G = 1{:}50$$
$$\text{depth at } D = \frac{1}{50} \times \frac{16}{1}$$
$$= \frac{16}{50}\text{ m}$$
$$= 320\text{ mm}$$
$$\therefore EL \text{ at } D = 3.180 + 0.320$$
$$= 3.500$$

Check:

$$D\text{ (total)} = 35\text{ m}$$
$$G = 1{:}50$$
$$\therefore F\text{ (total)} = \frac{1}{50} \times \frac{35}{1}$$
$$= \frac{35}{50}\text{ m}$$
$$= 700\text{ mm}$$
$$\therefore EL\text{ (total)} = 2.800 + 0.700$$
$$= 3.500$$

Excavation calculations

This section deals with the mathematics required to calculate soil volumes in trenching work. The drainage contractor needs to be able to determine the amount of excavation necessary in order to quote accurately for the job.

Excavated soil volume calculations are an extension of work already covered in the section dealing with the calculation of solids—in particular, the cuboid.

Examples

1 Trench with a uniform depth.
Calculate the volume of soil contained in a trench 20 m long, 1.5 m deep and 750 mm wide.

$$V = l \times w \times d$$

where V = volume (in m^3)
l = length of trench (in m)
w = width of trench (in m)
d = depth of trench (in m).

Then $V = 20 \times 1.5 \times 0.75$
answer = 22.5 m^3

2 Trench with non-uniform depth.
Calculate the volume of soil to be excavated from a trench 30 m long, 2 m wide and 6 m deep at the connection end and 2 m deep at the shallow end (see Figure 1.37).

$$V = \frac{l(A_1 + A_2)}{2}$$

where V = volume (in m^3)
l = length of trench (in m)
A_1 = area of trench cross-section at deep end (in m^2)
A_2 = area of trench cross-section at shallow end (in m^2)

FIG 1.37 A trench

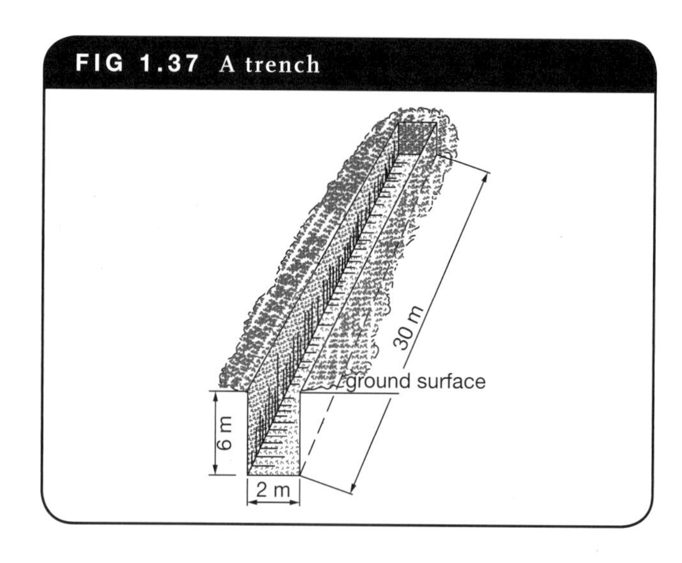

$$\text{Then } V = \frac{(30[6 \times 2] + [2 \times 2])}{2}$$
$$= 30\left(\frac{12 + 4}{2}\right)$$
answer = 240 m^3

Pipe calculations

Layout and measurement are an important part of pipework. This section aims to give the student the necessary basic skills in the application of formulae during pipework fabrication and assembly. The installation of pipe systems involves changes in direction by the use of approved fittings—elbows, tees, junctions and so on.

The most common techniques for changes of direction are the 45° and 90° offsets. For this reason, the examples in this section will be centred on the use of these angles.

The right triangle rule is used to assist calculation of pipe lengths. In pipe language, the three sides of the right triangle are labelled as:

1. offset (*O*)
2. advance (*A*)
3. travel (*T*).

(See Figure 1.38.)

FIG 1.38 The three sides of a right triangle

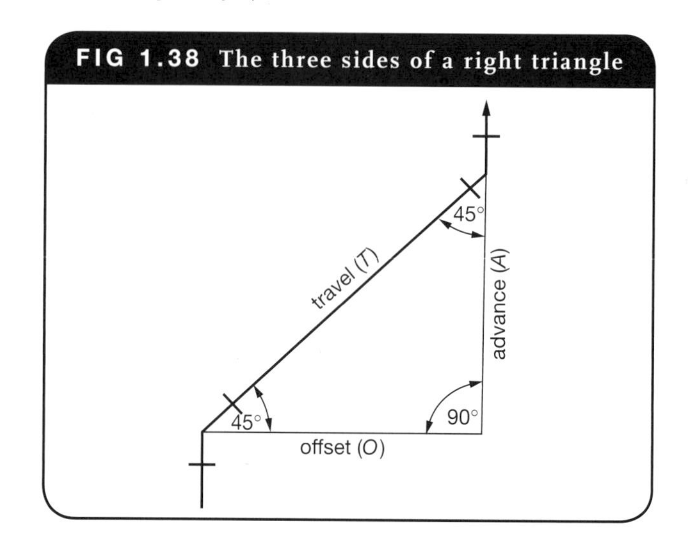

RULES FOR 45° RIGHT TRIANGLE

1. The triangle is built up of three sides, having two 45° angles and one 90° angle.
2. The lengths representing the offset (*O*) and advance (*A*) will form the sides of the 90° angle.
3. The offset and advance are equal in length.
4. To calculate travel length (*T*), use:

$$T = O \times 1.41$$
$$\text{or } T = A \times 1.41$$

5. To calculate offset (*O*) or advance (*A*), use:

$$O = \frac{T}{1.41}$$
$$\text{or } A = \frac{T}{1.41}$$

Note: 1.41 is a constant for a 45° triangle.

Examples

1 Calculate the length for travel (*T*) given that advance (*A*) is 1 m. Refer to Figure 1.39.

$$T = A \times 1.41$$
$$= 1 \times 1.41$$
$$\text{answer} = 1.41\text{m}$$

FIG 1.39 Calculating the length for travel

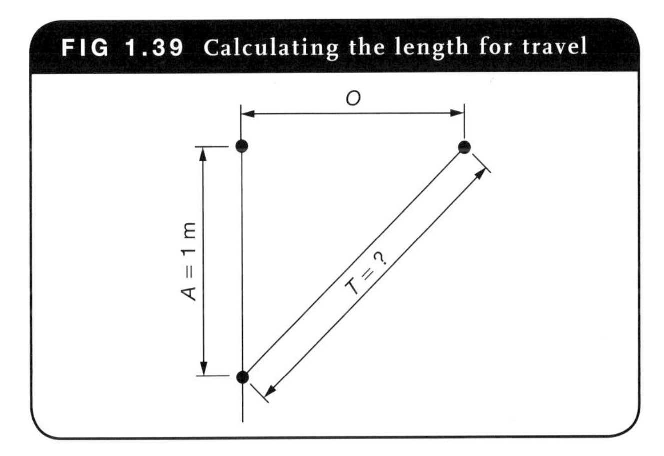

2 Calculate the lengths for *A* and *O*, given that *T* = 5 m. Refer to Figure 1.40.

FIG 1.40 Calculating lengths for A and O

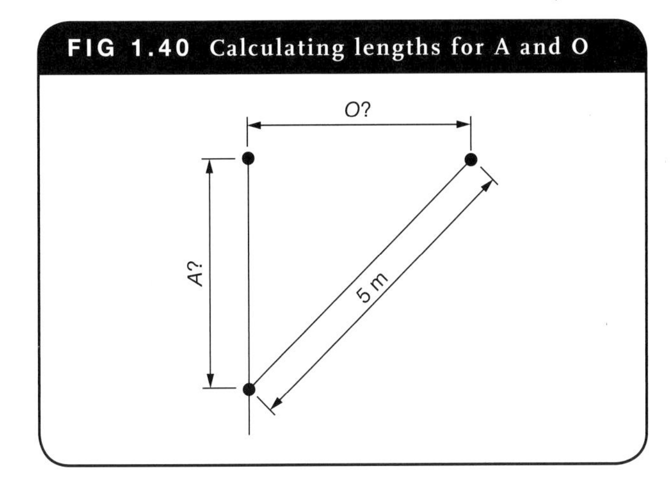

$$A = \frac{T}{1.41}$$
$$= \frac{5}{1.41}$$
$$A = 3.5 \text{ m}$$

A and *O* are the same.

$$\therefore A = 3.5 \text{ m}$$
$$\text{and } O = 3.5 \text{ m}$$

Offsets fall into two groups:

- simple offsets
- compound offsets.

The factors which contribute to simple offsets have already been covered in the previous examples.

The use of compound offsets, commonly referred to as 'rolling offsets', has the advantage of reducing the amount of piping and fittings required. In compound offset calculations, it is necessary to know two measurements in order to compute the third; e.g. to obtain travel (*T*) one must know both offset (*O*) and advance (*A*) distances.

The formula for a 90° rolling offset is:

$$T = \sqrt{O^2 + A^2}$$

Example

Figure 1.41 shows an isometric sketch for a simple offset, with centre-line (℄) measurements (*O* = 2 m, *A* = 1.5 m). Calculate *T* using 90° fittings to eliminate piping and fittings as shown in Figure 1.41.

FIG 1.41 Simple offset

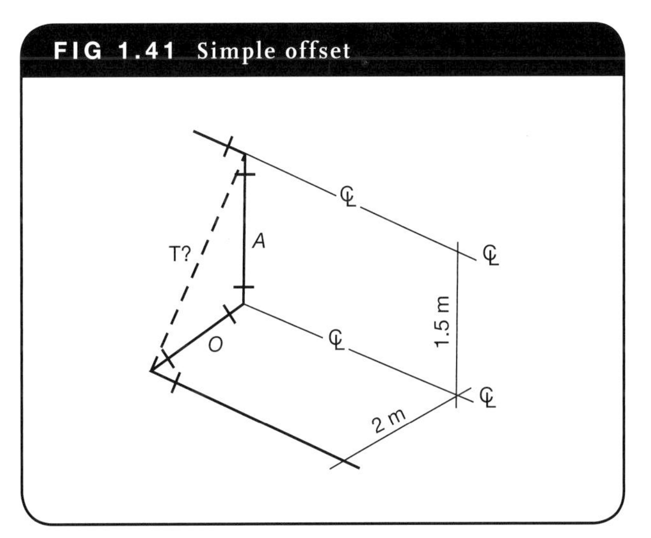

$$T = \sqrt{O^2 + A^2}$$
$$= \sqrt{2^2 + 1.5^2}$$
$$= \sqrt{6.25}$$

answer = 2.5 m (℄ measurement)

Pipe bending

It is essential for the student to become familiar with the mathematics required for dealing with pipe bending. The general characteristics and other considerations relating to pipe bending will be covered in Chapter 4. It is intended in this section to provide an understanding of the processes associated with pipe bending formulae.

All calculations used in pipe bending are based on the centre-line radius ($℄r$) of a bend, because the throat is compressed and the back (heel) stretches. In theory, we assume that the ($℄r$ remains unchanged during a bending operation.

Calculation for 'range' of bend (90°)

The following data must be known to enable a calculation to be made:

- angle of bend
- pipe diameter
- throat radius (Tr) of bend.

The 'throat radius' is usually given as the number of pipe diameters.

$$R = ℄r \times 1.6 \times \frac{ABT}{90}$$

where R = range of bend
℄ = centre-line radius
1.6 = constant
ABT = angle through which the pipe is bent

The range for a full circle = $2\pi ℄r$.

However, all pipe bends relate to a 90° bend (see Figure 1.42).

FIG 1.42 (a) Bend relates to a quarter circle, (b) showing throat radius, heel and centre-line radius

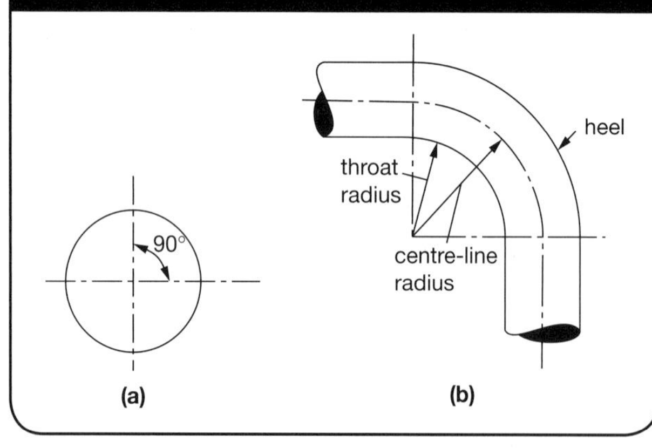

$$\text{Since a } 90° \text{ bend} = \frac{1}{4} \text{ circle}$$

$$90° \text{ bend} = \frac{2\pi ℄r}{4}$$
$$= \frac{2 \times 3.142 \times ℄r}{4}$$
$$= 1.6 \times ℄r$$

Example

Calculate the range of a bend given the following details:
90° bend
40 mm diameter pipe
throat radius 2 pipe diameters

Step 1 Determine the throat radius (Tr) by multiplying the diameter of the pipe by the number of pipe diameters (40 × 2 = 80 mm).

Step 2 Determine the centre-line radius.

$$℄r = Tr + \frac{1}{2} \text{pipe diameter}$$
$$= 80 + 20$$
$$= 100\text{mm}$$

Step 3 Calculate the range (R) of bend.

$$R = ℄r \times 1.6 \times \frac{ABT}{90}$$
$$= 100 \times 1.6 \times \frac{90}{90}$$
$$\text{answer} = 160 \text{ mm}$$

Range for bends (other than 90°)

All bends other than 90° may be calculated by comparing the angle through which it is bent with that of a 90° angle bend.

Example

Calculate the range of bend for a pipe having outside diameter 50 mm and throat radius 3 pipe diameters. The pipe is to be bent through 60°.

$$Tr = 50 \times 3$$
$$= 150 \text{ mm}$$
$$℄r = Tr + \frac{1}{2}.\text{pipe diameter}$$
$$= 150 + 25$$
$$= 175 \text{ mm}$$
$$R = ℄r \times 1.6 \times \frac{ABT}{90}$$
$$= 175 \times 1.6 \times \frac{60}{90}$$
$$\text{answer} = 187 \text{ mm}$$

Note: The range for a 45° bend would be half that of a 90° bend, all other factors being equal.

Offsets

Example

Calculate the length of pipe required to complete the offset in Figure 1.43.

section B: 60° bend
Tr = 2 pipe diameters
section D: 105° bend
Tr = 5 pipe diameters
pipe diameter = 20 mm

FIG 1.43 Calculating 'range' for a given offset

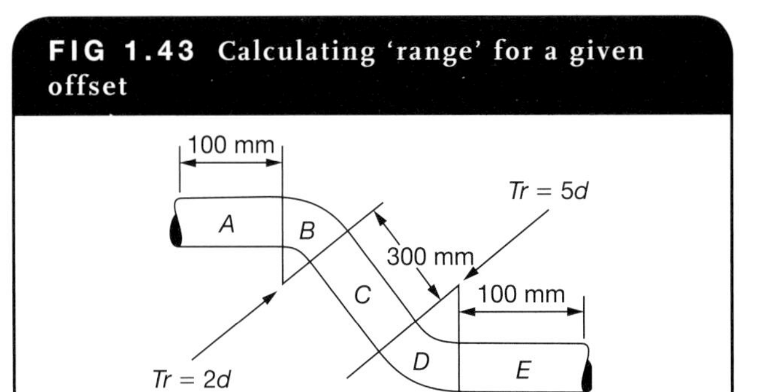

Range for bend *B*:

$$R = \text{℄}r \times 1.6 \times \frac{ABT}{90}$$
$$= 50 \times 1.6 \times \frac{60}{90}$$
$$= 53 \text{ mm}$$

Range for bend *D*:

$$R = \text{℄}r \times 1.6 \times \frac{ABT}{90}$$
$$= 110 \times 1.6 \times \frac{75}{90}$$
$$= 147 \text{ mm}$$
$$\therefore \text{total length} = A + B + C + D + E$$
$$= 100 + 53 + 300 + 147 + 100$$
$$\text{answer} = 700 \text{ mm}$$

MECHANICS OF WATER SUPPLY

This section deals with the hydraulics relevant to water plumbing. In particular, emphasis will be focused on natural physical forces and how they influence and assist in the operation of apparatus used in plumbing systems.

To understand the operation of the units associated with hot and cold water supply, it is essential to have a basic knowledge of the physical conditions which enable them to operate. The physical conditions considered in this section fall within the broad scope of the mechanics of water supply.

Natural forces

To enable such equipment as pumps, cisterns, siphons and flushometers to function, naturally occurring physical conditions need to be present. The two most important physical conditions affecting this area of work, atmospheric pressure (atm) and gravity, are considered in this section.

Atmospheric pressure (atm)

This is the pressure which the atmosphere exerts on every open surface and is equivalent to approximately 100 kPa (kilopascals) at sea level (1 atmosphere).

Note: 1 pound-force per square inch (lbf/in² or psi) = 6.89 kPa (approximately).

Gravity

Gravity is a natural force of attraction possessed by all bodies. All downward movement of free-falling bodies is due to gravity. From a plumbing viewpoint, it can therefore be seen that gravity is responsible for water flow. Thus, we are able to provide water reticulation services using the effect of gravity to produce water pressure and flow.

Velocity of water flow

The velocity of falling bodies is uniformly accelerated by the force of gravity and can be determined from the following formula:

$$V = \sqrt{2gh}$$

where V = velocity (in m/s)
g = gravitational constant (approximately 9.8 m/s)
h = height body falls (in m)

Example

Calculate the velocity that water will attain in a pipeline if the vertical height between the water source and free outlet is 25 m.

$$V = \sqrt{2gh}$$
$$= \sqrt{2 \times 9.8 \times 25}$$
$$\text{answer} = 22 \text{ m/s (approx.)}$$

Water hydraulics

Although the generally accepted definition of hydraulics deals with liquids in motion, it should include all conditions of liquids, whether in motion or at rest. A storage tank filled with water, for example, is subject to a 'hydrostatic' pressure. A water supply reticulation system is subject to the laws of hydraulics.

'Pressure' and 'head'

'Pressure' is defined as 'force per unit area' and is measured in kilopascals (kPa). 'Head' is the vertical height below a given datum level and is measured in millimetres (mm) or metres (m). The intensity of pressure exerted by water is directly proportional to the depth of the water. The head is therefore responsible for pressure (see Figure 1.44, overleaf).

Examples

1 In Figure 1.45, assume that h_1 is 1 m below the static water level in the tank. Calculate the pressure at h_3, which is 3 m below the given datum level.

Note: 1 m head is equivalent to 9.8 kPa.

Calculation:

1 m head = 9.8 kPa
$\therefore$ 3 m head = 9.8×3
answer = 29.4 kPa

FIG 1.44 Pressure is proportional to depth

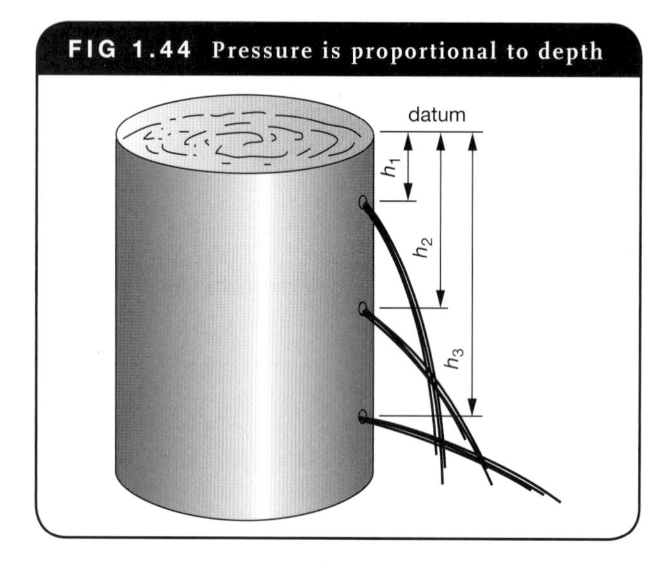

2 Refer to Figure 1.45 and calculate the minimum and maximum working pressures for flushometer valves at the levels indicated.

Calculation for minimum head:

1 m head = 9.8 kPa
$\therefore$ 2.5 m head = 9.8×2.5
answer = 24.5 kPa

Calculation for maximum head:

1 m head = 9.8 kPa
$\therefore$ 40 m head = 9.8×40
answer = 392 kPa

FIG 1.45 Conversion of 'head' to 'pressure'

datum
minimum head 2.5 m
maximum head 40 m

USEFUL FORMULAE AND CONVERSIONS

Temperature

To convert from °C to °F, use:

$$°F = \frac{9}{5}C + 32$$

To convert from °F to °C, use:

$$°C = \frac{5}{9}(F - 32)$$

comfort temperature = 72 °F or 22.2 °C

body temperature = 98.6 °F or 37 °C

domestic hot water temperature = 140 °F or 60 °C

Coefficient of linear expansion

$$E = Kl\,(T_1 - T_2)$$

where E = expansion
K = coefficient of expansion
l = length
T_1 = temperature when hot
T_2 = temperature when cold

Water pressure conversions

10 mm or 1 cm = 0.1 kPa (approx.)
100 mm or 10 cm = 1 kPa (approx.)
1000 mm or 100 cm or 1 m = 10 kPa (approx.)
1000 cm or 10 m = 100 kPa (approx.)
1 or 1 psi = 6.89 kPa (approx.)

Volume

cube = l^3
rectangle = $l \times w \times d$
cylinder = $\pi r^2 l$ or $\pi r^2 h$

TABLE 1.4 Coefficients of expansion per degree Celsius for some common substances

Zinc	0.000029
Lead	0.000029
Aluminium	0.000023
Tin	0.000022
Silver	0.000019
Brass	0.000018
Copper	0.000017
Steel	0.000013
Cast iron	0.000011
Glass	0.000009

FOR STUDENT RESEARCH

- Australian Standard. AS/NZS 3500: 2003 Glossary of Terms, Plumbing and Drainage

PLUMBER PROFILE 1.1

DAVID MARSH

Job Title: Great Southern Divisional Manager, Active Plumbing Pty Ltd, WA

David has more than 30 years' experience in the plumbing industry, with expertise in managing a plumbing division that services vast areas in southern Western Australia. He is particularly passionate about the development of apprentice plumbers, and his training has produced both the winner and third place apprentice in the MPA Apprentice of the Year awards in 2010 for the state of Western Australia.

How did you begin your career as a plumber?

I did my apprenticeship at Swan Districts Hospital, starting with drainage work and then moved into the cottage industry. I moved into contracting and eventually went from a senior estimator to assistant manager, and then into the managerial role I am in today.

Why did you decide to move into a managerial role?

I was sick of people cutting my pay! So I decided to grapple on the other side of the fence and worked my way up in the office.

What does a typical day at work look like?

I run a division in Albany in the great southern Western Australia and I manage about 35 people. My division is 400km south of Perth. I have to deal with different staff issues, contractors, quoting, dealing with accounts, all those sorts of things.

Is your job mainly office-based or do you travel as well?

I travel every second week. It's just part of the job. I have a young family so it's tough to balance that but I'm doing the job for my family as well, to put them in a better position.

What do you most enjoy about your job?

I like bringing young people into the industry and I like seeing apprentices succeeding in their trade, that's my favourite bit—getting these kids through their trade!

What is the biggest challenge you have faced over the course of your career?

The biggest challenge here in Western Australia is holding on to our trades and not losing them to the mining industry. The difference is about $80 000—that's how much more they can get up there as opposed to being a plumber working for a company. My favourite story is, we had this 20-year-old kid who we lost to the mining industry, who goes from a $50 000 job to a $140 000 job. Two weeks on, one week off, accommodation paid for and a maid coming in twice a week to clean the room. Insane isn't it? We can't compete at all.

How do you try to motivate the apprentice plumbers who work with you?

We try different things—we have an apprentice program and we are going to take all our people to Melbourne for a weekend to watch the footy. That's one of the things we do. We always try and do fun things, like have BBQs and try to create a good atmosphere, and get people to join not just the job but the lifestyle we've got down here. Up north it's all red dirt and rocks, but down here in the south coast of Western Australia it's all beautiful countryside and a beautiful place to live. It's a great lifestyle.

What are the biggest jobs you have worked on?

We have just finished off a sports complex, I have done 16 units with four stories and we are about to undertake a $700 000 job.

What was the most memorable experience you have had in your career?

Last year we took out the Apprentice of the Year for the state of Western Australia. Definitely my most memorable moment, as I've had five goes at that and never even placed, and then we took out both first and third. It was an absolutely awesome result.

How did you get involved in doing charity work at your company?

I couldn't live with myself if my own children didn't have the specialised equipment needed if they ever got ill, or any of our people who don't have the facilities to help their own kids. It has to be more than money; you have to give back to the community. We are currently doing a charity house which is called the Telethon Trek Home. We do all of that for free and our suppliers contribute with us. All the money that is made out of selling the charity home gets put back towards areas of need for the kids down here, for example, specialised hospital equipment. We also did the Albany Hospice House and the Royal Flying Doctors Service Charity Home.

Any advice to future plumbers?

Study hard and be the best you can be. The better they are at the job, the better they are going to get paid. They have to be committed to their trade. We are looking for apprentices who have a good head on their shoulders, good core values and who are hard workers.

CHAPTER **TWO**

Basic graphics

LEARNING OBJECTIVES

In this chapter you will learn about:

2.1 drawing equipment and layout
2.2 geometric concepts
2.3 orthogonal projection
2.4 isometric drawing
2.5 how to read and interpret plans
2.6 pattern development.

INTRODUCTION

Some of the most valuable skills in industry today are those used in communication. For any industry to be effective and economically viable it is vital that messages between architects, tradespeople, drafters, managers and customers are as clear and concise as possible. Professionally-prepared mechanical and building drawings provide a means of communication between designers and fabricators. Plumbers need to be able to interpret these drawings to obtain such information as the shape and dimensions of equipment.

This chapter is intended to serve as an introduction to the basic elements of drafting. It is by no means comprehensive, aiming only to provide the plumber with basic skills and understanding. It is divided into two sections: geometric concepts and building drawings.

Many principles are common to all graphic drawings. While these should always be followed, variations in skill and approach will permit an experienced person to develop an individual style. Figure 2.1 shows an example of the type of drawing required so that fabrication can take place.

FIG 2.1 Pipe detail: a typical technical drawing

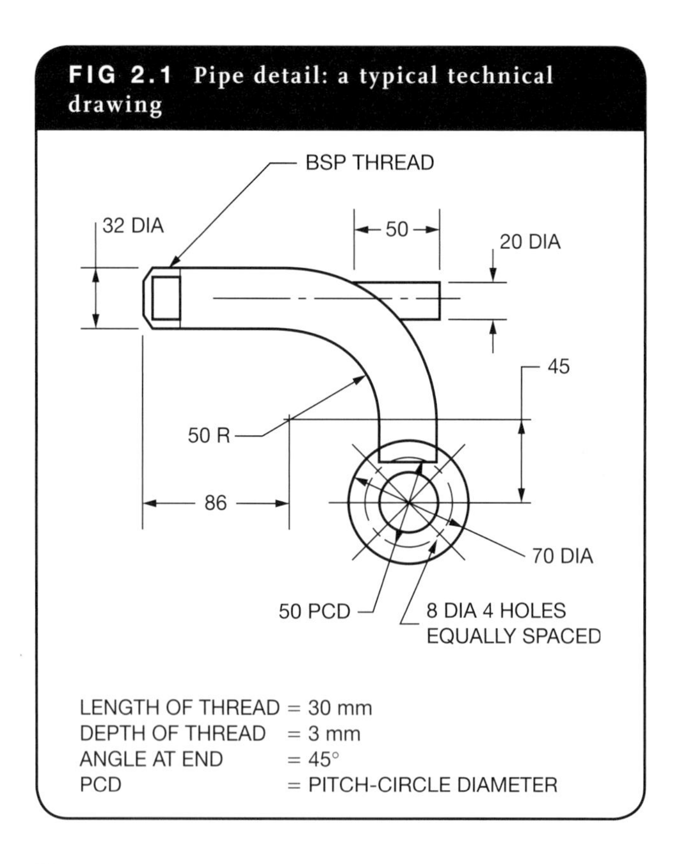

EQUIPMENT

Pencils

A medium hard pencil such as a 3H or 4H will produce light lines suitable for the initial development or construction work. For outlines, lettering and dimensioning a softer pencil such as an HB or 2B will provide a darker result, contrasting with the previous lines.

Scale ruler

All technical and building drawing measurements should be shown in millimetres. Using metric scale rulers will save a lot of time when drawings are being produced on a reduced or enlarged scale. Some common scales are 1:20, 1:50, 1:100.

Eraser

There are many erasers available today capable of producing excellent results for most applications. Most stationery retailers will be able to recommend a suitable eraser to suit individual requirements.

Drawing set

While not essential, the ancillary equipment often included in drawing sets makes their purchase desirable and permits a greater range of applications for the divider and compass. A compass is essential and can be purchased separately if the greater expense of a complete drawing set is not justified.

Drawing boards

These vary in size, but should be large enough to accommodate the drawing sheet to be used. The left-hand and right-hand edges of the board must be perfectly straight to allow accuracy when using a T square, and the board's surface should be smooth and flat. A backing sheet should always be placed under the drawing sheet and both sheets should be held in place with masking tape or fastening clips. The use of drawing pins to secure drawing sheets is not recommended.

T square

An accurate T square with a transparent working edge provides a good basis for any technical drawing. The T square can be used to produce horizontal, parallel lines, or by placing a set square against the working edge, accurate vertical or oblique lines. The plastic edge should be protected against damage at all times as every indentation will show up on the drawing sheet.

Set squares

The two most commonly used set squares are the 30° × 60° and the 45°. The clear plastic type with dimensions printed on the surface is the most suitable. Set square edges can easily be damaged so they require care both while being used and in storage. Regular cleaning will eliminate uneven linework and help to keep the drawing sheet clean. To obtain angles not available by the use of set squares a protractor is used.

Lettering guide

Any drafters who are regularly required to include lettering on drawings should use a lettering guide. This instrument allows parallel guide lines for lettering to be drawn quickly and accurately to a variety of dimensions. Instructions on how to use the instrument will accompany any good lettering guide.

FIG 2.2 Drawing board, T square, set square and drawing sheet held down with masking tape

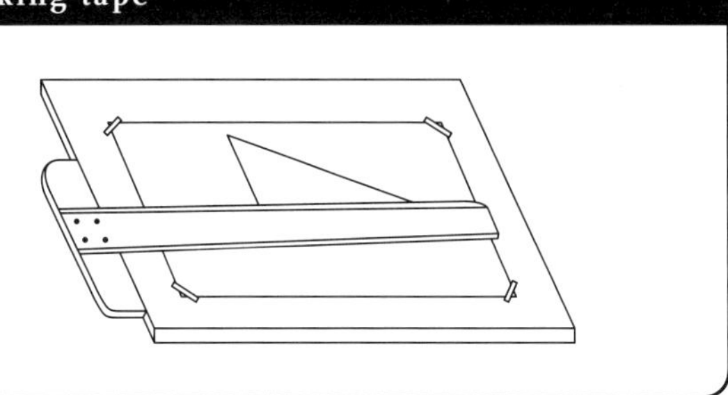

Paper: sheet sizes (rectangular)

Sheet sizes are related to each other. Generally, the smaller sheet measurement is doubled to form the next larger size of sheet. Each sheet will therefore have double the surface area of the next smaller size sheet.

- A4 = 210 mm × 297 mm
- A3 = 297 mm × 420 mm
- A2 = 420 mm × 594 mm
- A1 = 594 mm × 841 mm
- A0 = 841 mm × 1189 mm.

Therefore Al has an area equal to twice the area of A2, or four times the area of A3 or eight times the area of A4.

LAYOUT

Care should be taken with a drawing in relation to the position of the main object, dimensions, lettering and title block. The drawing should be centrally-positioned on the sheet with sufficient room between the drawing and the edge of the sheet to allow for dimensions and lettering to be included. The title block is usually located towards the bottom right-hand corner of a drawing sheet.

FIG 2.3 Drawing sheet showing location of title block with minimum detail

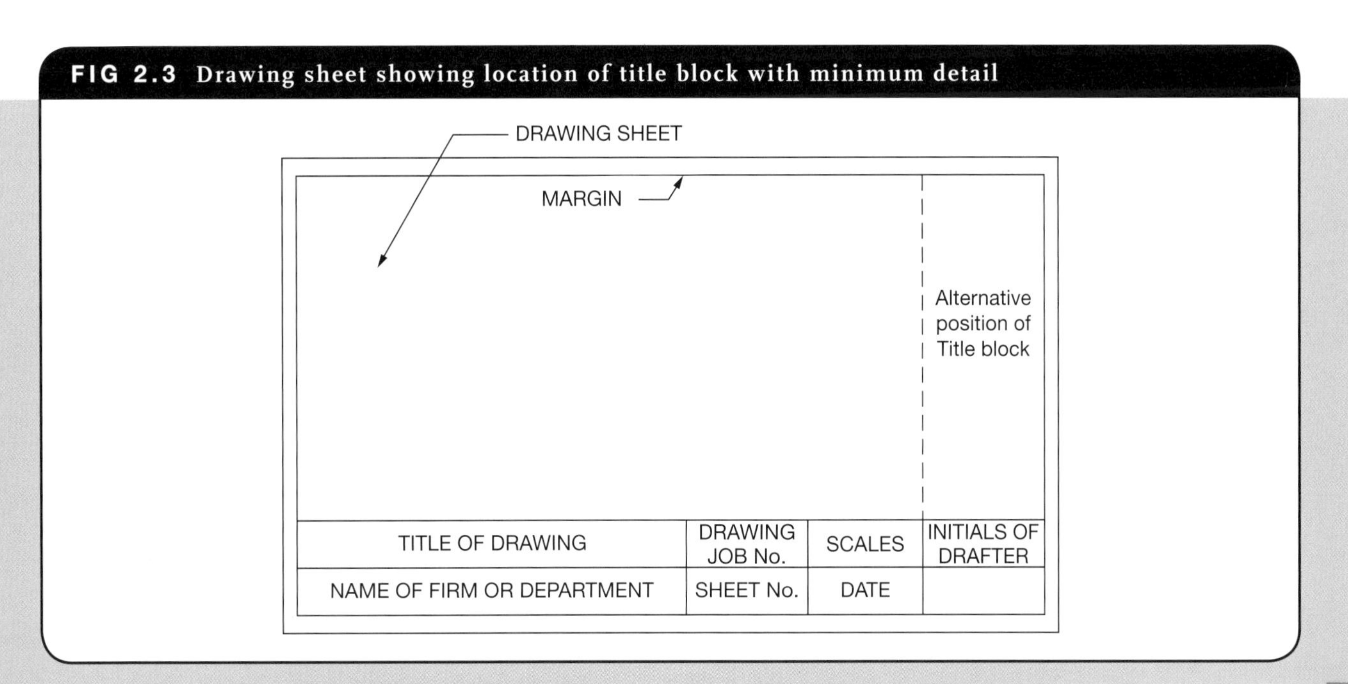

Sometimes the title block may extend right across the base of a drawing sheet or vertically down the right-hand side. The title block should contain the following information:

- title of drawing
- name of firm or department
- drawing job number
- sheet number
- scale
- initials of drafter and date
- date drawn or amended.

GEOMETRIC CONCEPTS

An understanding of basic geometric concepts will provide a good foundation for the development of graphic skills. Familiarity with geometric shapes or forms permit the easy recognition of solid forms from which geometric objects are made or developed.

Plane figures

Triangle: a three-sided figure (see Figure 2.4)

(a) equilateral, all sides equal, all angles equal

(b) isosceles, two sides equal, two angles equal

(c) scalene, no sides equal, no angles equal

(d) right-angled, two sides meet at 90°

(e) acute-angled, all angles less than 90°

(f) obtuse-angled, one angle greater than 90°

Quadrilateral: a four-sided figure (see Figure 2.5)

All quadrilaterals can be divided to form two triangles.

(a) square, all sides equal, all angles 90°

(b) rectangle, opposite sides equal, all angles 90°

(c) parallelogram, opposite sides and angles equal

(d) rhombus, all sides equal, opposite angles equal

(e) irregular quadrilateral, four sides unequal, four angles unequal

(f) trapezium, only two sides parallel

FIG 2.4 Triangles

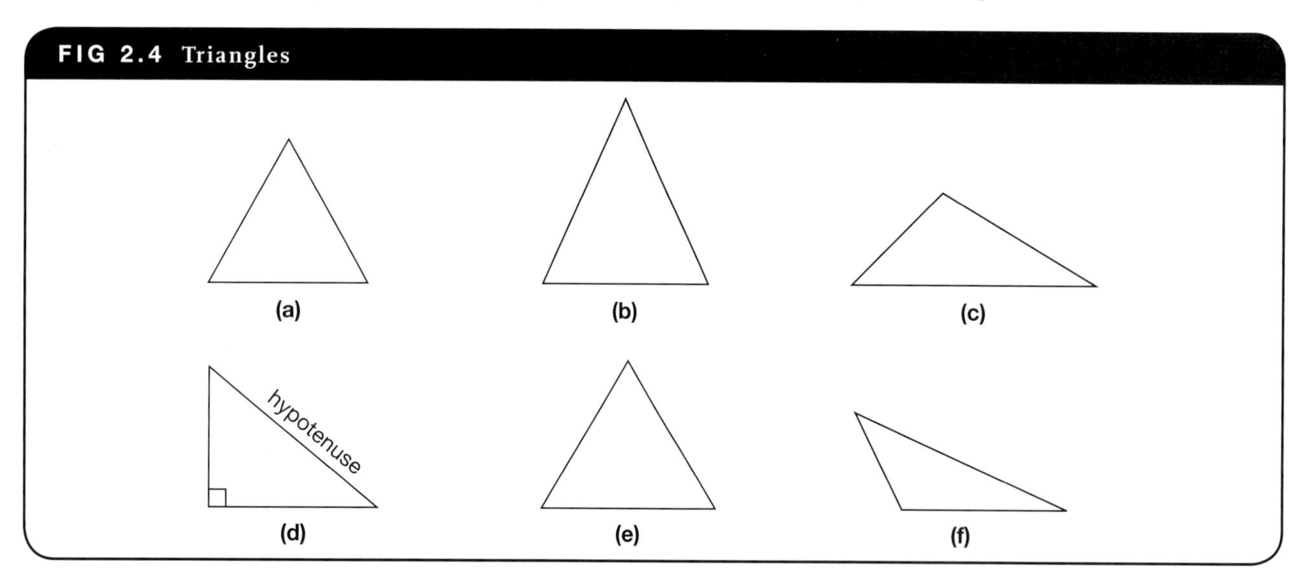

FIG 2.5 Quadrilaterals

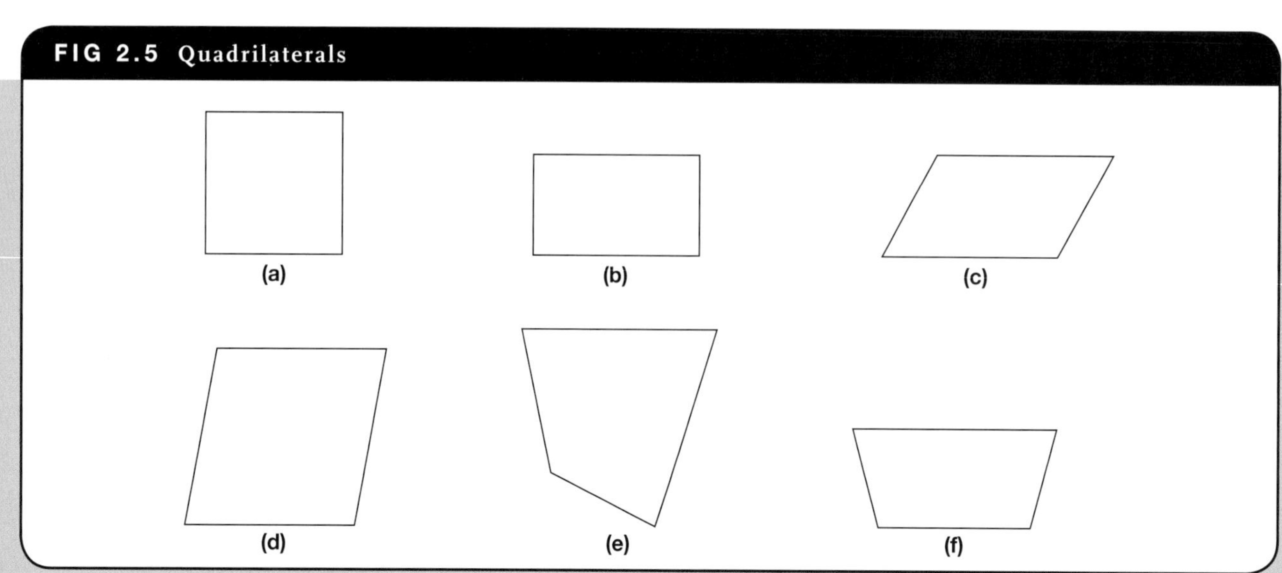

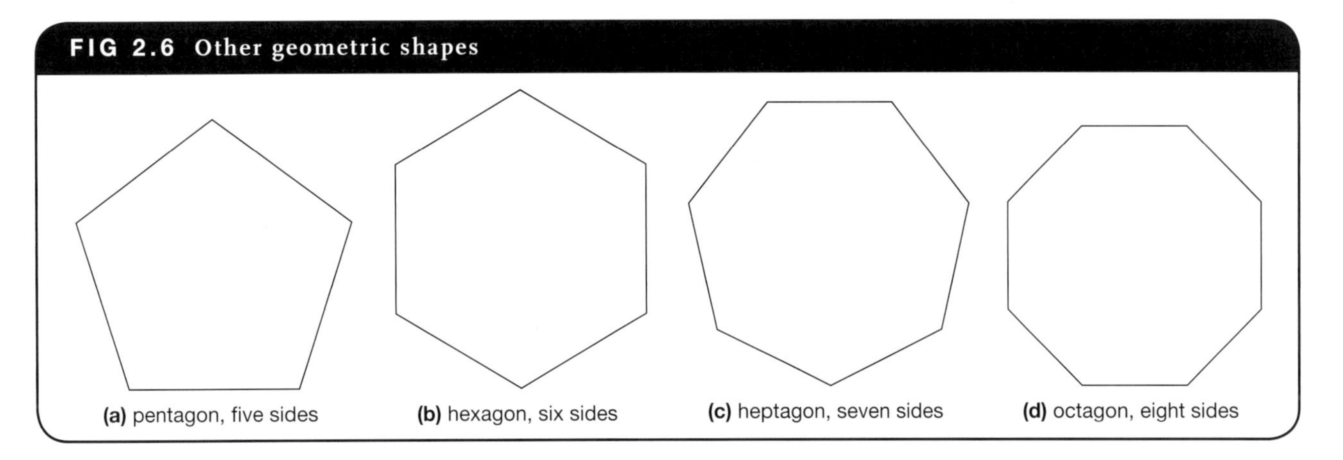

FIG 2.6 Other geometric shapes

Polygon

Polygons are those figures which have more sides than quadrilaterals. Regular polygons have angles and sides which are equal. Irregular polygons have one or more angles or sides which are unequal.

Circle

A circle is a perfectly round figure on a flat plane.

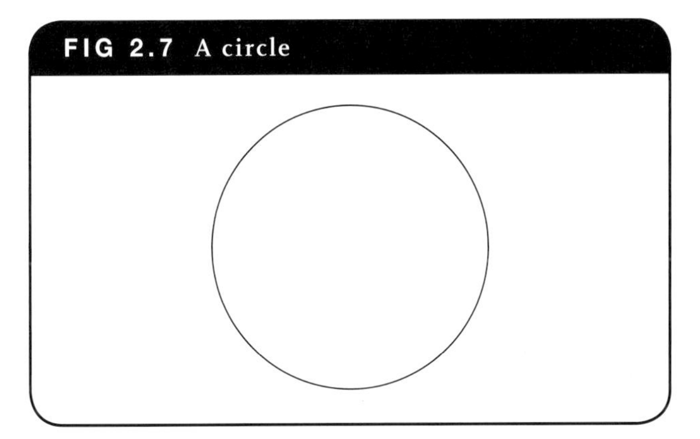

FIG 2.7 A circle

Tangents and other circle parts

A tangent is a straight line which touches the perimeter of a circle but does not enter the area of that circle. A chord is a line segment linking any two points on a circle.

Arc

An arc is part of the perimeter of a circle or part of a curve.

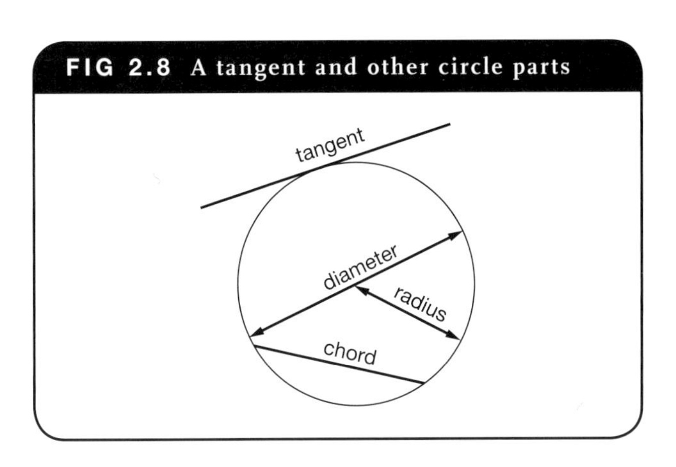

FIG 2.8 A tangent and other circle parts

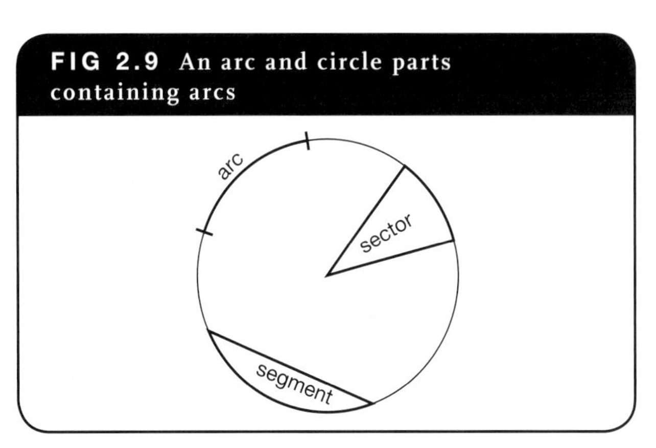

FIG 2.9 An arc and circle parts containing arcs

Concentric circles

Two or more circles having a common centre are concentric.

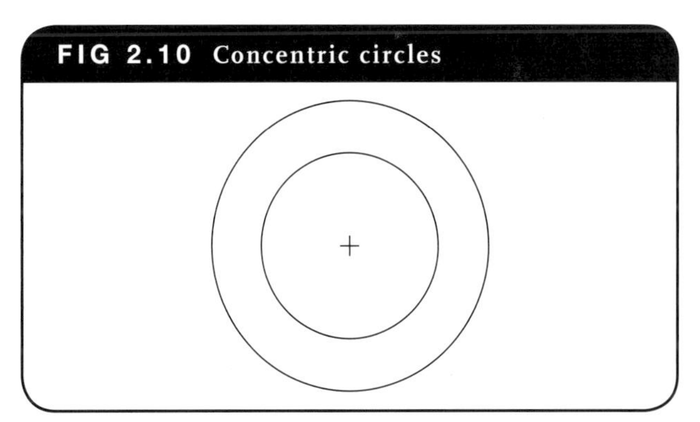

FIG 2.10 Concentric circles

Eccentric circles

Circles within each other but not having a common centre are eccentric.

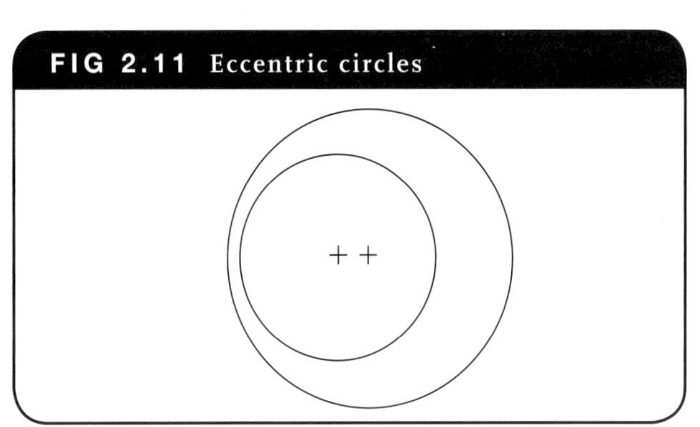

FIG 2.11 Eccentric circles

Ellipse

An ellipse is an elongated circle on a flat plane—a regular oval shape.

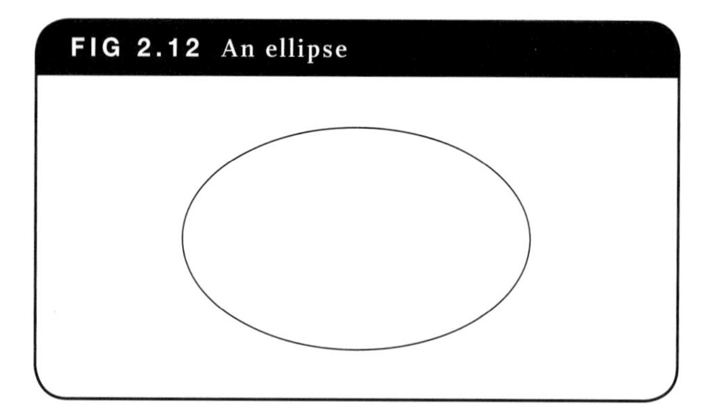

FIG 2.12 An ellipse

Solid forms

A solid is a three-dimensional object. It is from these forms that specific shapes are made.

Cube

This solid has six equal sides which are at right angles to each adjoining side. It can also be called a regular hexahedron.

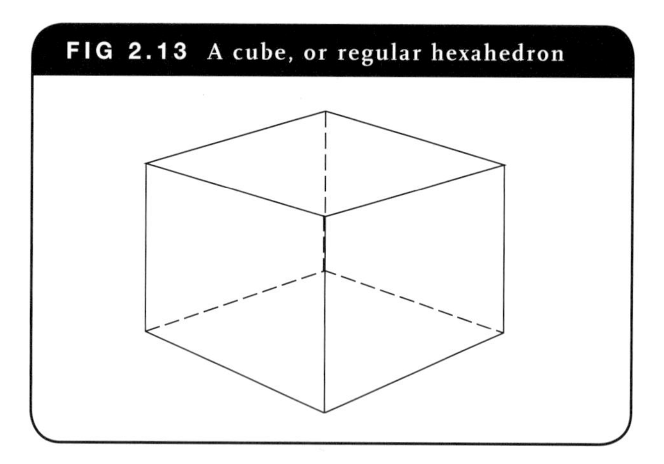

FIG 2.13 A cube, or regular hexahedron

Rectangular prism

This solid has four equal and flat rectangular sides and two equal and flat square ends.

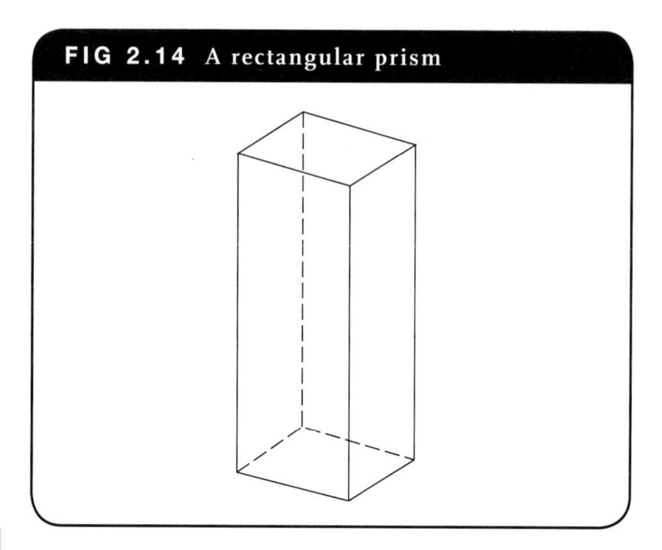

FIG 2.14 A rectangular prism

Triangular prism

This solid has three flat rectangular sides and two flat triangular ends.

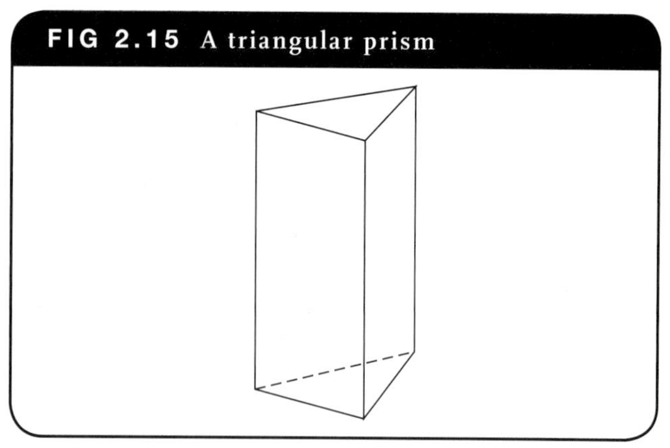

FIG 2.15 A triangular prism

Pyramids

Pyramids are classified according to the shape of their base. A square pyramid is a solid with a flat square base and four flat triangular sides of equal area which meet at a point above the centre of the base. A triangular pyramid is a solid with a flat triangular base and three flat triangular sides of equal area which meet at a point above the centre of the base.

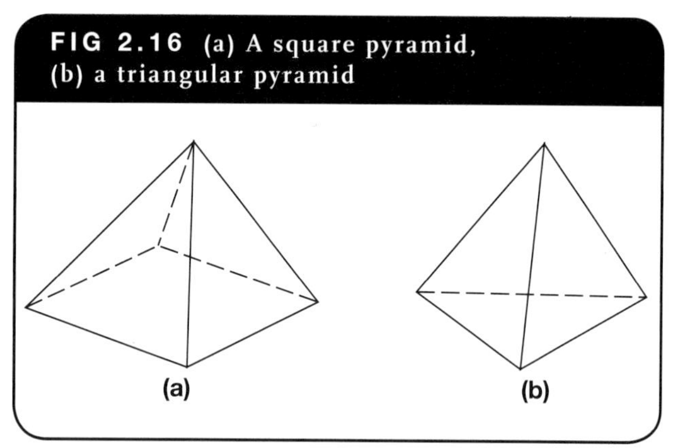

FIG 2.16 (a) A square pyramid, (b) a triangular pyramid

Cylinder

This is a roller-shaped form of constant diameter.

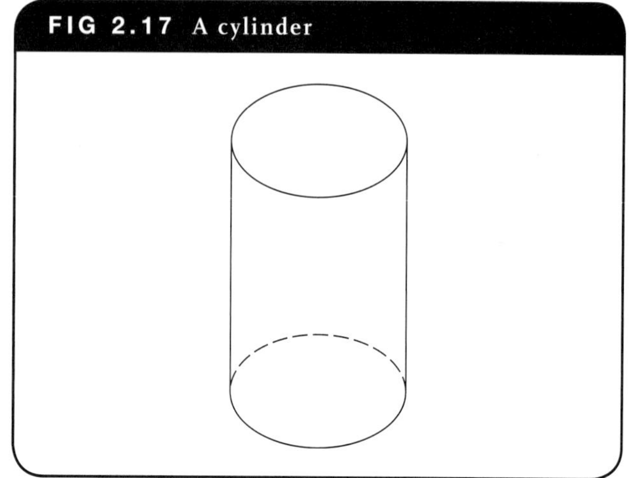

FIG 2.17 A cylinder

Cone

This solid has a flat circular base and a curved side rising to a point above the centre of the base.

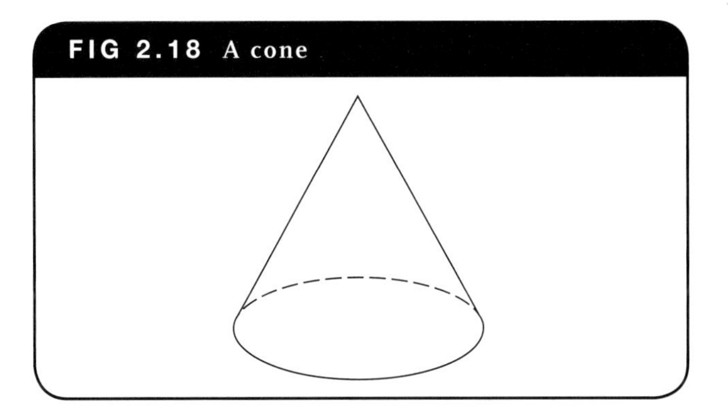

FIG 2.18 A cone

Sphere

This is a circular solid of constant diameter. All diameters taken from any point on the surface and passing through the centre are equal.

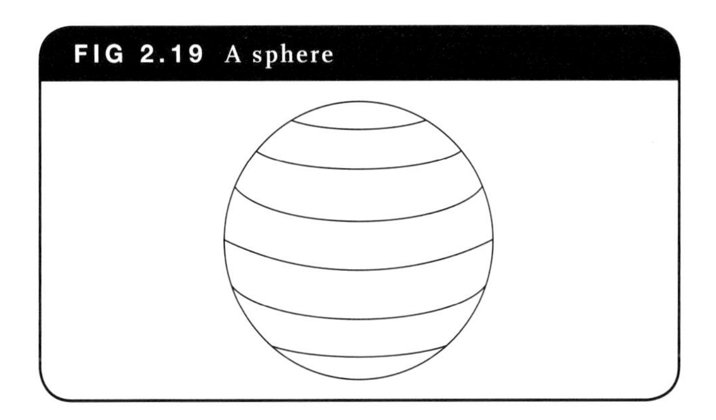

FIG 2.19 A sphere

Orthogonal or orthographic projection

The drawing of an object can be achieved in many different ways. A single view may show three dimensions, or several views may be drawn, showing the dimensions as seen from various positions.

If we viewed a cube from a position where only two dimensions were visible, we would be viewing the cube perpendicularly (at right angles) to the plane or surface which is visible. The planes adjacent to the surface we are viewing would also be perpendicular to the visible plane. As we draw each visible plane, we are drawing an orthogonal projection—that is, a view that is perpendicular to the visible surface plane, but which only shows two dimensions of an object.

Of the six views available, only three are usually employed. These are the top side, the front side and the right side views. This method of drawing an object is known as 'orthogonal projection'. 'Third angle projection' (Figure 2.22) is recommended as the standard method of orthogonal projection. In this method of drawing, the object is placed below or behind the plane on which the drawing is produced.

Isometric drawing

An isometric drawing (Figure 2.23) shows three sides of an object with lines being drawn to actual length or to an even scale. A T square and a 30° × 60° set square are required for isometric work. The base or axis lines are drawn at 30° to the horizontal while vertical lines remain the same. Lines which are not parallel to vertical lines or axis lines are called 'non-isometric' lines. Isometric drawings are not confined to rectangular objects. The flat circular end of an object will

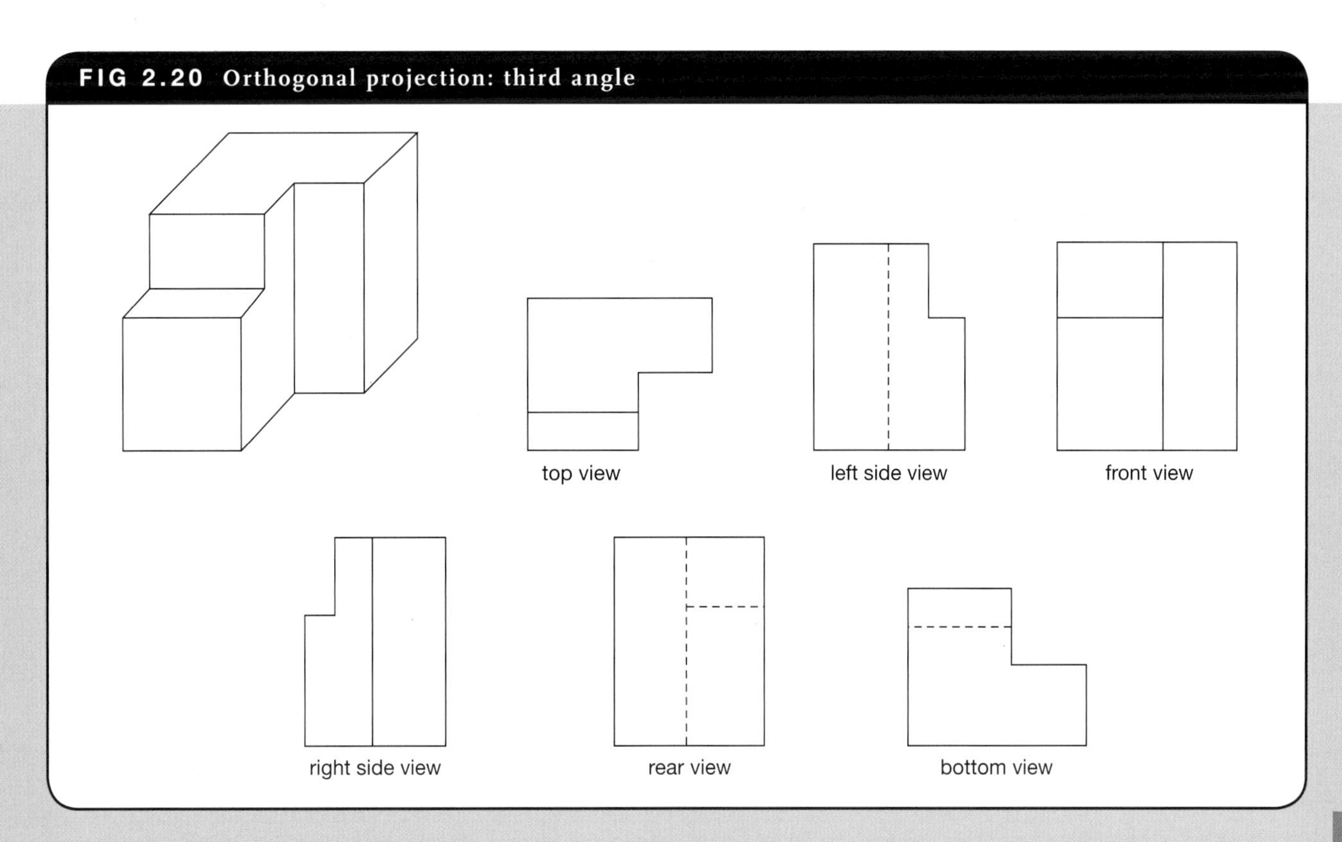

FIG 2.20 Orthogonal projection: third angle

FIG 2.21 Designation of views: (a) a cube showing two dimensions, (b) the same cube showing three dimensions, (c) six views of a cube

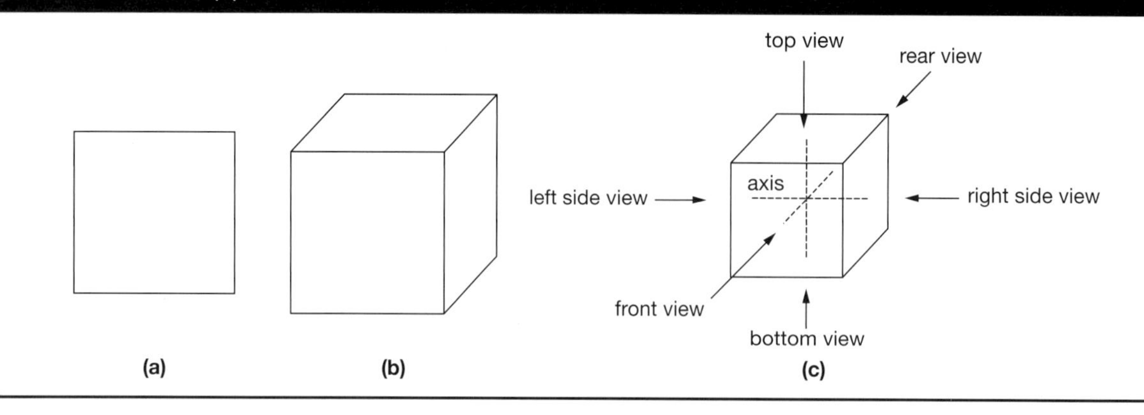

FIG 2.22 Orthogonal views: (a) objects positioned in a quadrant are projected perpendicular to an adjacent plane, (b) produced from third of angle projection

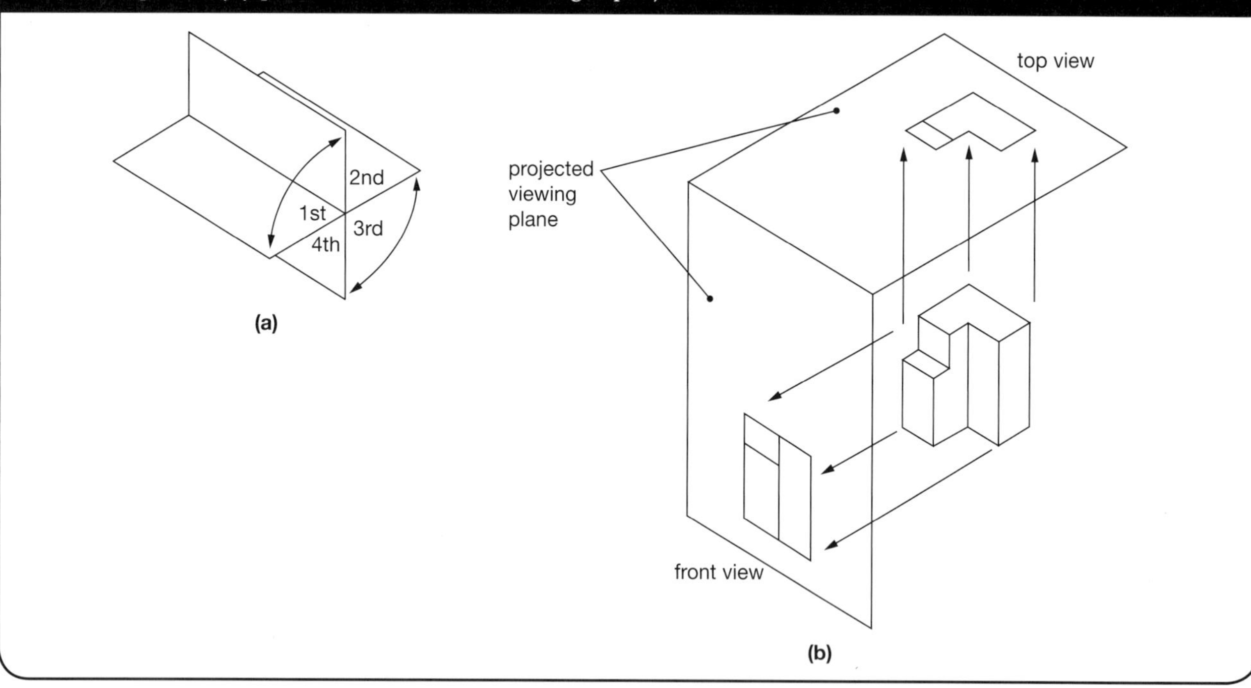

FIG 2.23 An object drawn in isometric projection

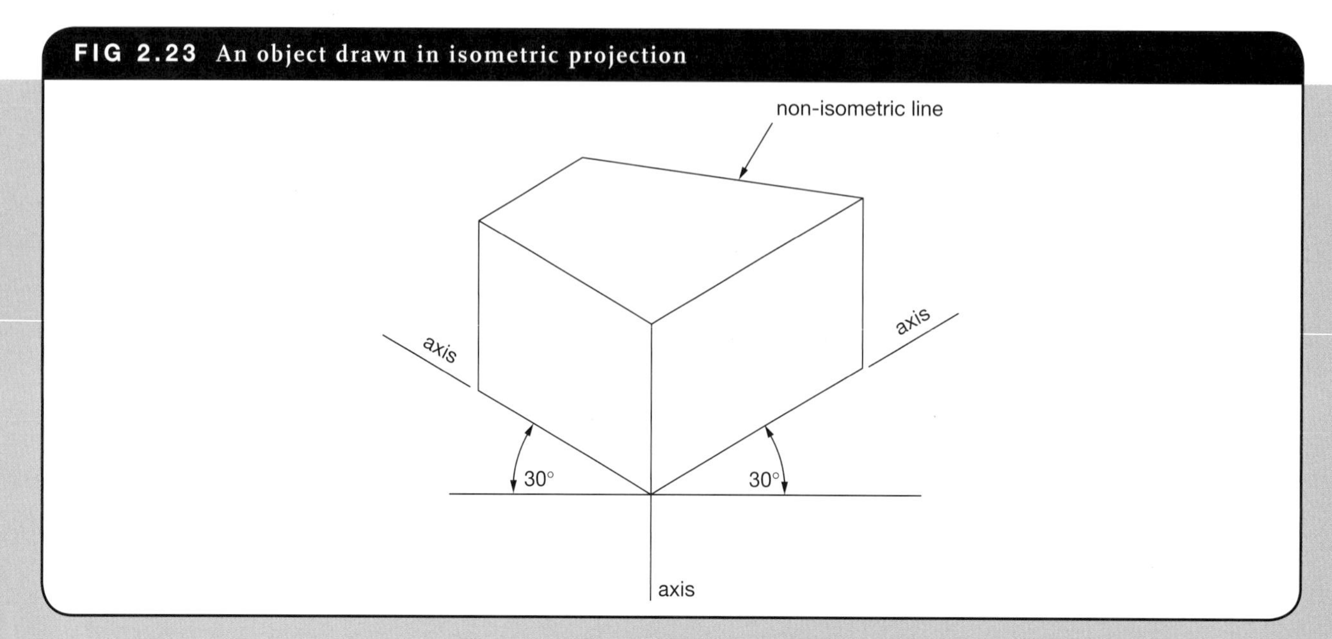

appear as an ellipse when viewed or drawn in isometric projection, as shown in Figure 2.24.

Pipelines are usually drawn before construction of a building for design purposes. Using an isometric drawing, the layout of a pipe system becomes immediately apparent. More accurate estimating of materials, location of penetration points and determination of levels and dimensions can all be obtained from an isometric drawing (Figures 2.25(a) and (b)).

FIG 2.24 Ellipse formed from isometric construction, using a compass to swing

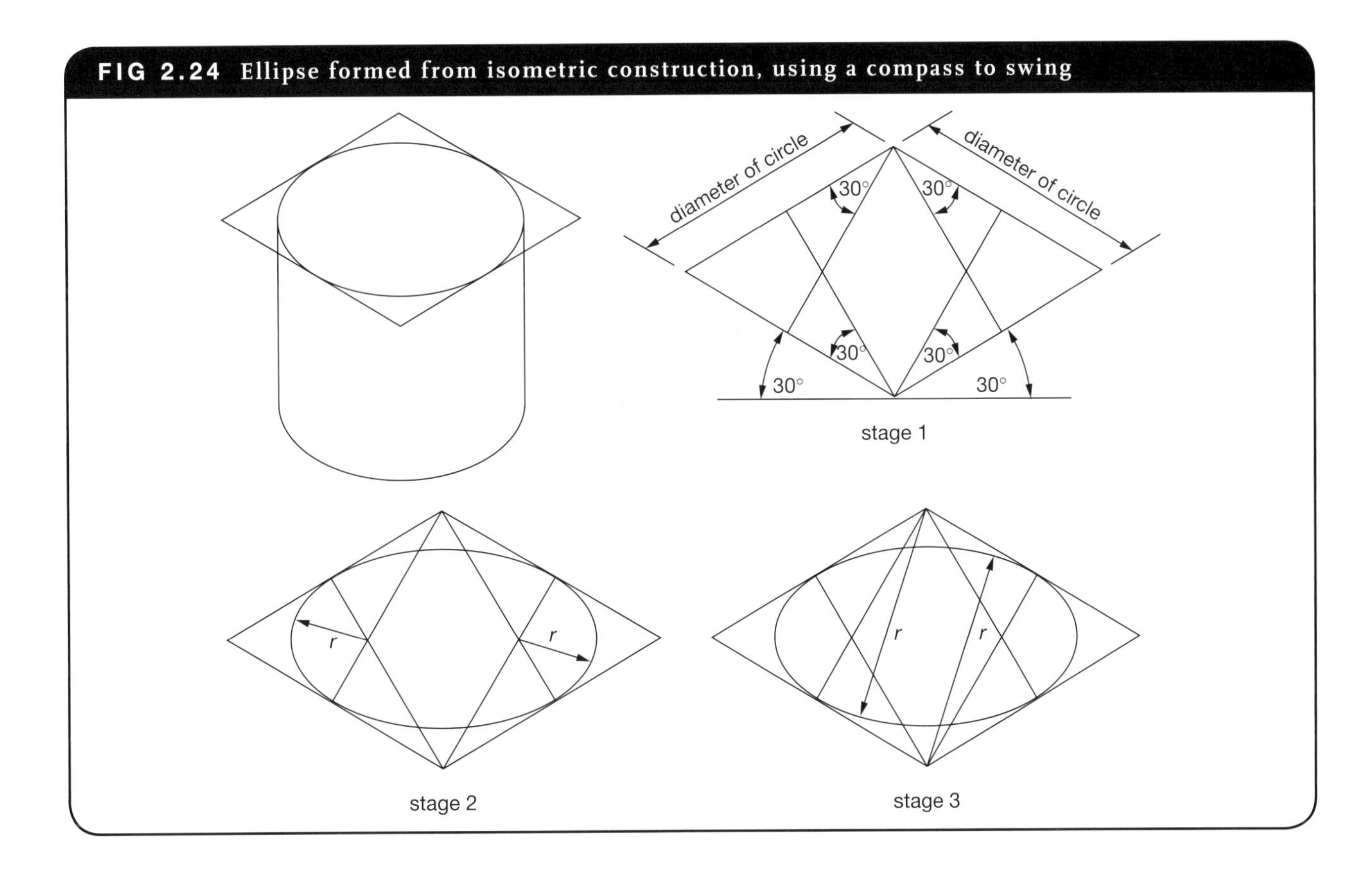

FIG 2.25 Isometric pipelines: (a) a simple system, (b) a more detailed isometric pipeline system

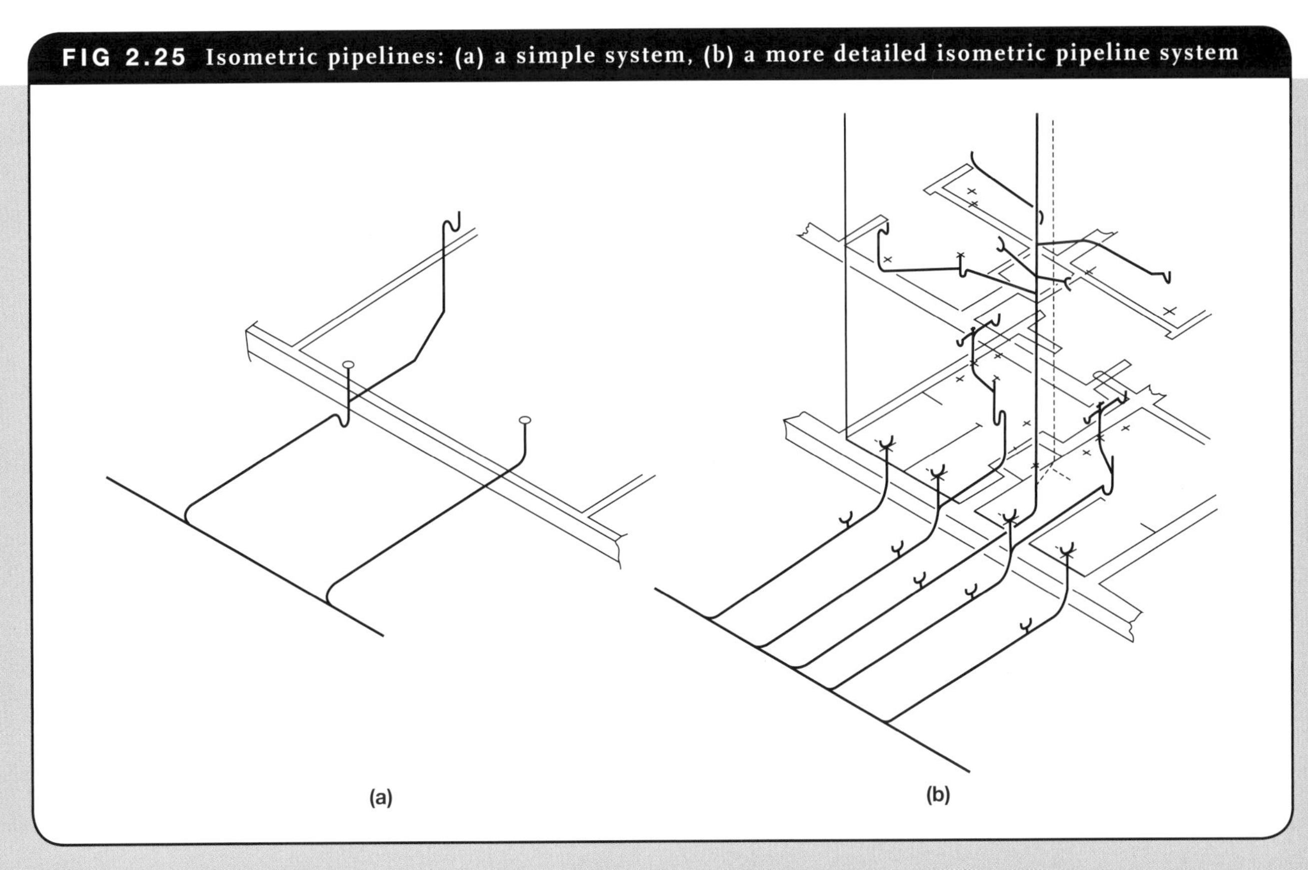

BUILDING DRAWINGS

As trainees progress in the plumbing industry, they should not only be able to interpret a building plan but should be able to add to or develop plans associated with plumbing work. Uniformity of drawing standards and techniques enables all personnel likely to use a plan to obtain accurate information easily.

In this section we will cover how to:

- identify the types of plans and drawings
- understand the function of plans and drawings
- identify commonly used symbols and abbreviations.

Drawings are required by the following:

- owners of the proposed building
- local authorities
- water and sewerage authorities
- engineers, mechanical, electrical, structural and plumbing consultants
- financiers
- costing estimators
- project officers
- trades people and subcontractors.

Types of drawings and their functions

In the construction industry there are two types of drawings used:

1. Pictorial drawings, which are used mainly by architects and designers to determine the completed appearance of the project
2. Working drawings that give an overhead view and instruction for the completed project. Working drawings enable estimators, builders and sub-contractors to perform their respective tasks. Most working drawings have a standard layout of plans, elevations and detail sections. This helps the people who need the information and where to look.

We will focus on the uses of working drawings here. A set of working drawings for the construction of a building consist of some of the following:

- site plan
- floor plan
- elevations
- sections.

General requirements for working drawings are that they are drawn to scale of 1:100 except sectional details, which may be drawn to 1:5, 1:10 or 1:20. Site plans to 1:500 and 1:200 are sometimes preferred by local authorities.

Site plan

The site plan should show the outline, location and dimensions of existing and proposed buildings in relation to site boundaries and to the street. It is an overhead view of the construction site and the home as it sits in reference to the boundaries of the allotment. Site plans should outline location of the utility services, retaining walls, setback requirements, easements, fences, location of driveways and walkways, and sometimes even topographical data (contours) that specifies the slope of the terrain (Figure 2.26).

Note: North should always be shown (Figure 2.27).

FIG 2.26 Contour lines

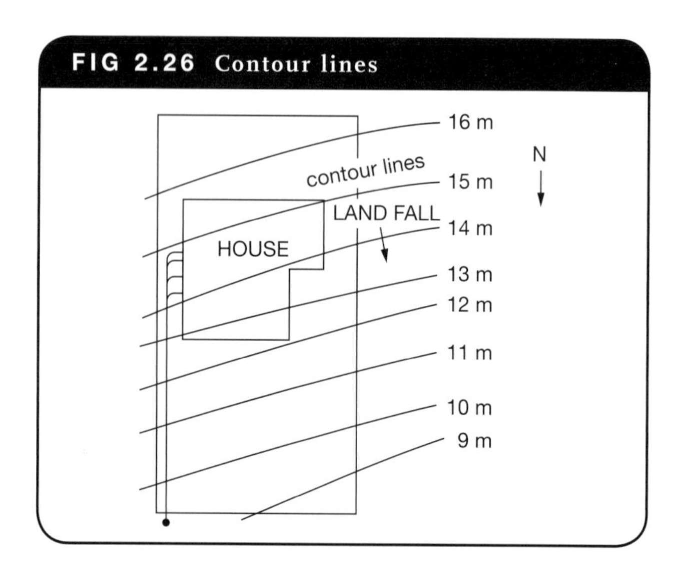

Floor plan

The 'plan' is viewed from above and is a horizontal section of a building. Plan views normally show wall thickness, rooms, windows, doorways, major fixtures and roof outlines. Floor plans also include details of fixtures like sinks, water heaters, gas appliances and so on. Floor plans include notes to specify finishes, construction methods and symbols for electrical items (see Figure 2.28).

Other information can be shown, but too much information on one plan results in a confusion of linework.

FIG 2.27 A site plan

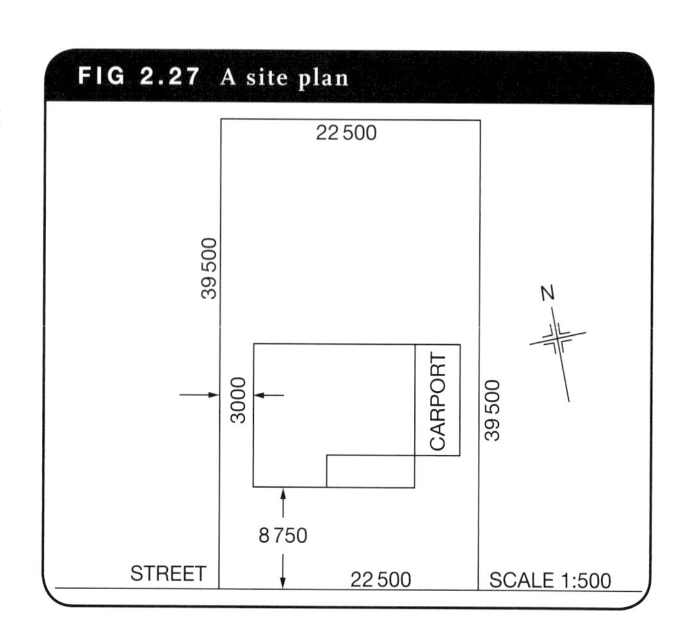

FIG 2.28 Floor plan

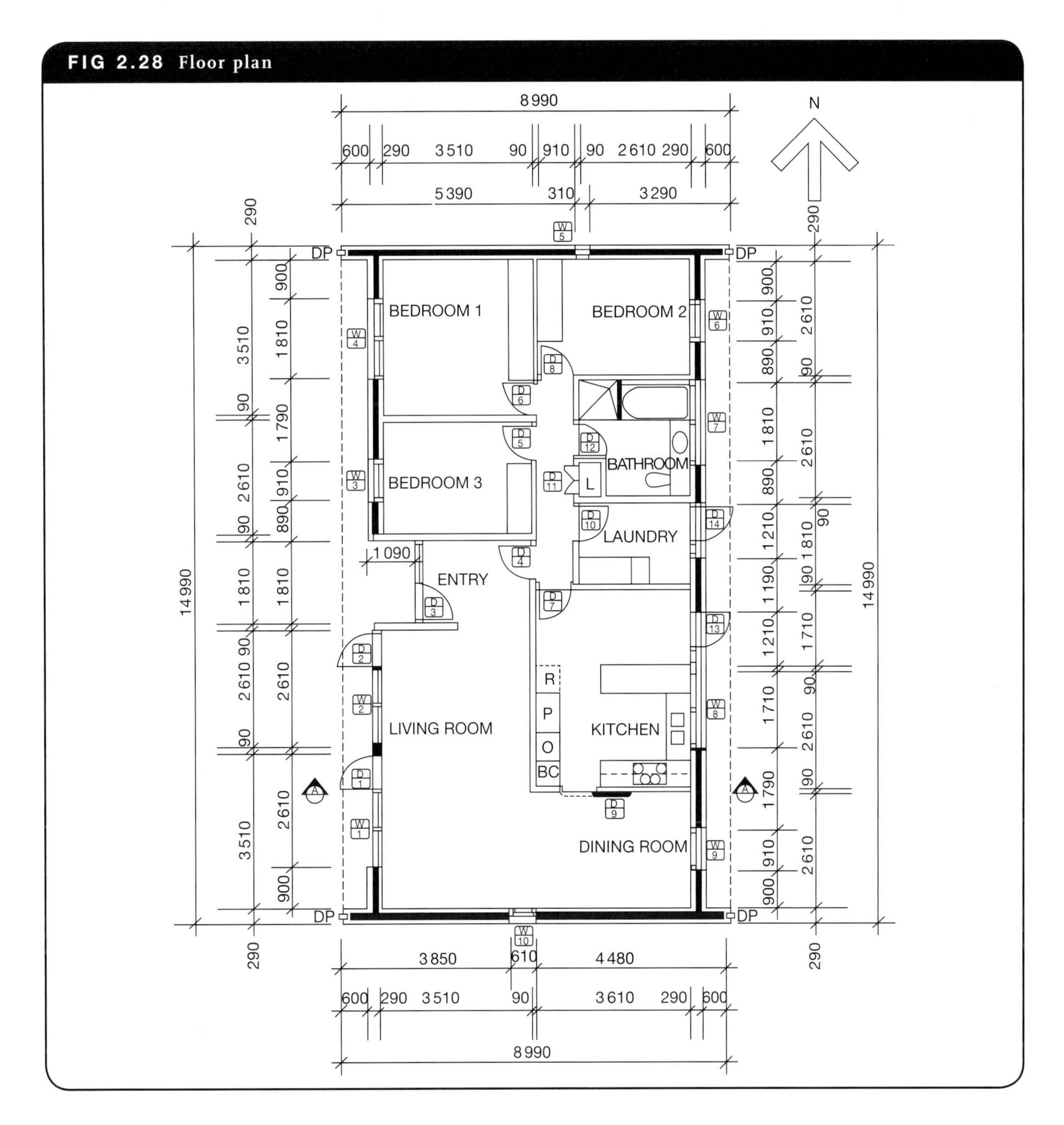

Elevations

Elevation drawings show the external appearance and size of a building as viewed from a position in front of that elevation (Figure 2.29 overleaf). Elevations are named by the pole direction which they face (for example, north elevation faces northwards).

Elevations specify:

- ridge heights
- roof shape and pitch (slope)
- windows and doors to external walls
- external walling (type and finish)
- roof overhang (eaves)
- finished ceiling level (FCL)
- finished floor level (FFL)
- handrail heights to verandahs
- position of meter box
- position of down pipes.

Sections

It is a common drawing practice to 'cut through' a building to show particular features of construction which are not visible from other views (Figure 2.30 overleaf). These plans are called 'sections'; they are usually drawn larger than elevations.

The section to be viewed is indicated on the floor plan by section lines which, by the aid of the arrow heads, indicate the direction in which the section is viewed.

FIG 2.29 Elevations

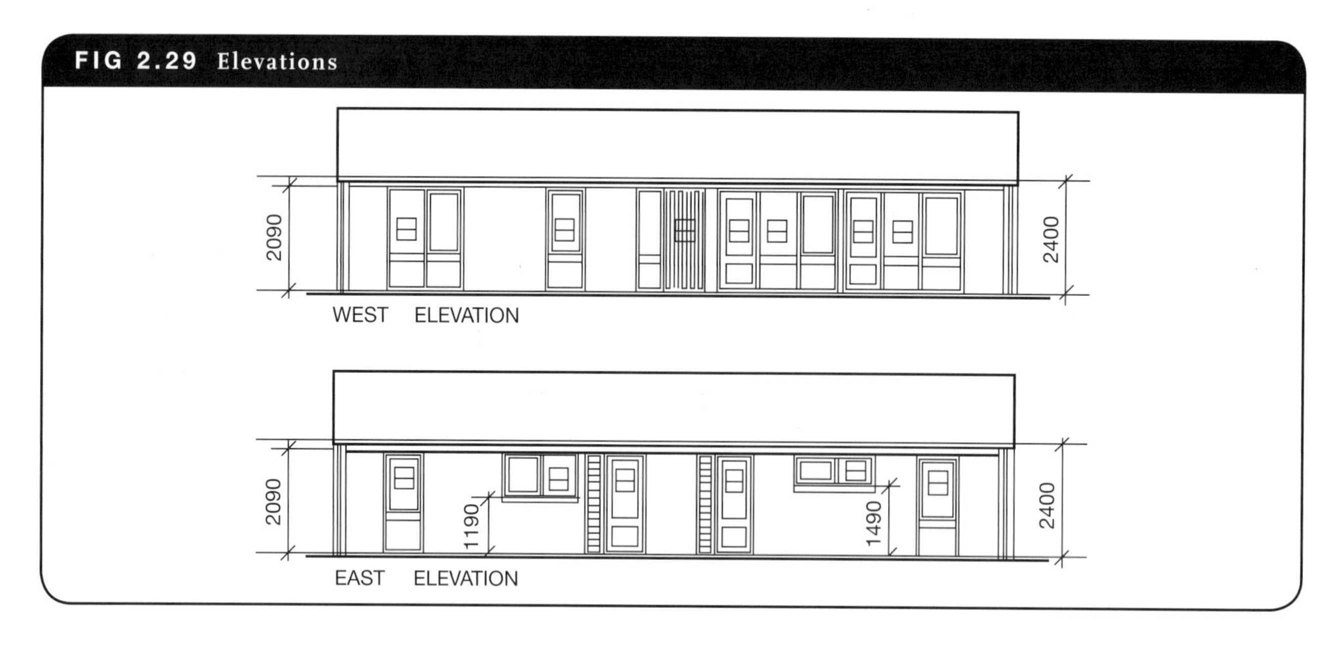

FIG 2.30 A cross-section A–A. See Figure 2.8

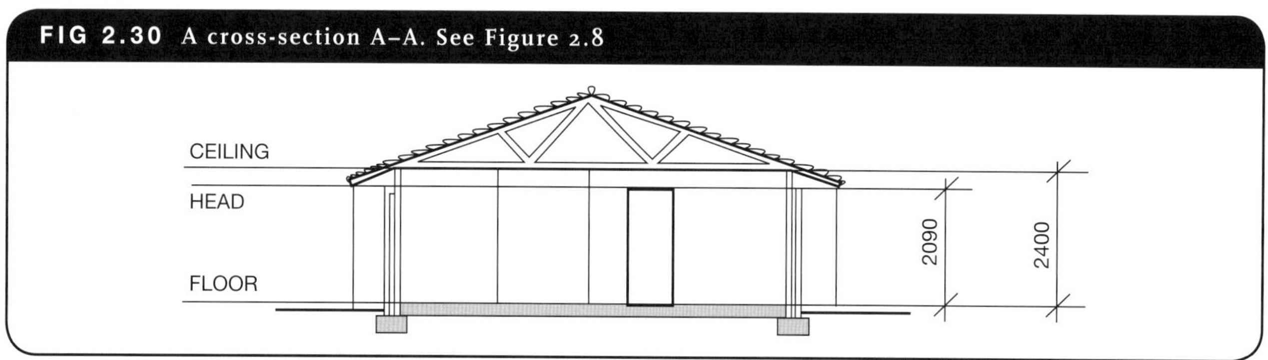

The sectional elevation shows:

- height above ground level (GL)
- ceiling height
- depth of footing
- roof construction and pitch.

Title block

The title block gives a professional appearance to printed plans and provides important document information. Title blocks are usually located at the bottom right-hand corner of the drawing (Figure 2.31).

Title block includes:

- plan name
- designer name
- project title
- building address
- date printed
- drawing number
- page number (when there is a set of drawings)
- signature of approving authority
- date of amendment
- scale.

FIG 2.31 Typical title block

Client: D.C. Brown		
Project: Proposed Child Care Centre		
Location: 21 Drive Road Happy View SA 5000		
Project Consultants: Black and Blue Consultants		
Architect: B.E. Sharp		
Drawn: A.D.	**Checked:** J.B.	**Issue:** A
Scale: 1:100	**Date:** March 2011	
Project No: BCA- 652	**Drawing No.** WD- 03	

It is very important to check the title block to ensure the latest version of the drawing is being used to make sure all amendments are noted.

Scales

Because a plan cannot be of the same size as the building, scale drawings have to be used.

Scaling is a drawing method used to enlarge or reduce a drawing in size while keeping the proportions of the drawing the same and fit them on a manageable, standard-size piece of paper for convenience.

Scales are generally expressed as ratios and the most common scales used in the construction industry for working drawings are:

- site plan 1:500, 1:200
- floor plan 1:100
- elevations 1:100
- sections 1:100
- details 1:5, 1:10, 1:20

Interpreting scales

If a building was drawn to full size, the scale would be represented as 1:1. Given this is not a manageable size, various scales are used to reduce the size of the drawings.

Note: Measurements on an architectural drawing take precedence over a scaled measurement.

Scales are represented like this 1:50

The first figure (1) is the actual size.

The second figure (50) is how many times it has been reduced, in this case 50 times (1/50th the size of the completed building).

Example

If a wall measures 10 mm on the drawing, the actual size is $10 \times 50 = 500$ mm (0.5 m).

If you do not have a scale rule at hand, this method can be used on site with an ordinary ruler or a tape measure.

Reading a scale rule

A scale rule is similar to an ordinary rule in having graduation lines marked at one millimetre intervals, longer lines at the five millimetre marks and longer lines at the ten millimetre marks. The ten millimetre marks also show the graduation number. On a scale rule the graduation number indicates the scaled-up value rather than the true length (Figure 2.32).

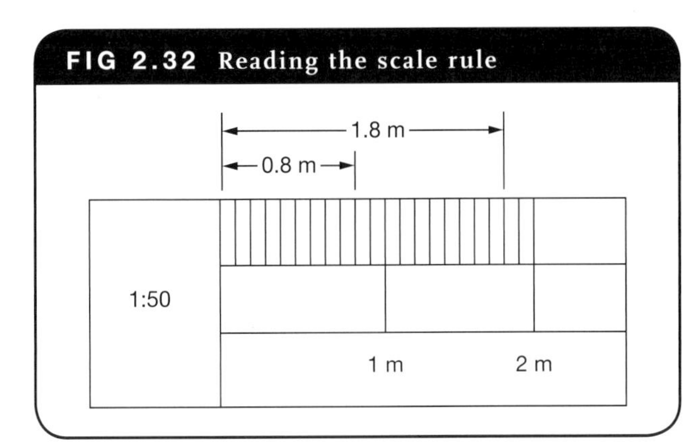

FIG 2.32 Reading the scale rule

Dimensions

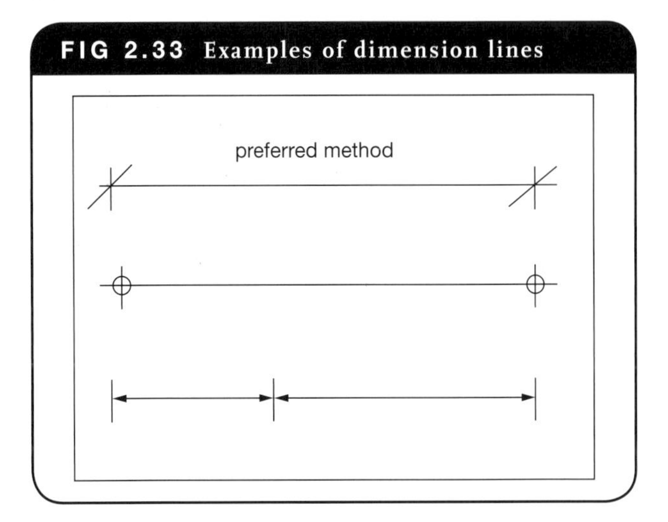

FIG 2.33 Examples of dimension lines

Legends and symbols

The legend is used to explain or define a symbol or special mark placed on a drawing (Figure 2.34).

The important thing is that you understand the meaning of the symbols on the drawing on which you are working.

FIG 2.34 Symbols for plumbing fixtures

Symbol	Fixture
	bath
	water closet
	basin
DF	drinking fountain
	bidet
	urinal
HWT	hot water tank

Construction symbols including hatching

Representations for various materials of construction for use on drawings are specified in national standards and should be adhered to wherever possible (Figure 2.35).

For further student resources, refer to page 155 for additional symbols and abbreviations.

FIG 2.35 Symbols of materials

Material	General location drawings (section) Scale: 1:50 or less	Large scale drawings (section)
brickwork		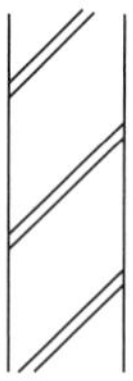
cement render, plaster	too fine to hatch	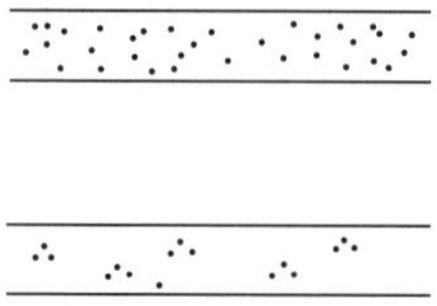
concrete	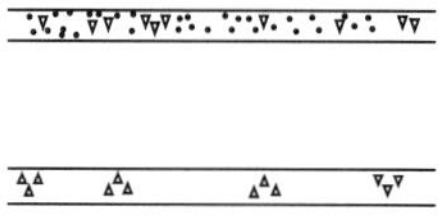	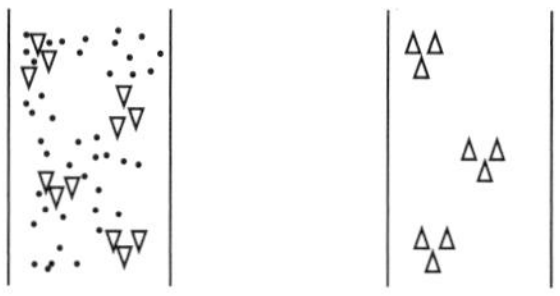
concrete block	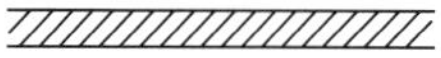	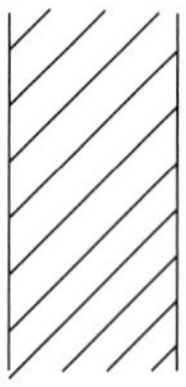
cut stone masonry		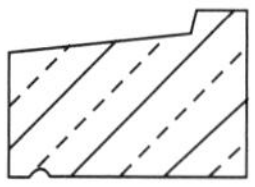
earth		
fill	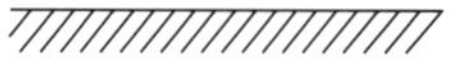	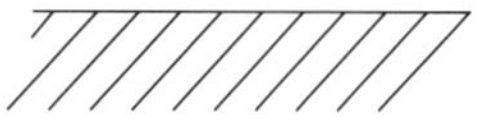
glass	too fine to hatch	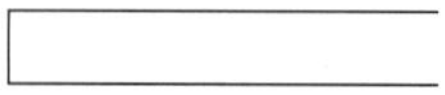
hardcore		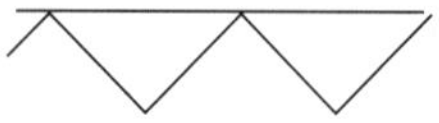
insulation		

FIG 2.35 Symbols of materials (*continued*)

Material	General location drawings (section) Scale: 1:50 or less	Large scale drawings (section)
partition block		
rock		
structural steel		
stud walls	(grey shading)	
timber	usually too fine to hatch	sawn dressed

Specification

A well-drawn construction drawing cannot adequately reveal all the aspects of a construction project. There are many features that cannot be shown graphically. For instance, how can anybody show on a drawing the quality of workmanship required for the installation of the plumbing equipment, or who is responsible for supplying the materials, except by extensive hand-lettered notes? The standard procedure then is to supplement construction drawings with written descriptions. These detailed written instructions, commonly called specifications (or specs), define and limit the materials and fabrication according to the intent of the engineer or the designer. The specifications are an important part of the project because they eliminate possible misinterpretation and ensure positive control of the construction.

A specification should be followed where there is a difference between the plan and the specification. Seek clarification from the designer. Plumbing specifications would include the type of:

- waste drainage system
- gas system
- hot water system
- sanitary fixtures
- tap ware.

Pattern development

Hollow objects such as boxes, cylinders, pipes and cones which are made of thin materials can be formed from flat metals. Plumbers often produce shapes from sheet metals, either on-site or in a workshop. The process of marking out a shape prior to cutting and folding is called 'pattern development'. Three methods are used to develop patterns geometrically:

- Parallel line development is used to mark out the pattern for such shapes as cylinders, boxes and prisms, which have parallel sides.
- Radial line development is used for cone-shaped objects which have sides converging to a single point called the 'apex'.
- Triangulation is used where the other two methods are unsuitable, such as for transition pieces, which may change from a square to a round shape.

Parallel line development

Example

Development of a truncated cylinder pattern (A truncated cylinder is one with part of its end cut off.)

1 Draw, in orthogonal projection, a plan and elevation view of a truncated cylinder (Figure 2.36).

2 To divide the circumference of the circle into equal parts: using the radius of the circle, step off and mark the circumference either side of each of the horizontal and vertical lines which pass through the circle. This will divide the circumference into 12 equal parts.

3 Number each point on the circumference and rule a vertical line through each point up to the top edge of the elevation. Number each point on top of the elevation using the same number used at the base of the vertical line on the circle. The vertical lines are parallel lines.

4 From the base of the elevation produce a horizontal line to the right. On this line, this is the base of the pattern, step off the 12 equal parts forming the circumference of the circle. Number the points from 1 to 12. The last point (point 13) is also the original number 1.

5 Rule a vertical line up from each of the 13 points. From the numbered points on top of the elevation produce horizontal lines across to the same numbered vertical line of the pattern. Connect the points on the pattern with a smooth curved line.

Radial line development

Example

Development of the pattern for a truncated cone

1 Draw the plan and elevation view of a truncated cone, one above the other (Figure 2.37).

2 Divide the circle into 12 equal parts, numbering each point on the circumference from 1 to 12. From each of these points produce a vertical line up to the base of the elevation. From the base of the elevation, the lines are continued up to the apex of the cone.

3 Where each line passes through the truncated part of the cone, produce a horizontal line to the side of the cone. Number each point on the side of the cone.

4 To produce a pattern, open a compass out to the length of the side of the cone, from the apex to the base line of the elevation. Draw an arc and divide the arc into 12 parts equal to the circumference of the base circle.

Number each point and produce a line from the apex of the arc to each number point. True length distances are then taken from the side of the elevation and transferred to the pattern. Number 1, for example, is measured from the apex of the elevation down to number 1 on the side of the cone. This distance is then marked on the pattern on line number 1, from the apex back towards the arc. All points are transferred to the pattern and then joined to form a smooth curve.

Triangulation

Triangulation is the process of determining the length of the third side of a right-angled triangle when given only the length of the other two sides. The unknown length becomes the hypotenuse. Base lines are obtained from the

FIG 2.36 Pattern development of a truncated cylinder using the parallel line method

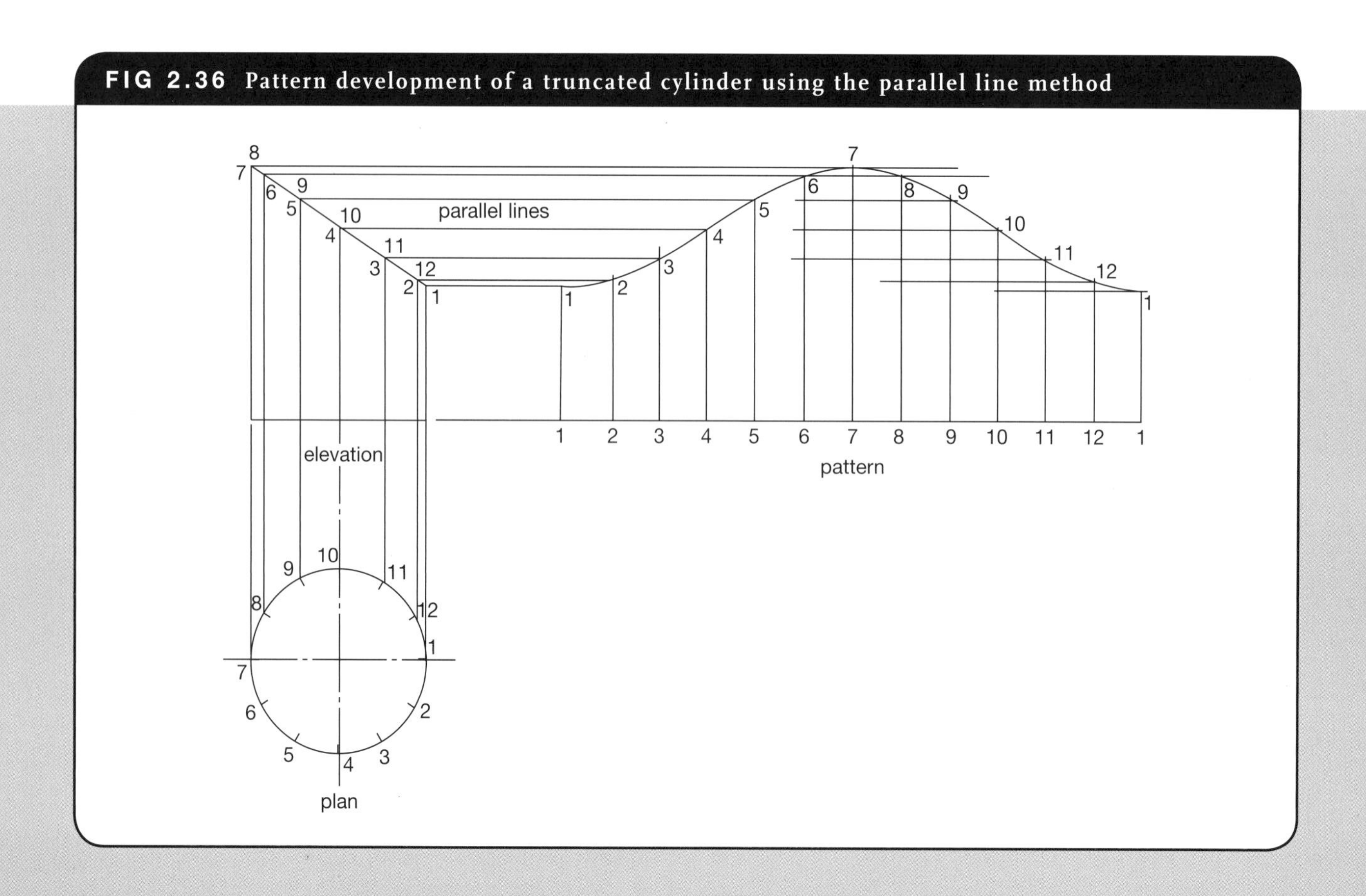

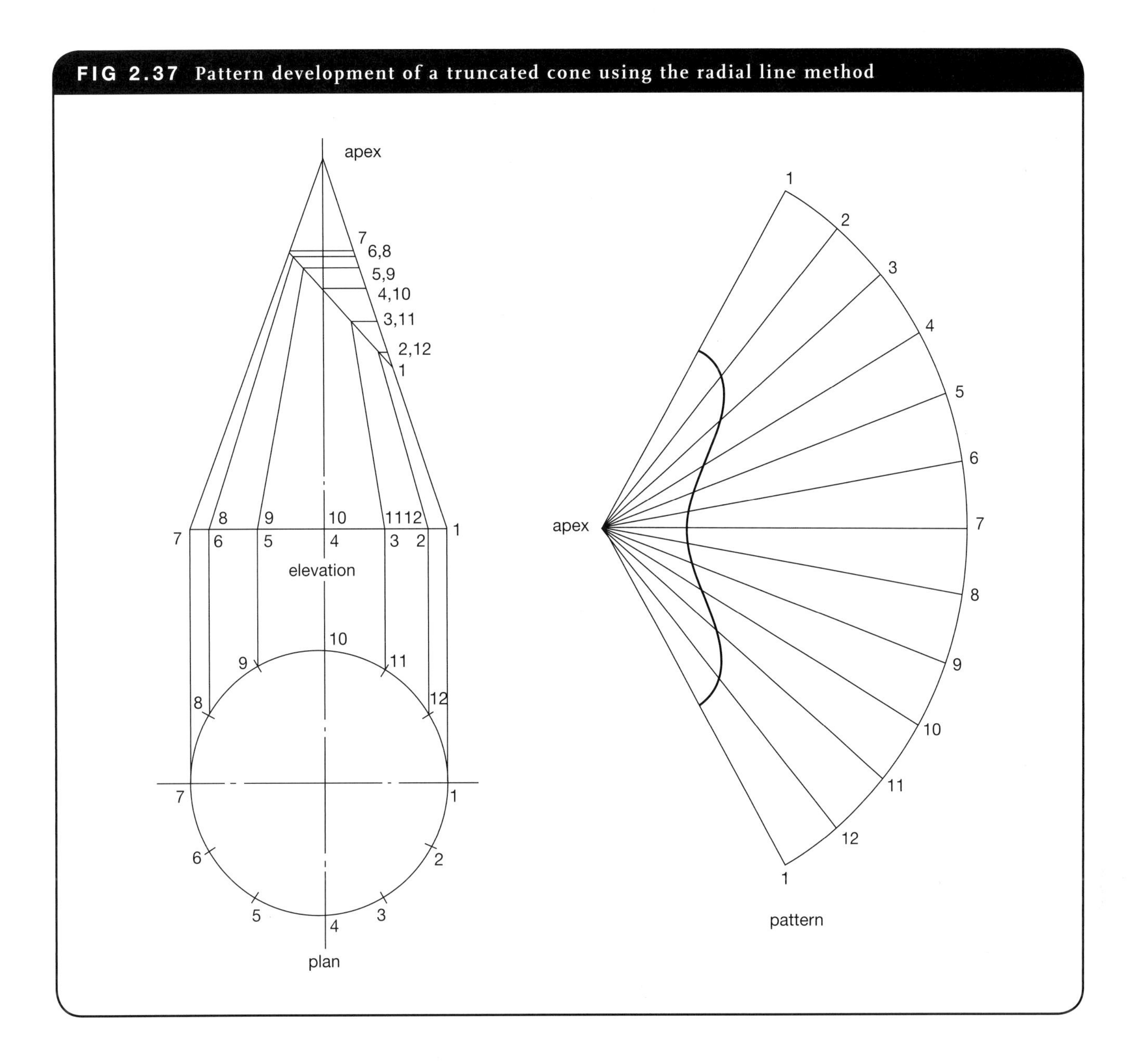

FIG 2.37 Pattern development of a truncated cone using the radial line method

plan view and heights are obtained from the elevation. The two known lengths are then joined at right angles on a true length diagram from which the length of the hypotenuse is determined.

Example

Development of the pattern for a 'square to round' transition piece:

1 Produce the orthogonal plan and elevation view of a transition piece of pipework (Figure 2.38).

2 Divide the circular top of the transition piece into twelve equal parts. Number each point on the circumference from 1 to 12. The base of the transition can be represented by the five letters A C, X, B, A. Seams are usually located on a short side of a model—Figure 2.38 (overleaf) shows the seam as being the line between point 5 and point 1.

Note: To develop a pattern, the unknown length between each number and the letter nearest to it on the base line must be determined. This is done by using a 'true length' diagram adjacent to the elevation.

3 A pattern is usually commenced opposite the seam. From the plan view, step off the distance 7-X and transfer it to the base of the true length diagram. The true length of 7-X can be taken from the top of the true length diagram down to point X. This distance becomes the starting line 7-X on the pattern. Draw a perpendicular line through × and mark points C and B, which are taken directly from the plan.

4 Draw a light line on the pattern between 7 and B, then between 7 and C. The distances between 7 and 6, and 7 and 5 are taken directly from the plan. Obtain this distance and draw an arc either side of point 7.

FIG 2.38 (a) Square to round transition piece, (b) pattern development using triangulation

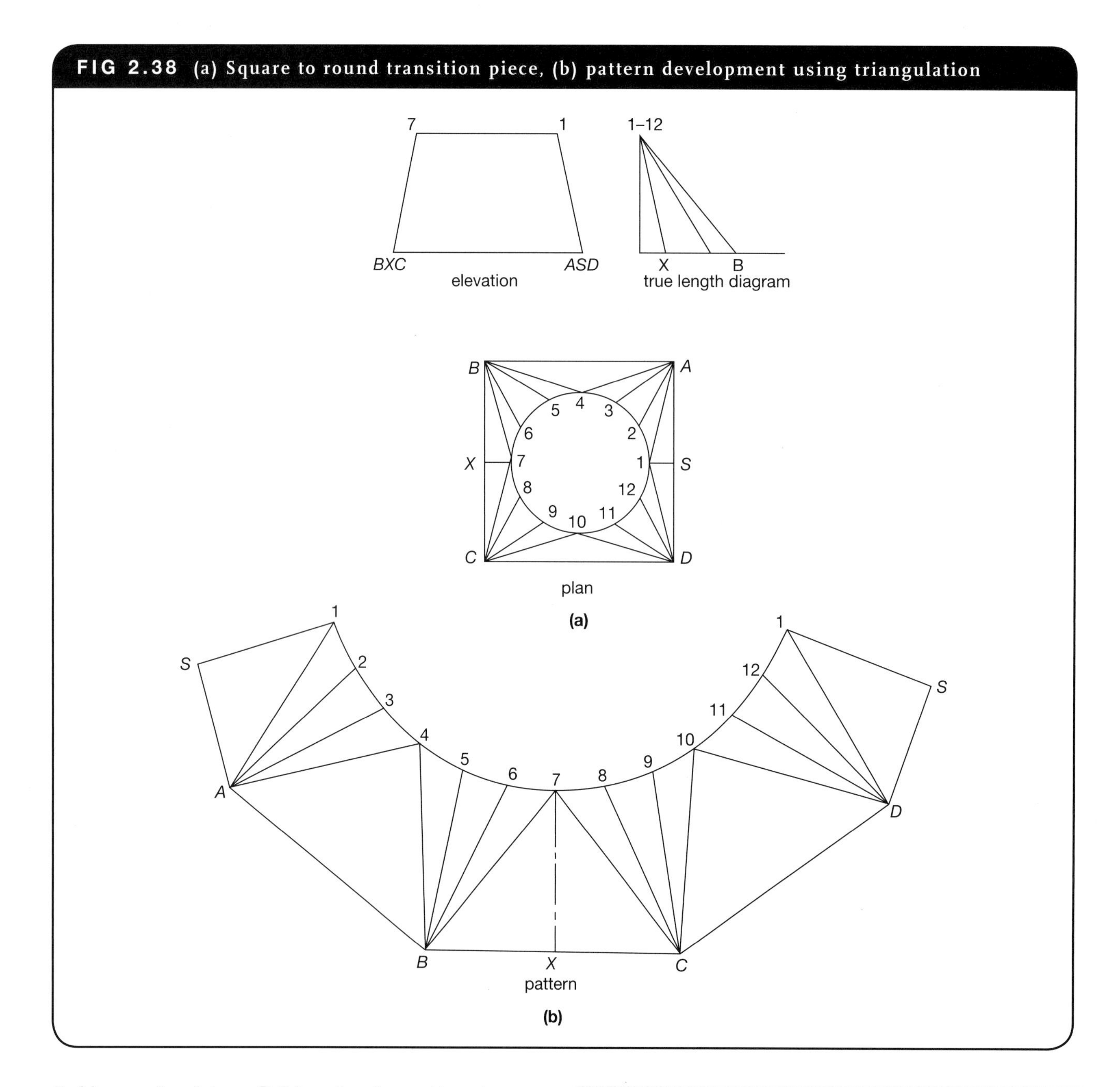

5 Measure the distance B-6 from the plan and transfer it to the base of the true length diagram. The true length of B-6 is the distance from the top of the true length diagram to point B; this distance is then transferred to the pattern. Place the compass point on B and draw an arc adjacent to 7 to locate point 6.

Note: There are only three different true lengths for the complete pattern. The numbers 1 to 12 are all equally spaced and taken directly from the plan. The distance between letters on the base line can be taken directly from the plan.

6 Point A is located on the pattern by using the true length diagram. Find the true length of 4-A and from point 4 draw an arc. Take the distance B-A from the plan and draw an arc from B to determine point A. The pattern starts and finishes on a seam line.

Note: Follow steps similar to those above to locate all the points.

FOR STUDENT RESEARCH

- AS 1100 Technical Drawing:
  - 1100.101 Part 101: General principles (1992),
  - 1100.201 Part 201: Mechanical engineering drawing (1992),
  - 1100.301 Part 301: Architectural drawing (1985),
  - 1100.401 Part 401: Engineering survey and engineering survey design drawing (1984),
  - 1100.501 Part 501: Structural engineering drawing (1985)
- AS/NZS 3500: 2003 Plumbing and Drainage. Glossary of Terms

PLUMBER PROFILE 2.1

ELERI DEAR

Job Title: Apprentice plumber at Select Solutions, Victoria

Eleri Dear was recently awarded NMIT's Encouragement Award for being the runner up of NMIT's 2010 1st Year Apprentice of the Year. As an apprentice plumber she has dabbled in almost all plumbing jobs and wants to one day have her own all-female plumbing business.

How did you get into plumbing?

They were advertising for girl plumbers and I gave it a go and I stuck to it. I was doing horse training before for ten years as I had never been the girl to do an office job, and needed a career change. I looked at all the different trades out there but a plumber just gets to do a lot more than a carpenter or electrician does, there is a lot more variety of work you do. You are never bored when you work!

What is the most difficult aspect of your job?

That's a hard one as everything is pretty tough. The hardest part for me is just the physical work. You get used to it though. I was thinking about doing more training to keep up with the boys but before you know it you just get stronger. From when I first started, I can tell I am a lot stronger than I used to be.

What is your most memorable experience?

I was runner-up for Apprentice of the Year award at the end of last year. Also just doing a reline of a house on my own—you get really excited when you do it all on your own!

What has been the worst job you have done so far?

I was doing a septic tank and it all poured out and I fell into it. I stunk of it all day. You expect the worst in plumbing when you're working with toilets.

What do you most enjoy about your job?

Just working outdoors, except when it's 2 °C like today, which is not the best, but every day is different. I have never been bored once. As much as it's hard to work in the cold sometimes you just know tomorrow is not going to be the same as today.

What is it like working in a male-dominated industry?

It's not as bad as everyone thinks. As long as you get along with guys reasonably well and deal with the common habits of burping and farting and just being a normal bloke, it's not that bad. They don't pick on you, or don't act extra nice to you and feel sorry for you. I get along with all the guys so I wasn't concerned about it: I could always stand up for myself.

What kind of job would you want to do in the future?

Probably your household jobs, old people's homes and that kind of thing. There is a lot of opportunity there as a female as I think they trust you more. It's still really early in my apprenticeship and I have a lot to learn, so I haven't pinpointed exactly what I want to do but I probably want to start an all-girls business. It hasn't been done and it is something I would want to look into to give other girls a go. A lot of girls are probably worried about working with so many guys as well.

Do you have any words of advice for future female plumbers?

Just give it a go if you are thinking about it. If you can't handle the boys you are in the wrong industry.

PART 2 TRADE SKILL ESSENTIALS

CHAPTER **THREE**

Cutting and drilling

LEARNING OBJECTIVES

In this chapter you will learn about:

- **3.1 cutting with hand tools**
- **3.2 how to sharpen drills**
- **3.3 the types of drilling machines**
- **3.4 tapping and threading.**

INTRODUCTION

Many of the metal fabrication and assembly processes the plumber is expected to perform require the use of various cutting tools. Since tools were first used technological innovation has constantly improved the equipment—from the simple stone tools of prehistory to the sophisticated new tools and equipment used by today's workforce. The main purpose of all tools is to make work fast, easy, comfortable and safe.

This section deals with the hand tools used for cutting in current plumbing practice, but also includes information on the latest machine-cutting tools and techniques used for the wide range of approved plumbing materials.

HAND CUTTING

Hacksaws

The hacksaw is used to cut most types of metal, irrespective of the shape of its section, but is not used on flat sheet metals. The hacksaw is a bow-shaped frame with a handle. The blade is suspended across the bow and attached to steel pins at each end. A variety of designs is available for different applications (Figure 3.1).

When the blade is placed in position, with the cutting teeth facing away from the handle (Figure 3.2), the adjustable wing nut at one end of the bow is tightened until the blade is in tension.

Most hacksaw frames are adjustable to accommodate blades of different lengths, usually 250 mm and 300 mm. A 'junior' hacksaw is often useful (Figure 3.1(b)) since it can cut in awkward places more readily than larger hacksaws. Some brands of hacksaw are provided with a tension lever (Figure 3.1(a)) and these eliminate the problem of blade twist and wobble.

Blade selection and use

Hacksaw blades are made with various grades of cut so that they can be used for different cutting operations. Soft materials, for example, should be cut with a coarse blade (fewer teeth) and harder materials cut with a finer blade (more teeth).

FIG 3.1 Various types of hacksaw; (a) tension lever, (b) junior hacksaw, (c) pistol grip hacksaw (d) metawork hacksaw

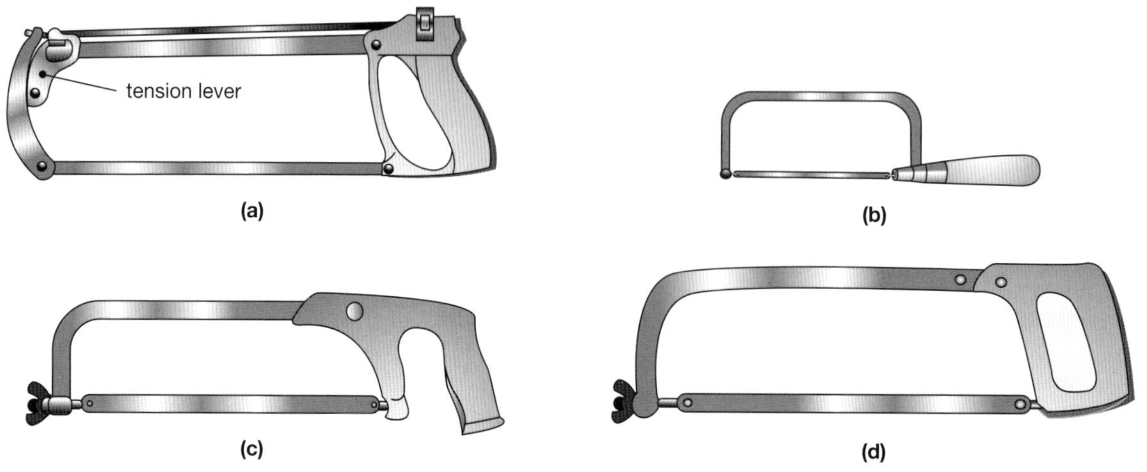

The grades are measured in two ways:

- pitch (P) measured in millimetres
- number of teeth (N) per 25 millimetres (t/25 mm) of blade length.

Hacksaw blades are usually made in standard pitches of 14, 18, 24 and 32t/25 mm of blade length. Table 3.1 sets out the recommended grade of blade for various uses. Selection of the blade should be based upon the material to be cut.

FIG 3.2 Placing a blade into a hacksaw frame

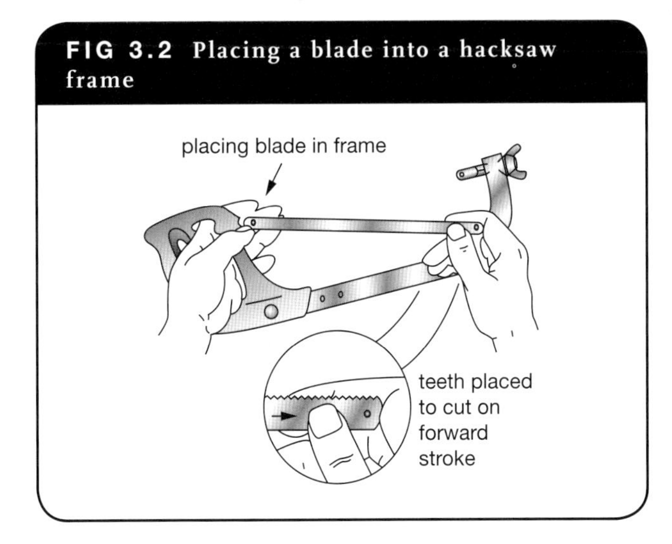

TABLE 3.1 Recommended grades

P (mm)	N (t/25 mm)	Example of material
1.8	14	Soft steel
1.4	18	Steel
1.0	24	Copper tubing
0.8	32	Thin sheet metal

Blades with 18t/25 mm are recommended for general use, but it should be noted that there is no such thing as an 'all-purpose' blade.

High-speed steel blades

High-speed steel blades are designed for cutting where blade life is important. Because high-speed blades are more durable than general purpose blades, they are used for more specialised tasks such as cutting alloy steel (stainless steel) and other hard metals. They are, however, more expensive than other blades.

Low-tungsten steel blades

These are designed to cut materials such as mild steel, brass, aluminium, copper and other similar soft metals.

Flexible blades

These are recommended for general use especially where the material to be cut cannot be held firmly or is in an awkward position.

Plastic pipes such as UPVC, polyethylene and polypropylene may be cut with either a 24t/25 mm or 32t/25 mm blade. A mitre box should be used to produce a square cut.

Blade checklist

Before a hacksaw is used, the blade should be checked for the following:

- Condition: check the teeth for excessive wear
- Direction of blade (fitting): check to see if the teeth are pointing away from the handle—this will enable the hacksaw to cut on the forward stroke
- Tension: the blade may be checked for tension by flicking it with the finger. This should produce a low-pitched ring when in proper tension

CUTTING WITH A HACKSAW

The following rules should be remembered when making a hacksaw cut. Refer to Figure 3.3, overleaf.

- Make sure the hacksaw is held at an acute angle to the work surface. This prevents stripping of the blade teeth, especially on thin materials (Figure 3.3).
- The cut is commenced by placing the nail of the left thumb (for a right-handed person) on the metal to guide the blade. Apply a light pressure until a groove appears after a few strokes.
- Use slow steady strokes. This will reduce excessive wear and overheating and prolong the life of the blade.
- Three teeth of the blade should be in contact with the material to be cut at all times.

These points should be kept in mind irrespective of the type of section of the material being cut.

Conservation of blades

Breakages make blade replacement necessary much more frequently than normal wear. Blades are often broken because excessive pressure is applied when cutting through a small surface. The concentration of pressure at a single point on the blade causes it to buckle and break. 'Cramping' and 'binding' also cause many breakages that could be avoided by cutting with straight-line strokes. Care must be taken not to tilt or cant the saw frame. Likewise, insecurely held work is a frequent cause of blade breakage. Loose work will result in a sudden binding and distortion of the blade and will practically guarantee a broken blade.

These problems can be eliminated provided the list of rules is kept in mind during all cutting operations. A forward, cutting stroke at no more than one stroke per second will help to maximise the life of the hacksaw blade.

FIG 3.3 Correct cutting angle

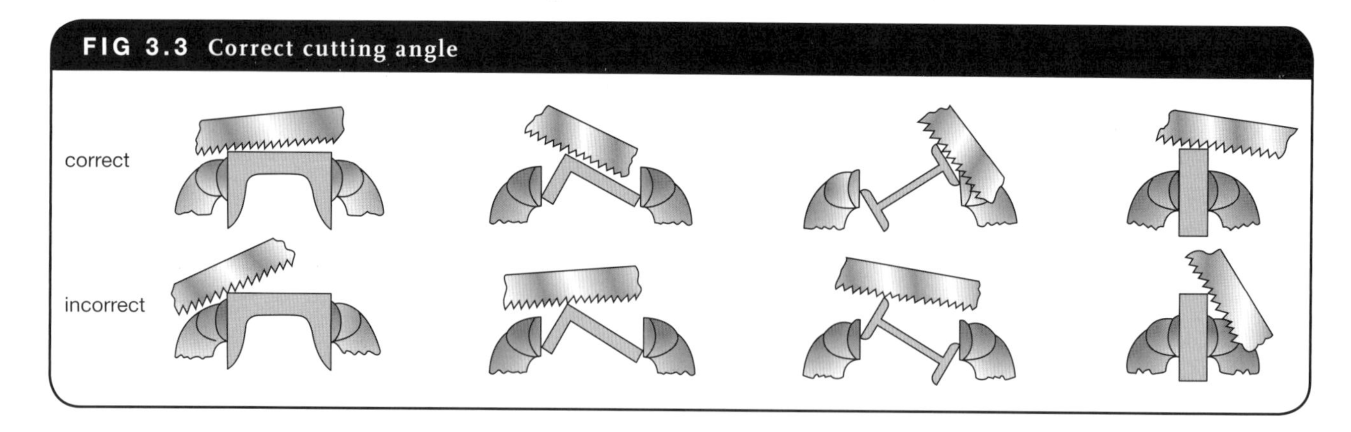

TIN SNIPS

As part of the diverse range of work, especially in areas of work relating to roof plumbing, the plumber often needs to use snips.

There are a great many types of snips currently available to cut sheet metal. Having a wide range to choose from enables the plumber to select the correct size and shape for any particular job. Snips are obtainable in sizes up to 400 mm in length. The most commonly used types are (a) straight snips, (b) curved snips, and (c) universal combination snips. Figure 3.4 shows these and other basic types.

Straight snips

These are sometimes referred to as 'tin snips' and are used for making straight cuts and large external curves (Figure 3.4(a)).

Curved snips

As the name implies, these are designed with curved blades and are used to cut around inside curves (Figure 3.4(b)).

Universal combination snips

These are sometimes referred to as 'gilbows'; the blades are designed for universal cutting of straight lines, or internal and external cutting of contours (Figure 3.4(c)). They are available for right-handed and left-handed operators and are easily identifiable as the top blade is either on the right-hand side or the left-hand side respectively.

Jewellers snips

These are a miniature version of the straight and curved snips (Figure 3.4(d)). Jewellers snips are used for working with light-gauge metals. Curved jewellers snips have a particular application when cutting small internal curves.

Aviation snips

As shown in Figure 3.4(e) these are heavy-duty snips for cutting hardened sheet metals up to 20 gauge, such as stainless steel. The blades are similar in design to combination snips, but have very fine serrations on the cutting edges. Double leverage, achieved by a linkage between the handle and jaws, substantially increases the initial pressure applied by the operator making cuts in the heaviest capacity-rated material.

The jaws are hot drop-forged from special steel, heat treated and tempered to withstand heavy and continuous cutting. Machine serrations on the outside face of cutting edges grip materials and facilitate smooth, positive cuts.

Aviation snips are available for right-hand and left-hand curve cutting and in combination design for straight and irregular cuts. They are popular because of their versatility.

FIG 3.4 Various types of cutting snips; (a) straight snips, (b) curved snips, (c) universal combination snips, (d) jewellers snips, (e) aviation snips

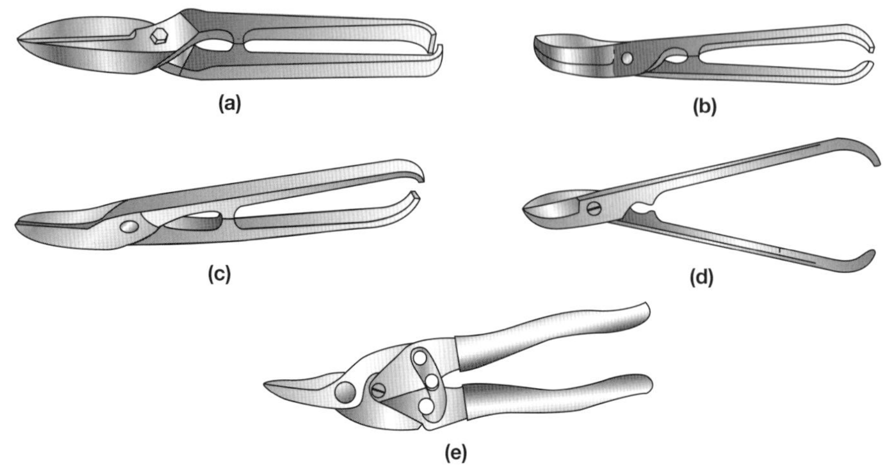

(a) (b)

(c) (d)

(e)

Design features

Conventional snips are forged from tool steel and the blades are twisted slightly giving a 'hollow ground' appearance. The cutting face is ground to an angle of 85°–87° (Figure 3.5) to prevent the edge of the cut from burring the metal.

Straight snips have thin blades which are strong only on a vertical plane. They are therefore suitable for straight cuts and external curves where surplus waste has to be removed.

Universal snips have short, thick blades which will withstand the twisting of the snips when being used on irregular curved cuts. Their design also provides an advantage when cutting thicker materials.

CUTTING WITH TIN SNIPS

- When making a cut, place the metal in a horizontal position and rest it flat on the cutting face of the blade.
- To obtain maximum leverage, the grip should be as close as possible to the end of the handles. Figure 3.6 shows the correct grip when cutting metal. The scissor action is provided by the finger movement indicated by the arrows.
- To produce maximum cutting force, the hand should be as far away as possible from the pivot (fulcrum). The metal must therefore be positioned close to the pivot.
- The full length of the cutting blade should not engage the metal when cutting with snips. This action forms serrations and produces a poor finish along the edge of the metal.
- Never extend the arms of a pair of snips to cut a greater thickness of metal than that for which they were designed. This careless action will result in deflecting the cutting blades and bending or shearing the fulcrum bolt.
- Blade clearances are very important and should be set to suit the material being cut.
- Correct clearance is essential to obtain the best shearing results.
- As an approximate guide, the blade clearance should not exceed 10 per cent of metal thickness and must be varied to suit the particular material.

FIG 3.5 Angle of cutting face

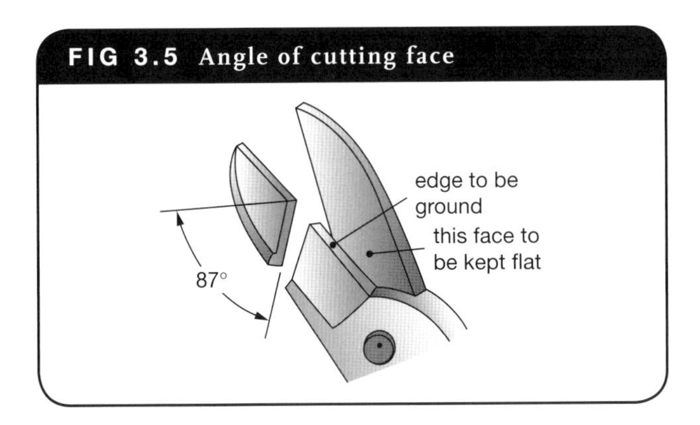

FIG 3.6 Correct leverage for snips

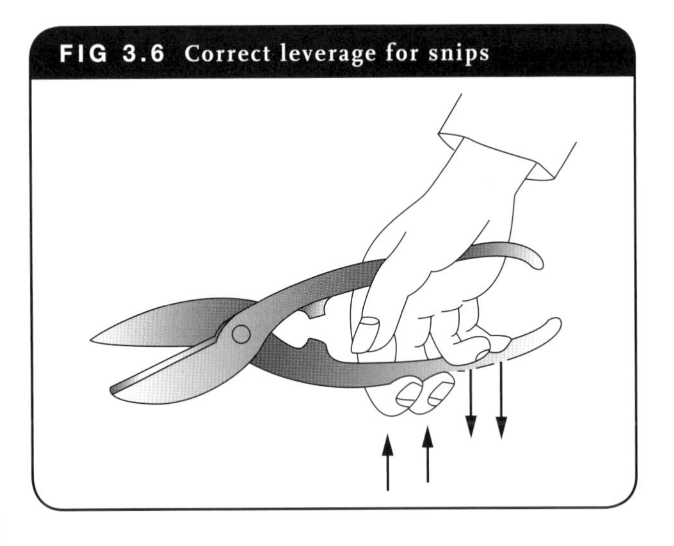

TUBE AND PIPE CUTTERS

Plumbers and pipefitters are frequently involved in small-bore copper and stainless steel tubing installations. Tube cutters with attached reamers are often used for cutting tubing (Figure 3.7).

FIG 3.7 A typical tube cutter

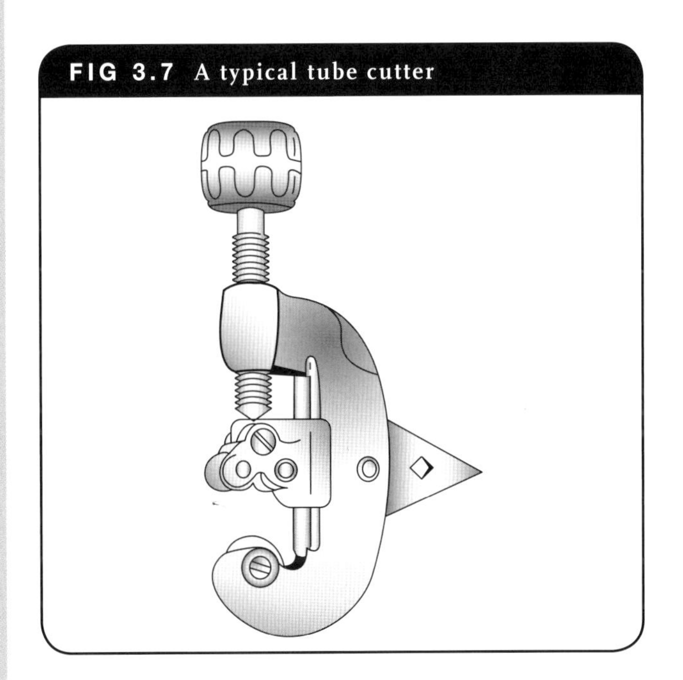

Copper tube cutters

Copper tube cutters are available in sizes ranging from 3 mm OD to 150 mm OD. Cutting is achieved by rotating the unit while gradually tightening the spindle to deepen the cut until the pipe is severed (Figure 3.8). Since copper is a soft material, it can be readily crushed or bruised if fixed in a vice, so the tube is hand held. The tube cutter is slipped over the end of the tube and the handle is turned to bring the cutting wheel against the copper. Excessive tightening of the cutting wheel against the tube may distort it. After the tube has been cut, the inside burr which forms must be removed. This is achieved by using the attached reamer as shown in Figure 3.9. The tube cutter is also useful when cutting small-bore chromium-plated tube.

FIG 3.8 A typical tube or pipe cutter, showing the method of use

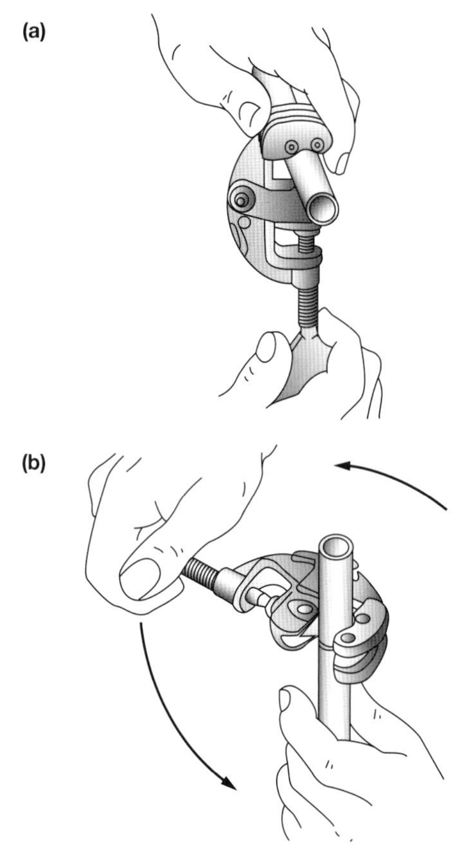

(a)

(b)

FIG 3.9 Using a reamer to remove the inside burr

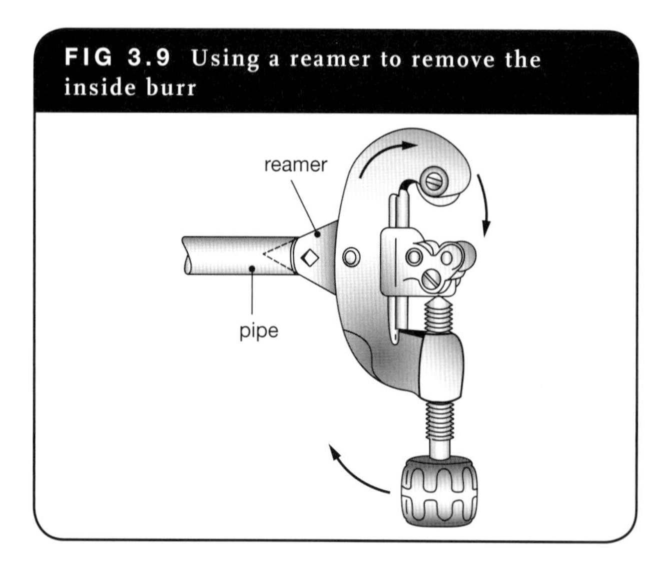

Rotary pipe cutters

The most common types of hand cutters fall into two categories:

- single-wheel
- multiple-wheel.

FIG 3.10 Single-wheel pipe cutter

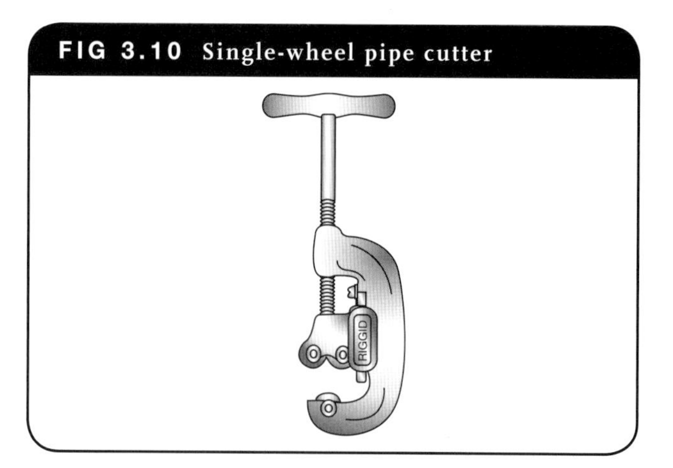

The single-wheel pipe cutter, as shown in Figure 3.10, consists of one cutting wheel and two rollers. It allows fast, clean, square cuts to be made by either hand or power.

PROCEDURE FOR MANUAL PIPE CUTTING

1. Before cutting the pipe ensure that it is securely fixed in the vice. (But remember copper pipe may bend or otherwise be damaged under pressure from the vice.)
2. Place the cutter over the pipe and ensure that the cutting wheel rests on the mark to be cut.
3. Tighten the handle until the cutting wheel commences to bite into the cutting mark on the pipe.
4. Rotate the pipe cutter one complete turn and examine the pipe to see that a circumferential line has been made completely around it.
5. Tighten the handle a little and make another complete turn.
6. Check to see that the cutting wheel is following the original groove.
7. Continue to rotate the cutter and gradually increase the tension on the handle until the pipe is completely cut through.

FIG 3.11 Power die and cutter

MACHINE CUTTING (COMBINATION POWER DIE AND CUTTER)

For workshop and site work where large quantities of pipe have to be cut, and especially for large-diameter pipework, machines are necessary. They are designed and built to achieve the maximum saving of time and labour, the physical effort required being reduced to a minimum.

Power-driven machines are constructed and arranged in such a way that once the pipe is cut, the dies are opened by the simple motion of a lever for the pipe to pass through without being removed from the machine. They are able to handle pipes with a wide range of diameters.

Soil pipe cutter

This type of cutter (Figure 3.12) is suitable for cutting a variety of pipe materials, but is mainly used by the plumber for cutting cast iron (CI) and vitrified clay (VC) pipes. It is particularly useful for cutting pipes in trenches or other restricted spaces because, unlike rotary pipe cutters, it does not require rotating to produce a cut.

The cutter provides an excellent mechanical advantage (the ratio is 200:1) and therefore requires little energy to operate. The cutter body incorporates a ratchet and the pumping action provided by the handle tightens the multi-wheeled chain against the pipe. This action sets up pressure points around the circumference of the pipe which increase until the pipe is severed. The hand guard protects the operator's knuckles at the time of the cut.

FIG 3.12 Chain-type soil pipe cutter

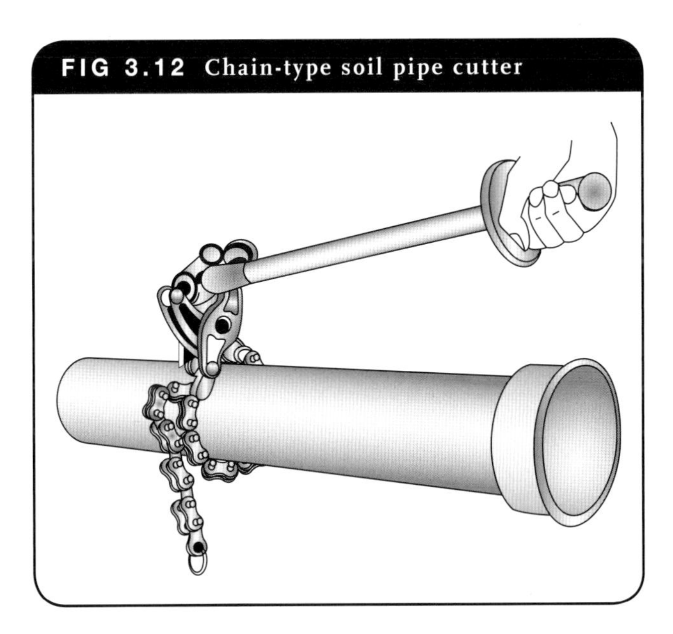

ANGLE GRINDERS

Angle grinders are used throughout the plumbing industry for cutting bricks, concrete, steel, cast iron, vitrified clay and slate.

The angle grinder is one of the most dangerous tools in any workplace. The spinning disc rotates at high speed and can cut through metal bars and pipes. Before using an angle grinder always inspect for cracks or flaws in the wheel and if there is any visible damage, replace it with a new wheel.

FIG 3.13 Angle grinder

USING ANGLE GRINDERS SAFELY

Eye injuries, hearing damage and other physical injuries can happen when using angle grinders. To avoid injury, use the correct personal protective equipment (PPE).

1. Always grip the grinder firmly with both hands and ensure you are positioned in a well-balanced stance.
2. When starting the angle grinder allow it to run up to its full speed. Check for any vibration and ensure the wheel is running 'true'.
3. Do not over-reach when using an angle grinder and avoid using them above shoulder height.
4. When grinding always keep the protective guard in place.
5. When you have finished the task, switch off the angle grinder and wait for the wheel to stop rotating before placing it down.

WARNING: **Never force the cutting wheel to stop.**

OPERATION OF DRILLS

Efficient drilling operations require a number of procedures to be followed, the most important of which is the selection of the correct drill for the application. Figures 3.23 to 3.28 inclusive show some of the types of drills available.

Most drills have a point angle of 118°. Although this angle suits most materials, the performance can be improved by repointing the drill to suit a specific material. (See Figure 3.18 for point geometry.) The speed and feed rate of the drill is restricted by the hardness of the material. Initially, moderate speed and feed are used, increasing one or both to achieve maximum production for a reasonable drill life before resharpening becomes necessary. The work piece must be held securely and supported as closely as possible to the drill. Always clamp the work whenever possible and never hold by hand. When using a straight-shank drill, hold it firmly in the chuck as slippage will frequently cause drill breakage. When drilling by hand, apply constant pressure. Do not allow the drill to dwell as this will cause dulling of the face edge.

The flutes of the drill must always be kept clear. Clogging of the drill flutes prevents sufficient lubricant from reaching the drill point.

The drill flutes may be cleared by occasionally withdrawing the drill after penetrating about two or three drill diameters in depth. This can lead to excessive heat, which will dull or blunt the cutting lip.

Drill nomenclature

Body

The portion of the drill extending from the extreme cutting end to the start of the shank.

Drill diameter

The measurement across the cylindrical lands at the outer corners of the drill.

Flutes

The grooves in the body of the drill which provide lips, permit the removal of chips and allow cutting fluid to reach the lips.

Flute length

The axial length from the outer corners of the cutting lips to the extreme back end of the flutes.

Overall length

The distance between two planes normal to the drill axis at the extreme ends of the cutting diameter and shank respectively.

Point angle

The included angle between the projections of the lines joining the outer corners and the corresponding chisel edge corners on a plane parallel to one (or both) of these lines and the drill axis.

Sharpening of drills

Most drilling problems are due to improper sharpening of the drill. For general purpose drilling, the point angle is 118°. The angle must be equal on both sides of the drill axis (59°) and the cutting edge lip must be the same length on both sides (Figure 3.15). If the two cutting edges are not equal in length and/or the angles are not equal, the drill will cut an oversize hole and breakage may occur. The drill must be sharpened with a lip clearance of between 12° and 15° (Figure 3.16). If the lip clearance is excessive, the strength of the cutting edge is reduced and may result in fracture of the cutting edge.

FIG 3.15 Point angle

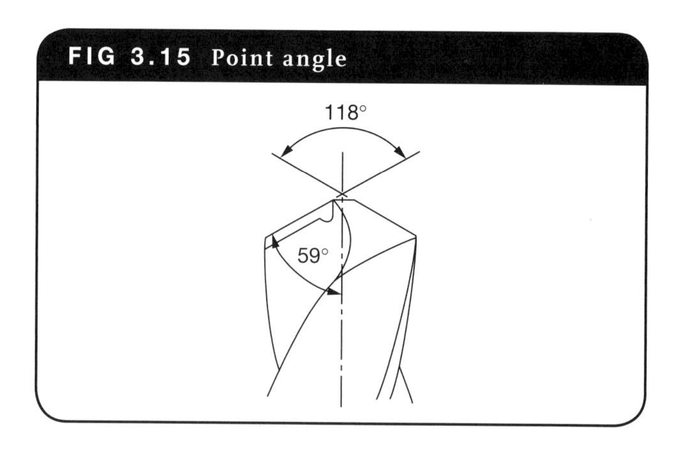

FIG 3.16 Lip clearance

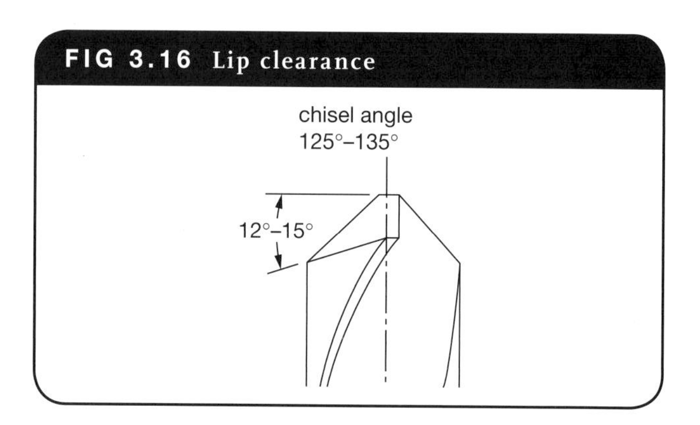

Whenever possible, drills should be sharpened on a drill-pointing machine. However, this is not always possible and the following hints will help to ensure good results when hand-sharpening drills (Figure 3.17).

FIG 3.14 Drill nomenclature

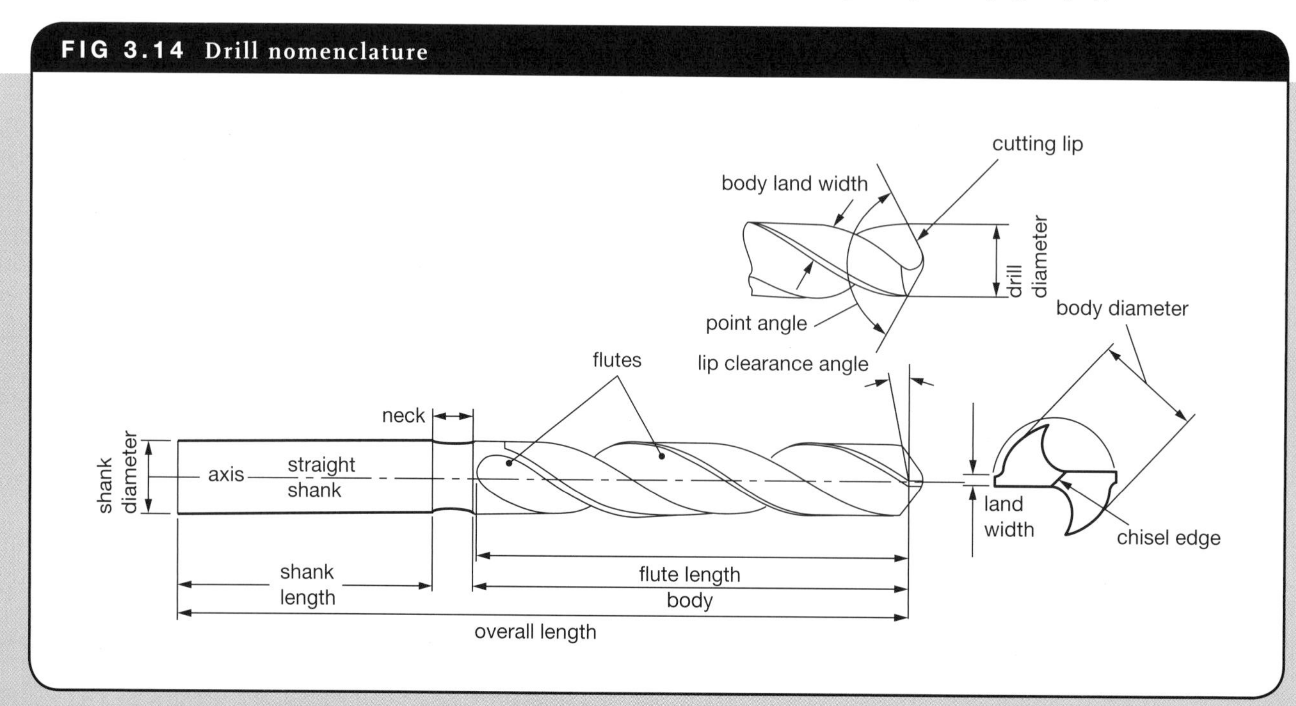

FIG 3.17 Hand-sharpening drills

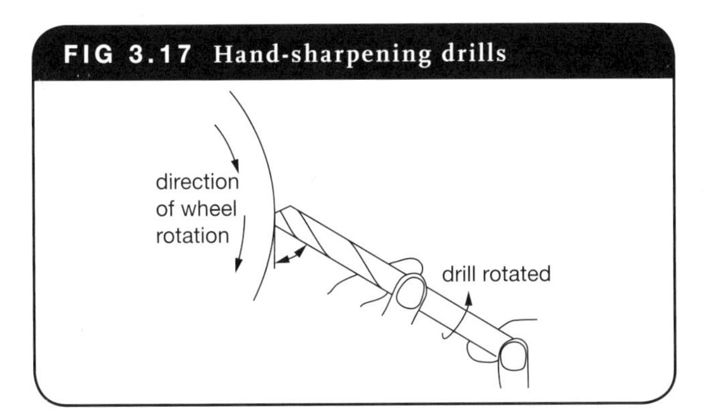

GUIDELINES FOR HAND-SHARPENING DRILLS

- The thumb and forefinger of the left hand are used as a pivot and the back of the drill is held with the thumb and forefinger of the right hand. The drill is then rotated in a clockwise direction, gently pushing the drill into the wheel.
- The intense local heat generated in grinding operations frequently results in surface cracking due to uneven thermal expansion and contraction. Grinding pressure should therefore be moderate when repointing drills and the use of a free-cutting wheel and the availability of a copious supply of water are desirable.
- When water is not available, grinding can be done by taking very light cuts. If excess heat is generated the drill should be left to cool in the air. Grinding cracks that are invisible to the naked eye can enlarge under working conditions and lead quickly to drill breakdown.

Most high-quality drills are pointed (sharpened) on precision automatic grinding machines when manufactured. As noted above, the standard point geometry has an included angle of 118° with a lip relief angle of 12° to 15°; this is suitable for most materials. However, it is sometimes necessary to repoint the drill to better suit a particular material. Figure 3.22 (overleaf) shows the point angles suitable for use with a variety of different materials.

Types of drill bits

There are many general-purpose and specialist drills available to the plumber. The types most likely to be encountered are listed below.

Jobber drill: straight shank

This is designed to give optimum performance in a wide range of materials (Figure 3.18).

Stub drill: straight shank

These are shorter in both overall length and flute length than the jobber drill and therefore have a greater rigidity (Figure 3.19). Major applications of the stub drill are for drilling sheet metal, shallow holes, hard metal (e.g. stainless steel) and for drilling out broken studs.

Masonry drill: tungsten carbide tipped

These are specially designed for use in masonry materials (Figure 3.20). They are suitable for use in portable electric drills, hand drills and drill presses. A specially designed

FIG 3.18 The jobber drill: straight shank

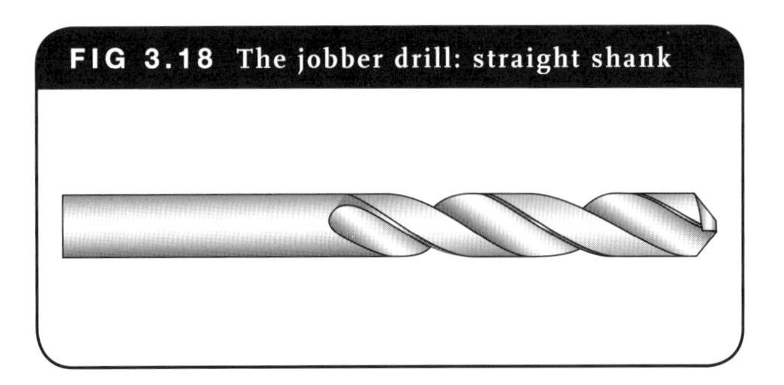

FIG 3.19 The stub drill: straight shank

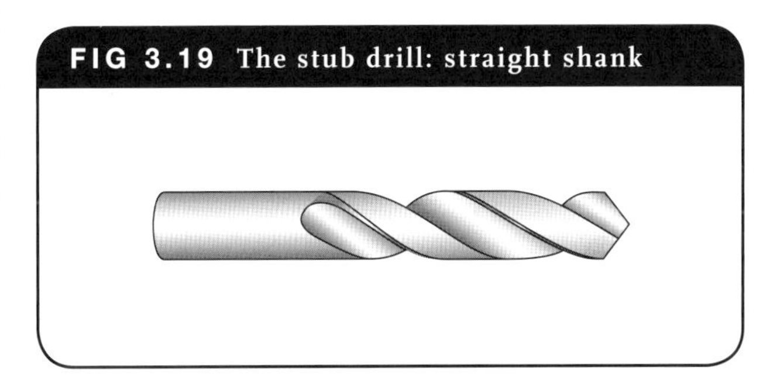

FIG 3.20 The masonry drill: tungsten carbide tipped

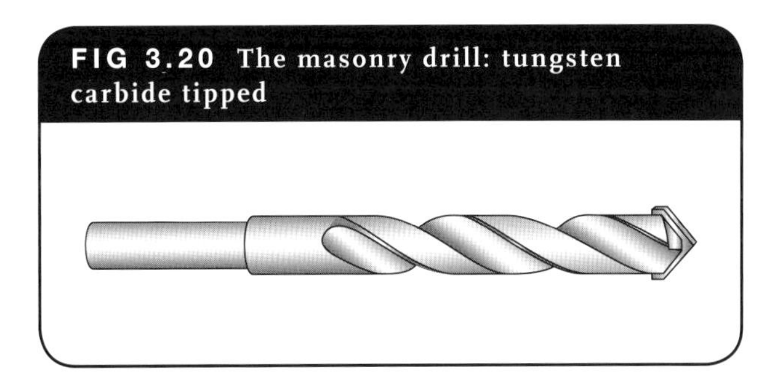

FIG 3.21 Panel drills: (a) single ended, (b) double ended

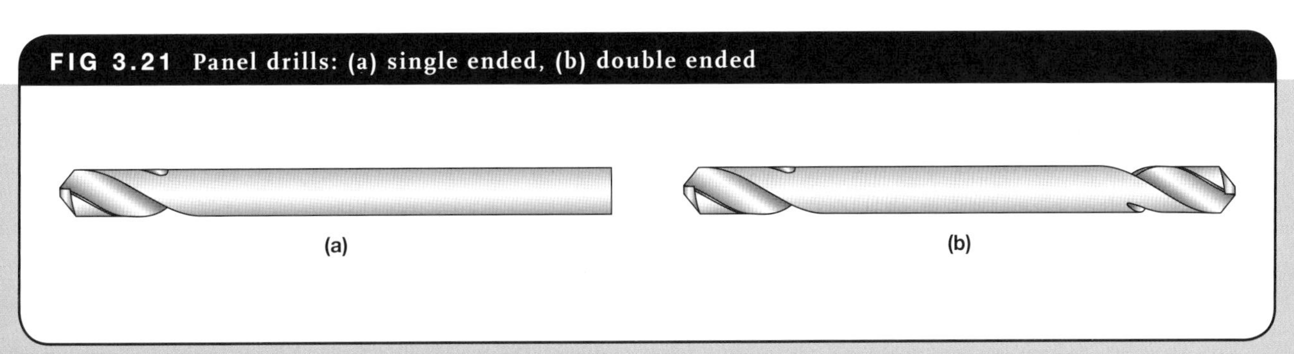

drill is available for use in rotary-impact portable drilling machines.

Panel drills

These bits have been developed to drill holes for rivets in flat and curved panels (Figure 3.21). These drills are recommended only for shallow holes no deeper than 1.25 times the drill diameter. When using double-ended drills, ensure that the chuck jaws are fully tightened. This will prevent the drill from cutting the inside of the chuck with the other point.

Types of drilling machines

Apart from the fixed drill stand, which is mainly used in a workshop environment, most electric drills are of the variable-speed, forward and reverse, hand-held type.

Three general types are used:

- light-duty drill (Figure 3.23)
- medium-duty or light industrial drill (Figure 3.24)
- heavy-duty drill (Figure 3.25).

FIG 3.22 Drill pointing table

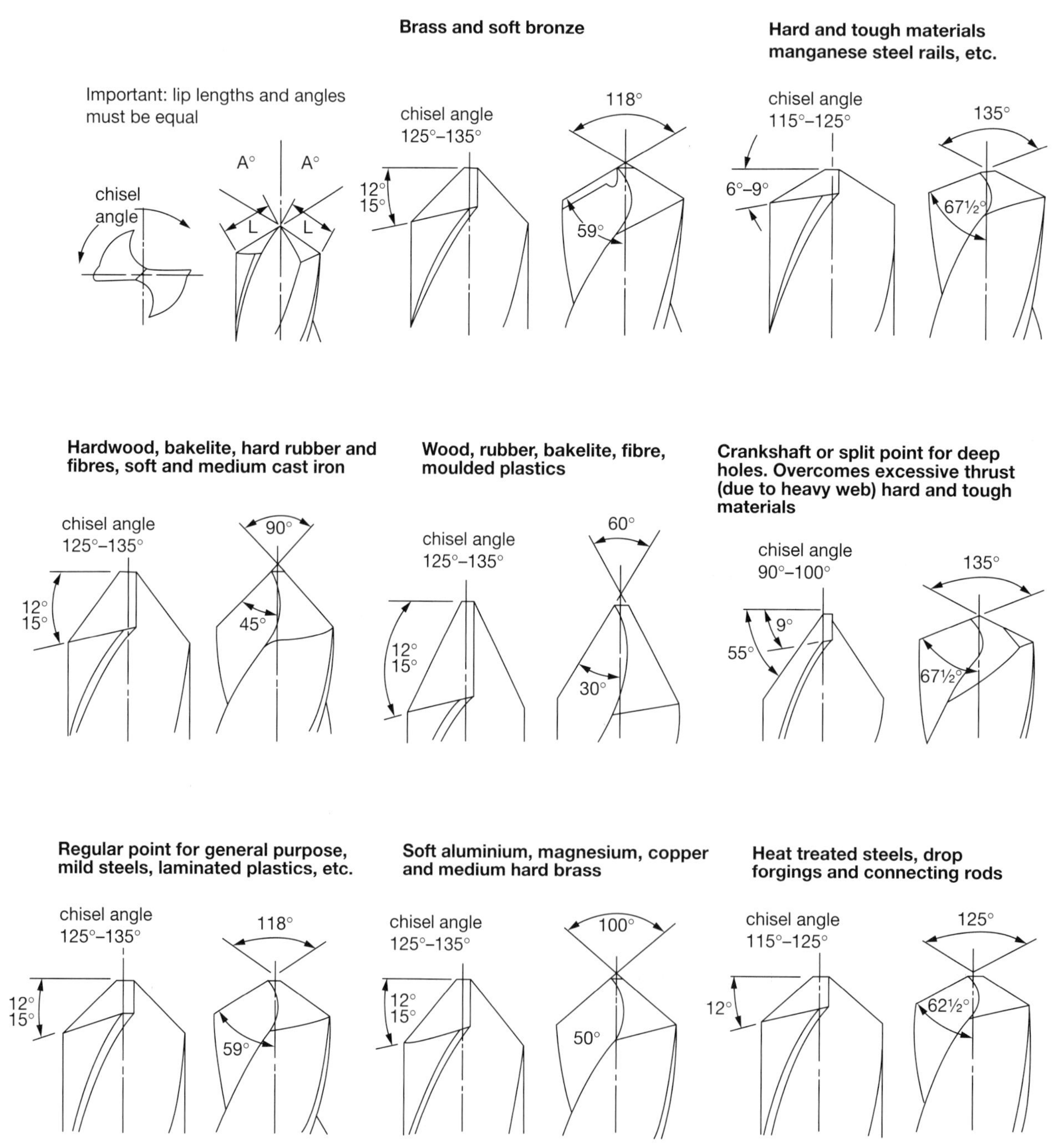

Important: lip lengths and angles must be equal

chisel angle

A° A° L L

Brass and soft bronze

chisel angle 125°–135°

12° 15°

118° 59°

Hard and tough materials manganese steel rails, etc.

chisel angle 115°–125°

6°–9°

135° 67½°

Hardwood, bakelite, hard rubber and fibres, soft and medium cast iron

chisel angle 125°–135°

12° 15°

90° 45°

Wood, rubber, bakelite, fibre, moulded plastics

chisel angle 125°–135°

12° 15°

60° 30°

Crankshaft or split point for deep holes. Overcomes excessive thrust (due to heavy web) hard and tough materials

chisel angle 90°–100°

9° 55°

135° 67½°

Regular point for general purpose, mild steels, laminated plastics, etc.

chisel angle 125°–135°

12° 15°

118° 59°

Soft aluminium, magnesium, copper and medium hard brass

chisel angle 125°–135°

12° 15°

100° 50°

Heat treated steels, drop forgings and connecting rods

chisel angle 115°–125°

12°

125° 62½°

FIG 3.23 Light-duty hand drill

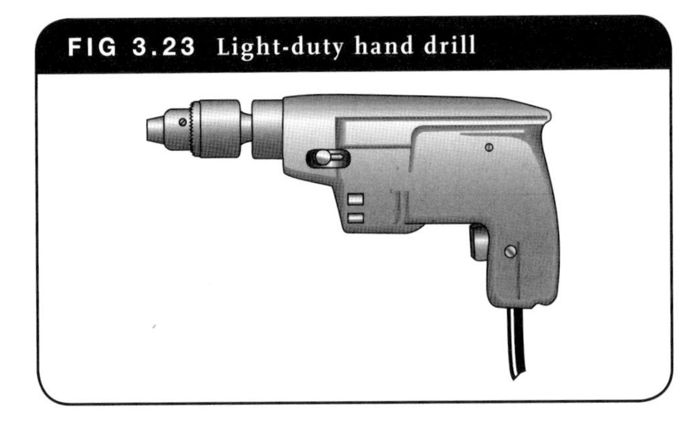

FIG 3.24 Medium-duty hand drill

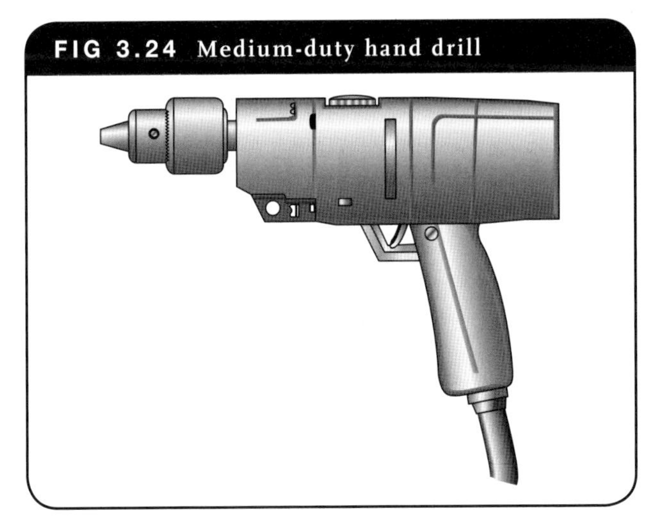

FIG 3.25 Heavy-duty hand drill

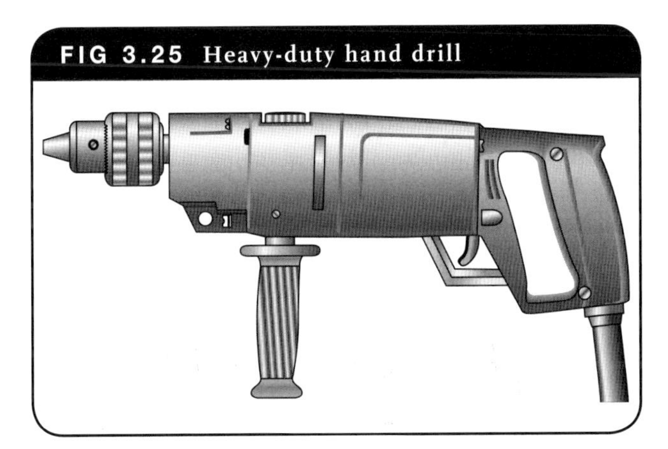

FIG 3.26 Impact drill

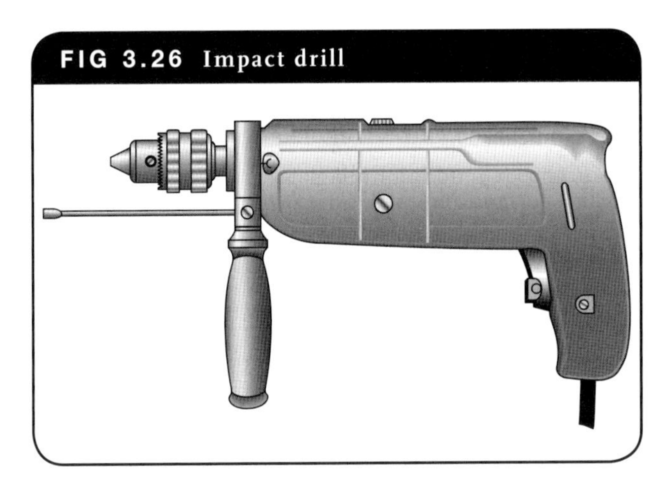

FIG 3.27 Reversible/variable-speed drill

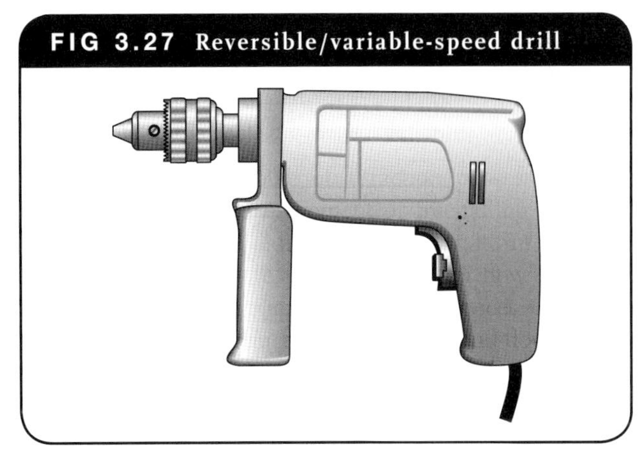

FIG 3.28 Battery-operated (cordless) drill

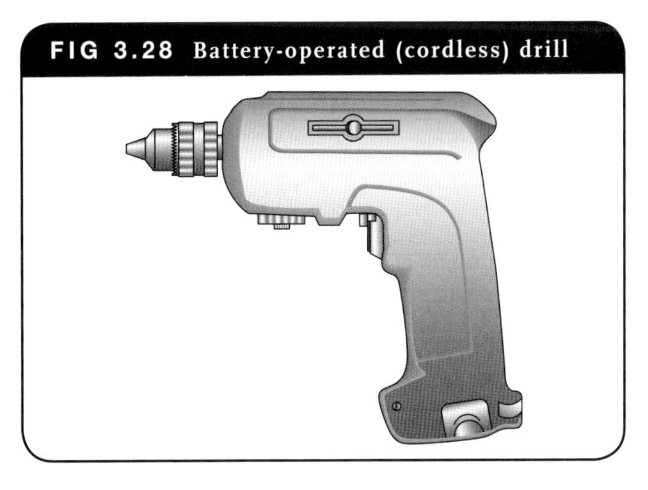

Drills usually fall into one of the following categories:

- impact drill (Figure 3.26)
- reversible variable-speed drill (Figure 3.27)
- battery-operated (cordless) drill (Figure 3.28).

Impact drill

The correct handling of an impact drill is most important as the tungsten carbide drill bits used are easily damaged. Most drill chucks have a self-centralising operation; however, the percussion action of the impact drill and the drill bit may cause the chuck to loosen. To assist in overcoming this, the drill chuck key should be used in more than one of the tightening holes; this assists the self-centralising mechanism to lock the drill bit in position.

When commencing a hole using an impact drill, care must be taken to avoid 'spin off' damaging the adjacent surface. A light pressure should be maintained on the drill. If possible, the 'impact mechanism' should be disengaged until the hole is started. The drill is then stopped, the 'impact' engaged and the drilling continued.

When using a tungsten carbide drill bit, a constant (but not heavy) pressure should be maintained. The drill bit should be removed frequently and the drilling dust blown out of the hole. This will reduce the friction on the drill shank and assist in reducing the heat build-up, thus increasing the life of the drill bit. When drilling large holes in concrete and masonry, the use of a pilot (smaller) drill to commence the hole is suggested. This allows the larger drill bit to penetrate more easily.

Reversible/variable-speed drill

The most common uses for reversible/variable-speed drills are in fixing of metal decking and wall panels. The method of fixing is frequently by self-drilling, self-tapping fasteners. The drill must be capable of removing the fasteners as well as installing them. A clutch system is commonly employed with reversing drills to avoid the torque of the drill stripping the thread, jumping off the fastener and breaking the fastener's head when it is tightened. In practice, the operator will adjust the clutch pressure to suit the material being fixed and the material being fixed into (e.g. timber or steel purling).

The variable-speed drill is operated by increasing or decreasing the pressure on the trigger on the drill. The advantages of these drills over single-speed and dual-speed drills are:

- When commencing any drilling operation, 'spin off' is easily avoided.
- When drilling holes in glazed surfaces (e.g. wall and floor tiles), damage to the tile surface is minimised.
- When fixing wall and roof panels, and decking (using self-tapping fasteners), the gradual increase in speed allows for quicker and more efficient fastening.

Battery-operated (cordless) drill

Battery-operated drills/screwdrivers are frequently used in fixing fastenings (e.g. pipe saddles) by screwing self-tapping screws into timber (see 'Anchors', Chapter 5). Battery drills are usually powered by rechargeable nickel cadmium (NiCad) batteries. The number of holes that can be drilled or screws that can be fixed depends entirely on the condition of the batteries. In practice it is advisable to 'flatten' (i.e. fully discharge) the battery pack prior to recharging. This will prolong the life of the batteries and give longer use between charges. The length of time taken to recharge the batteries will be specified by the manufacturer, but is generally one to three hours. It is not advisable to charge the batteries for longer than the recommended time.

Drill press

A drill press is preferable to a hand drill when the location and orientation of the hole must be controlled accurately. A drill press is composed of a base that supports a column; the column in turn supports a table (Figure 3.29). Work can be supported on the table with a vice or hold-down clamps, or the table can be swiveled out of the way to allow tall work to be supported directly on the base. Height of the table can be adjusted. The column also supports a head containing a motor. The motor turns the spindle at a speed controlled by a variable-speed control dial. The spindle holds a drill chuck to hold the cutting tools (drill bits, centre drills, deburring tools etc.).

WARNING: **Securing of the work piece is vital to ensure operator safety.**

FIG 3.29 Drill press

Care and maintenance

As with any electrical appliance, extreme care must be taken when drills are being used. Wet and damp conditions should be avoided—any appliance with a frayed or damaged electrical cord should not be used. The drill case should be examined for damage and the operation of any switch (i.e. power, hammer, reversing) should be checked prior to using the drill.

Eye protection should be worn at **all** times when using a drill. This is of particular importance when using a hammer drill.

TAPPING AND THREADING: TAPS AND DIES

The tools and equipment used to successfully tap and thread various materials are: taps for internal threads, called 'tapping'; and dies for external threads, called 'threading' (Figure 3.30).

Using taps and bar tap wrench

Taps are normally used in sets of three, consisting of taper, intermediate and bottoming, each tap having a different thread lead (Figure 3.32). The taper tap should always be used to start the thread, followed by the intermediate and then the bottoming tap. Through holes can, however, be completed with the taper tap.

From the tapping drill charts, select the correct drill size (Tables 3.3 and 3.4 overleaf). Clamp the work piece in a vice in an upright position, select the correct type of cutting fluid (Table 3.2 overleaf) and apply it to the tap. Hold the tap wrench securely with both hands close to the tap (Figure 3.32(a) and place the tap in the hole. Apply a light pressure and start to turn the tap in a clockwise direction in the hole (b). Turn the tap until a pressure of resistance is felt. Now turn the tap in an anticlockwise direction for between a quarter and half a turn. This breaks the chip and allows the teeth of the tap to cut without clogging and possibly snapping. During this process, ensure that the tap is held in an upright position perpendicular to the surface of the work.

FIG 3.30 Taps and dies: (a) hand tap, (b) adjustable button die, (c) die nut, (d) block die

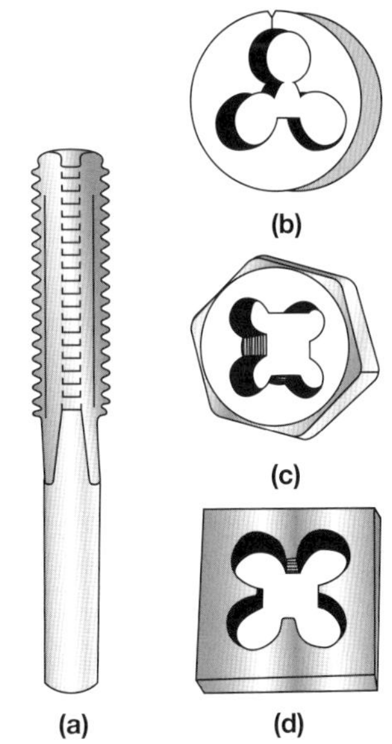

(b)

(c)

(a) (d)

FIG 3.31 Set of taps: (a) taper 7–9 thread lead, (b) intermediate 3–5 thread lead, (c) bottoming 1–1½

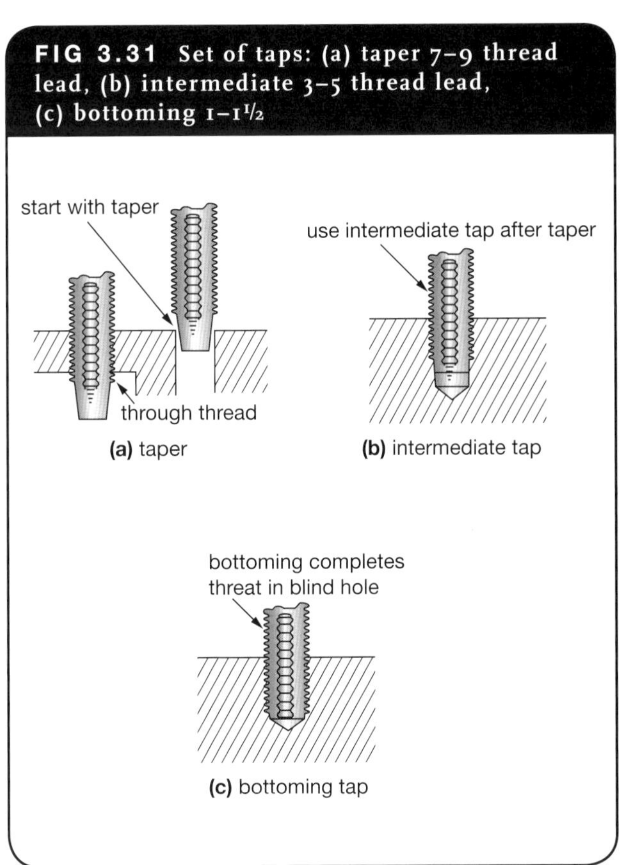

FIG 3.32 (a) Correct position of the hands when starting the tap in a hole, (b) correct position of the hands when the tap has commenced to cut and is drawing itself into the hole

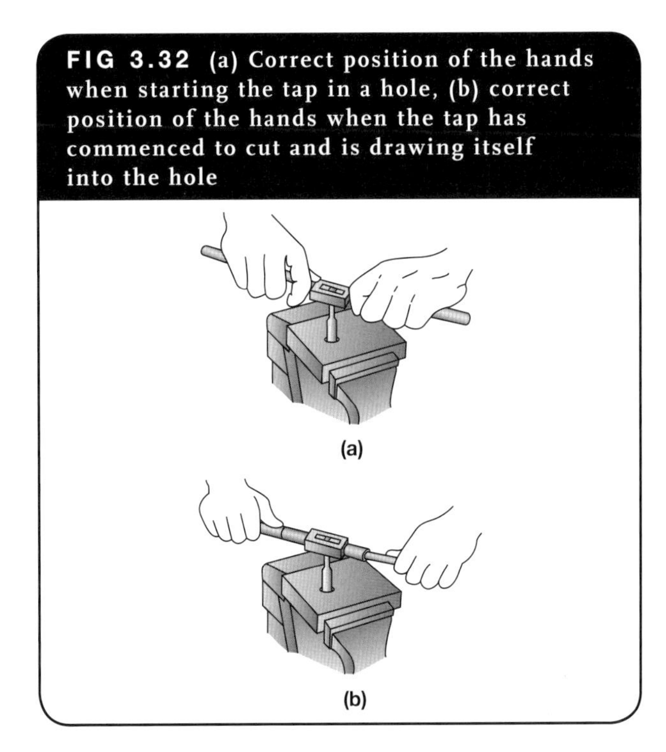

FIG 3.33 (a) Correct position of the hands when starting the die on a work piece, (b) correct position of hands when the die has commenced to cut and is drawing itself on to the work piece

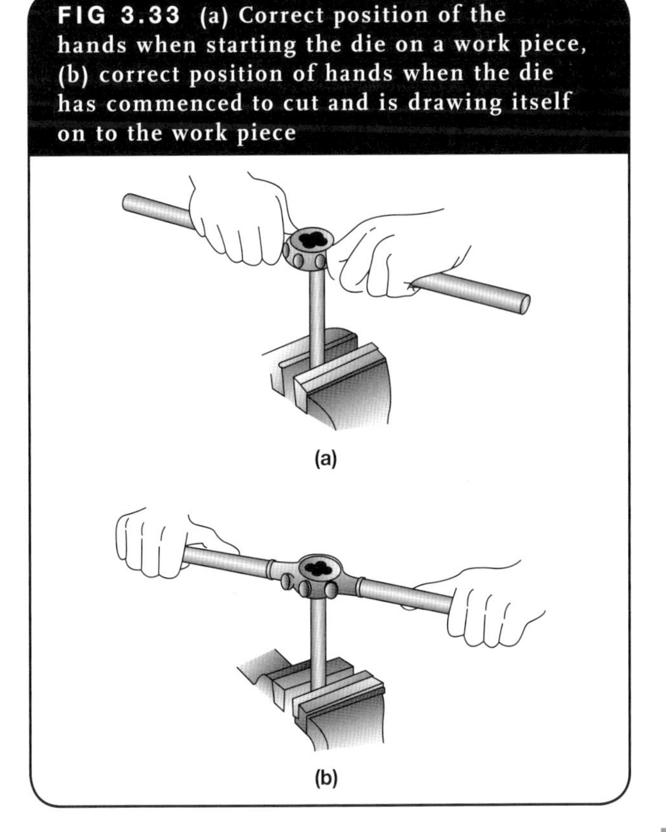

TABLE 3.2 Cutting fluid for taps and dies

Material	Cutting fluid
Steel	Neatsfoot oil or sulphur-based oil
Stainless steel	Neatsfoot oil or sulphur-based oil
Brass and copper	Kerosene and lard oil
Cast iron	Kerosene or soluble oil
Aluminium	Kerosene and light mineral oil
Plastics	Soap solution

Once a thread has commenced to cut, the tap will draw itself into the work. Continued downward pressure is therefore not necessary. However, when chip pressure is felt, the tap should be backed off a quarter to half a turn to allow the chips to break. Never force a tap as it can become tightly wedged in the hole and further pressure will cause it to break.

Using stocks and dies

To assist the die to start, a small bevel may be ground or filed into the rod or pipe to be threaded. In the case of water pipes this practice is rarely carried out. Secure the work piece in a vice, either vertically or horizontally, depending upon the length of the piece. Select the correct cutting fluid (Table 3.2), grasp the die holder with both hands close to the die and place the die head onto the job (Figure 3.33(a)). Press down firmly and, at the same time, rotate the die holder in a clockwise direction, reversing the direction when chip resistance is felt, as when tapping a thread. When the thread has started, pressure on the die is no longer necessary as the die will pull itself on as the thread cuts.

TABLE 3.3 Metric thread—ISO (coarse) 60° thread form

		Tapping drill sizes			
Size	Pitch	Preferred		Alternative	
(mm)	(mm)	(mm)	(inch)	(mm)	(inch)
2.0	0.40	1.65		1.6	1/16
2.5	0.45	2.1		2.05	
3.0	0.50	2.55		2.5	
3.5	0.60	2.95		2.9	
4.0	0.70	3.4		3.3	
4.5	0.75	3.8		3.7	
5.0	0.80	4.3	11/64	4.2	
6.0	1.00	5.1	13/64	5.0	
7.0	1.00	6.1		6.0	15/64
8.0	1.25	6.9		6.8	17/64
9.0	1.25	7.9	5/16	7.8	
10.0	1.50	8.6	11/32	8.5	
11.0	1.50			9.5	3/8
12.0	1.75		13/32	10.2	
14.0	2.00	12.2	31/64	12.0	15/32
16.0	2.00	14.25	9/16	14.0	35/64
18.0	2.50	15.75		15.5	39/64
20.0	2.50		45/64	17.5	11/16
22.0	2.50		25/32	19.5	49/64
24.0	3.00			21.0	53/64
27.0	3.00		61/64	24.0	15/16
30.0	3.50			26.5	1.3/64
33.0	3.50		1.11/64	29.5	1.5/32
36.0	4.00		1.17/64	32.0	1.1/4

Note: Metric ISO (coarse) replaces the following thread forms: BSW, UNC

TABLE 3.4 Metric thread—ISO (fine)

		Tapping drill size			
Size	Pitch	Preferred		Alternative	
(mm)	(mm)	(mm)	(inch)	(mm)	(inch)
8.0	1.00	7.1	9/32	7.0	
10.0	1.25	8.9		8.8	11/32
12.0	1.50		27/64	10.5	
14.0	1.50		1/2	12.5	
16.0	1.50	14.75	37/64	14.5	
18.0	1.50		21/32	16.5	
20.0	1.50		47/64	18.5	
22.0	1.50		13/16	20.5	
24.0	2.00		7/8	22.0	

Note: Metric ISO (fine) replaces the following thread forms: BA, BSF, UNF

Threading mild steel pipe

The threads on mild steel pipes are cut with the dies held in die holders commonly called 'stocks' (Figure 3.34). The stocks are rotated by means of detachable handles which

FIG 3.34 Hand-operated stock and block die

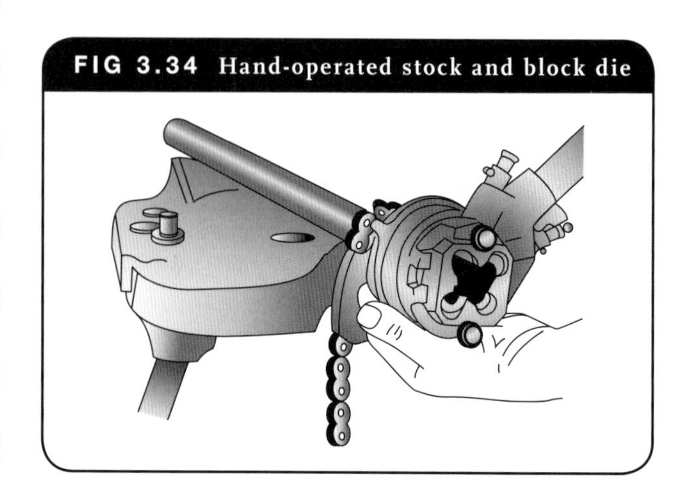

may be fixed in relation to the dies or of the ratchet type which permits the handles to be rotated independently of the dies; the ratchet allows rotation of the dies in a confined position.

The block die (Figure 3.30(d)) cuts a full thread in a single operation; the die is tapered to provide a non-parallel thread. This is to overcome any variation in pipe diameter. A separate block die is required for each pipe size. The sizes commercially available range from 7 mm to 50 mm.

Types of stocks

Stocks, as noted above, fall into two general categories, 'fixed' and 'ratchet'. Fixed stocks are available for both large and small diameter pipes. The handles screw into the body of the stocks. A guide is fitted to the back of the die head to ensure correct centring of the work and that the thread is parallel to the side of the pipe. The use of fixed stocks is limited, their main use being confined to open positions and threading pipes in vices.

Ratchet stocks are more versatile; the ratchet action allowing threading to be carried out in confined positions where a complete revolution of the handle is not possible.

CARE AND USE OF STOCKS AND DIES

When using stocks and dies on water pipe and steel the following points should be observed.

- Remove burrs from the end of the pipe before starting the thread.
- Select the correct lubricant; this assists in making a clean cut, assists in dissipating the heat generated and reduces friction.
- Keep dies clean and free from chips.
- Replace chipped dies.
- Protect the dies from rust and mechanical damage when not in use.

WARNING: The waste from the cutting process is known as 'swaif'. Care should be excercised in its clean-up as it is sharp and could lead to hand or eye injuries.

FOR STUDENT RESEARCH

Australian Standard

- AS 1722.2: 1992 Pipe threads of Whitworth form, Part 2: Fastening pipe threads

PLUMBER PROFILE 3.1

SEAN ANDREAS

Sean Andreas

Job Title: Co-owner of Plumbers Today & Gas Today, Queensland, WorldSkills Regional and National judge, WorldSkills trainer

Sean has been in the plumbing industry since he was 13 and has worked in the industry for over 30 years. He has been heavily involved in the WorldSkills competitions as a competitor, then as a judge and also training international competitors. He has travelled all over Australia and the world for WorldSkills and also has his own plumbing and gasfitting business that he set up with his brother in south Queensland. Sean is a recipient of the Plumbing Services Award.

When and how did you get into WorldSkills competitions?

As a second year apprentice—I was around 19 years old—I was recommended to compete at regional competitions in Toowoomba, won the gold medal there and went to compete nationally and there were 26 competitors there. I won gold regionally again in 1989, receiving a total of three gold medals and a bronze in regional, national and international competitions.

Currently I am not involved in any active way with WorldSkills but over my 22 years of participation I have been a regional/national/international competitor, regional/national/international judge/expert, regional/national/international project designer, regional/national/international trainer, international chief judge/expert and generally assisting in a volunteer capacity for competition set-up, operation and pull down.

What have you most enjoyed about your involvement in WorldSkills competitions?

Skill and personal/professional development, friendships with people all over the globe, seeing and being in places I never dreamed of, contributing to others' development and experience, being part of something that is totally positive.

Where have you travelled to as a competitor?

With the second round I went to national, and from that point went to Adelaide, Sydney, Perth, Canberra and onwards from there to the UK. I also went to work in Germany and Austria. So I competed internationally and then went on to do work and travel for around three months in Germany and Austria. I was doing hot-water pipe lines and general water systems over there. Over the years I have been all over Australia, Korea, Switzerland, The Isle of Man, United Kingdom, Germany, Austria, Belgium, Luxemburg and France.

How did you start up your own business?

I commenced my own business and started building it when I returned from my travels in Germany. I have had the business 22 years now; that's a bit scary! The most rewarding thing about it I think is the chance to play a part in developing tradespeople's skills and their personal development, and working with the community.

What has been your most challenging job so far?

Delivering contract work in remote locations: I had to deal with weather conditions, remote access and remote deliveries.

What was your worst job?

I think it was a major construction project with very tight timeframes, workforce restraints, weather constraints, trampling over each other to get stuff done, and demanding site managers. I got through it by working through the hard times and making the best of it, and being able to visualise the end result, even though it was difficult.

Could you describe one of your most memorable experiences as a plumber?

A great highlight as a plumber was receiving an international medal for second place from Margaret Thatcher in Birmingham in 1989.

What would you say has been one of the most rewarding things about your job?

A big part is the satisfaction of seeing a job well done but also contributing to people's professional and skill development.

I enjoy estimating and completing a job. Also, even though it was difficult, working in remote locations was rewarding. The lifestyle was different, things like four-wheel driving. It was just out of Cooktown. We were in the north end of Queensland for around six months, visiting new locations, going fishing out of hours—the lifestyle was just so different. Also that I have been all over the world—it has been quite a journey.

What would you advise plumbing students?

To listen and learn and apply what you do with a view to excellence both in skill and interaction with people.

Forming and bending

LEARNING OBJECTIVES

In this chapter you will learn about:

- **4.1 types of joints in sheet metal**
- **4.2 types of sheet metal machinery**
- **4.3 pipe and tube bending**
- **4.4 how to form copper tube**
- **4.5 junction forming of upper tube.**

INTRODUCTION

The plumber must be capable of fabricating sheet metal into various shapes and sections of flashing and ducting. In sheet metal the most commonly used joints are the 'grooved seam', the 'Pittsburgh lock' and the 'cleat joint'.

JOINTS IN SHEET METAL

The grooved seam

The grooved seam involves folding over the edges of the material to be joined in a folding machine, with one edge turned up and the other turned down (Figure 4.1). When interlocking the grooved seam, the folded edges should lie parallel to each other (Figure 4.2). The joint should be positioned on a solid base and the groover selected. Before commencing to groove seam the joint, centre punch each end of the joint to hold the seam in place. The grooves should be approximately 2 mm wider than the fold; this allows for the metal thickness to be accommodated during grooving. Place the groover over the folds (Figure 4.3) and clench the groove. The completed grooved seam appears as shown in Figure 4.4. The grooved seam is commonly used as a longitudinal seam on round, square and rectangular work, or for joining flat materials together.

FIG 4.1 Joint folded in preparation for grooving

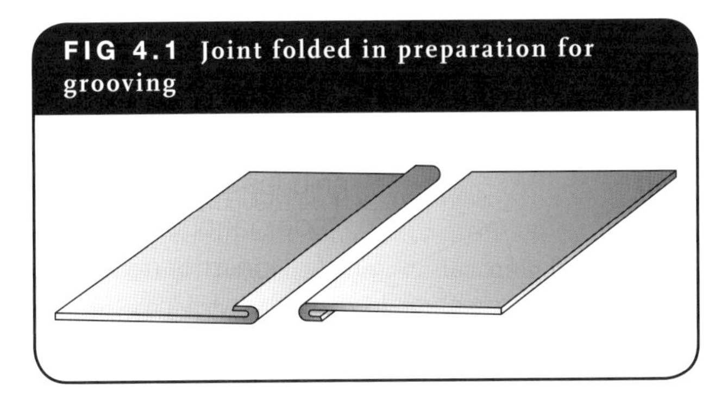

FIG 4.2 Joint interlocked in preparation for grooving

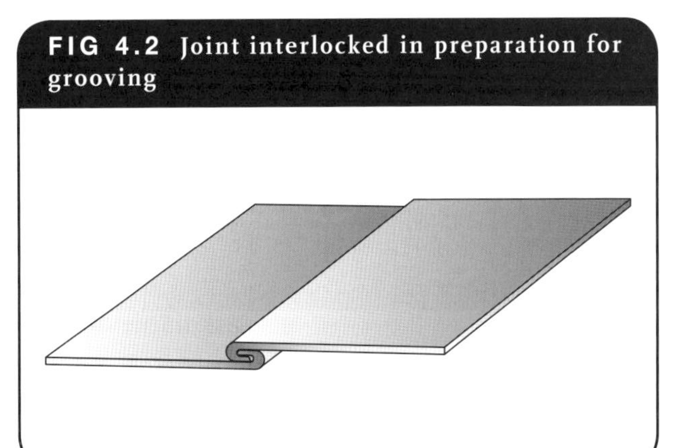

FIG 4.3 Clenching the grooved seam

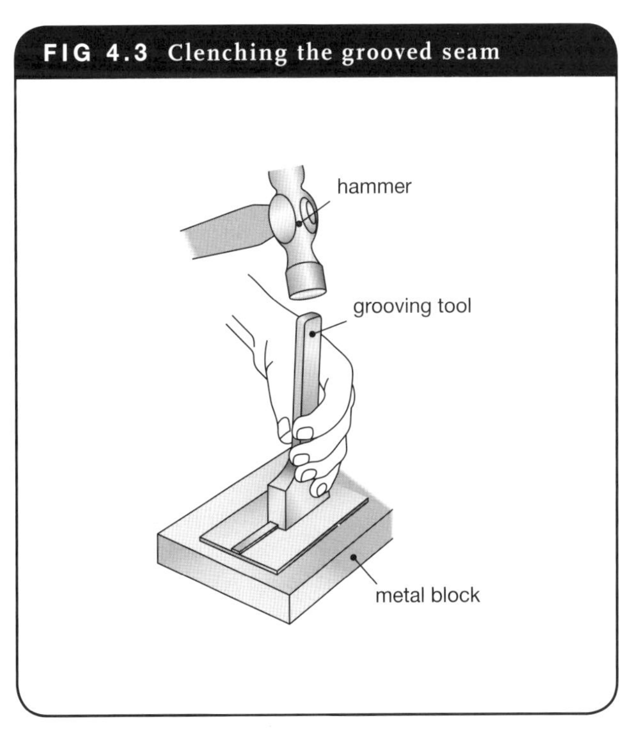

FIG 4.4 Final grooved seam

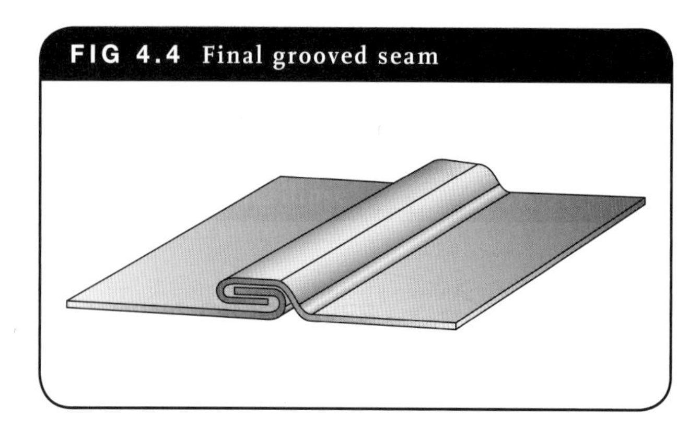

The Pittsburgh lock

The Pittsburgh lock seam is a roll-formed joint, usually made by passing the metal through a lock-forming machine. This machine has several rollers which gradually form the shape of the joint as the metal passes through them. The joint consists of two components, the 'lock form' and the 'male flange' (Figure 4.5). The standing edge of the lock form section is dressed over the male flange to secure the joint with a mallet (Figure 4.6).

FIG 4.5 (a) Lock form, (b) male flange components of Pittsburgh lock seam

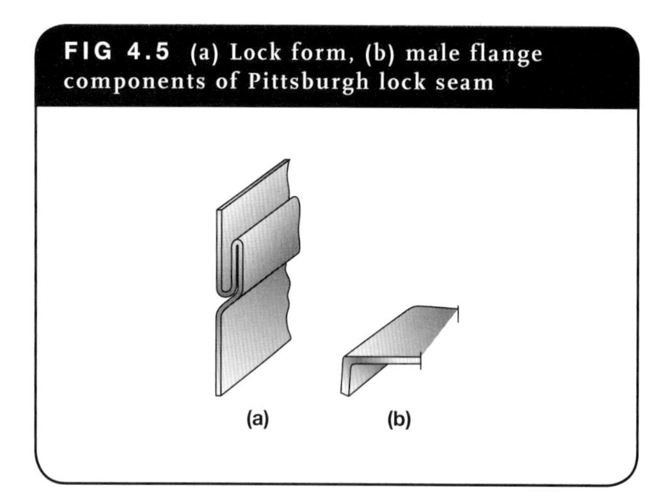

This joint is used as a longitudinal corner seam. It can only be used where one side of the material forming the joint has a straight edge. This limits its use to rectangular straight-section work.

This type of joint is now done in a factory environment using a specialist machine.

FIG 4.6 Completed Pittsburgh lock joint

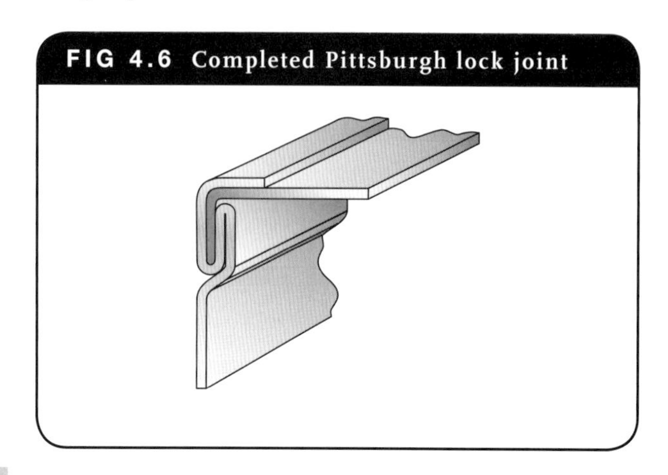

The cleat joint

The cleat consists of a strip of metal with two folds, both in the same direction, which drives over the folds made on the edges of the sections being joined (Figure 4.7).

A tongue is left on one end of the cleat and is folded over prior to final fitting (Figure 4.8). The cleated joint is used extensively in air-conditioning work, on square and rectangular ducts. It is also used to join wall panelling, particularly where ease of dismantling is needed.

FIG 4.7 Cleat and edge treatment of ducting

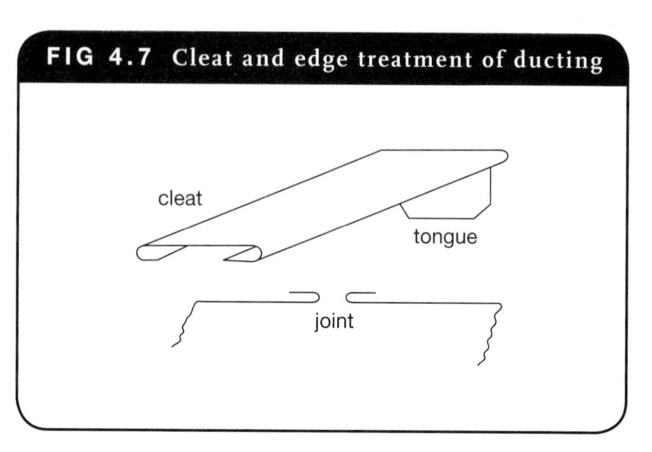

FIG 4.8 Fitting of cleats to a joint

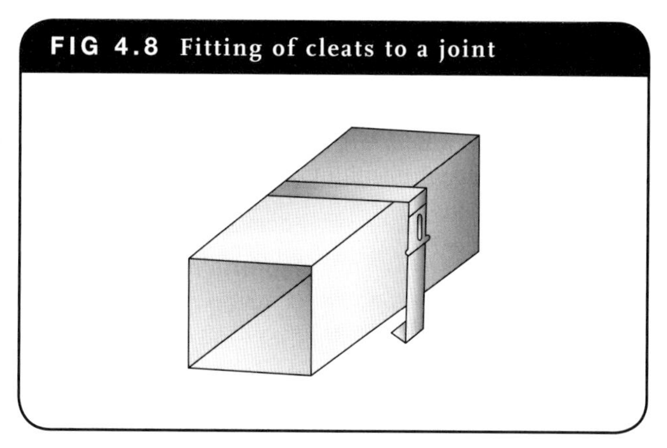

SHEET METAL MACHINES

The joints described above in the section on forming require the use of sheet metal-working machines. Some of the machines commonly found in plumbing workshops are: the cramp folder, the pan break folder, the guillotine, the lockformer, and rollers.

The cramp folder

This machine is used for general longitudinal folding of all sheet metal. The bending flap is carefully balanced and is made so that various bending bars can be used. Stops are provided at the front and rear of the machine, thus allowing for repetition work to be done with accuracy. Stops are also provided to allow various angle folds to be repeated. To allow for various thicknesses of material to be folded, an adjustment nut is situated on each end of the clamping bar and the front bending flap is also lowered to adjust for metal thickness (Figure 4.9).

FIG 4.9 (a) Cramp folder and bender, (b) detail of blade and folding edge

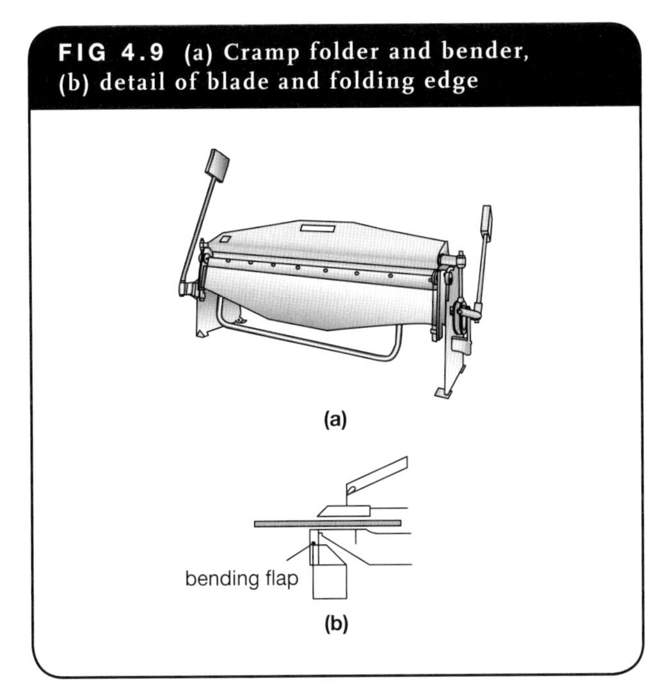

The pan break folder

This folder is more versatile than the cramp folder and is one of the most important folding machines in the workshop. The machine (Figure 4.10(a)) resembles and has similar adjustments to the cramp folder, but the top clamping blade is in sections of different widths (Figure 4.10(b)). This feature enables the forming of boxes and trays of varying depths and widths.

FIG 4.10 (a) Pan break folder, (b) detail of blade section at folding edge

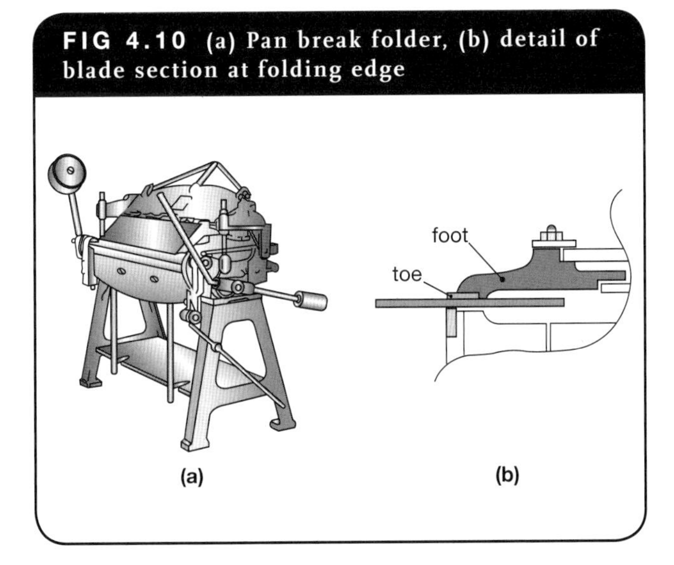

The guillotine

The guillotine is a shearing machine that is used to accurately produce straight cuts in sheet metal. Guillotines are available in both foot-operated and power-operated models (Figures 4.11(a) and (b)). The operation of these machines requires the utmost care and consideration both for the operator's safety and the machine's efficiency.

FIG 4.11 (a) Foot-operated guillotine, (b) power-operated guillotine

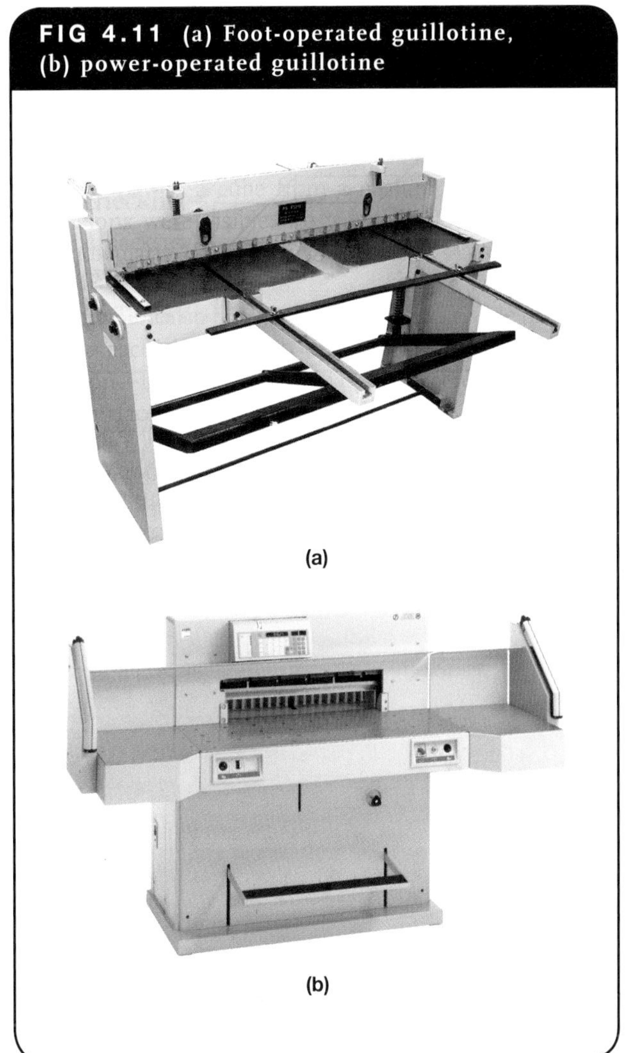

The lockformer

Cold-rolled forming machines (Figure 4.12(a)) are used to form special profiles for joints such as the Pittsburgh section, the cleat section and the flange section. The machine head

FIG 4.12 (a) Lockformer

consists of a series of hardened rotating formers shaped to gradually form the metal into the desired shape. The main section made by this machine is called a Pittsburgh lock, and the machine is therefore often misnamed a 'Pittsburgh machine'. In fact, the machine is more versatile than this and many shapes can be formed.

Rollers

Rollers are used in the plumbing industry for curving metal to the desired shape. Curving rollers are used for general curving of sheet metal. They consist of three rollers (Figure 4.12(b)) mounted in end frames. The front two rollers are joined to and activated by a crank handle; the third roller is situated behind the work. This roller is adjustable and is the pressure roller for curving the metal to the required shape.

FIG 4.12 (b) Curving rollers

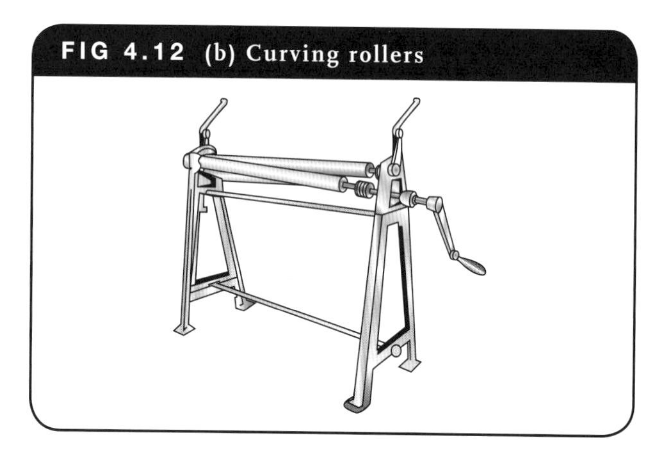

PIPE AND TUBE BENDING

For a proper understanding of the various tools, machines and techniques employed in pipe and tube bending, it is desirable that students should appreciate clearly what changes occur to the pipe itself during bending—distortion. One essential bending requirement is that the metal must have sufficient ductility for it to deform into the shape of the bend without seriously weakening the wall. The most common tendency is for the pipe to flatten or buckle along the inside of the curve. Distortion usually takes place because the throat of the pipe compresses, the heel stretches and the sides bulge outwards. In most cases the pipe walls must be supported either internally or externally in order to prevent the stresses set up from causing undue deformation. The pipe walls must remain parallel if the true round section is to be maintained in the bend. In Figure 4.13, the original length of die tube O–O remains unaltered along the centre-line of the pipe bend O–O. The inside of the bend is shortened and compressed while the back is lengthened due to stretching. A collapse of the pipe bend will result unless precautions are taken to prevent it. When bending manually, it is necessary to support the pipe walls during the bending operation. Support can be provided by using specially designed, coiled internal or external springs.

Hand bending (copper tube)

Hand bending may be carried out with tubes of any diameter and gauge to any throat radius commencing with approximately 1.5 pipe diameters. With larger diameter tubes, it may be necessary to use levers to assist the effort of pulling the pipe to the required angle.

Sand filler

Sand used in copper tube and steel pipe bending must be dry and free from impurities. It must be dry because once in the tube with both ends sealed, the application of any heat would cause steam to form, which could burst the tube. Hot sand bending is used on tubes requiring a small throat radius.

LOADING PROCEDURE (COPPER TUBE/SAND LOADED)

1. Warm the tube to evaporate any moisture that may be present inside the tube.
2. Insert a wooden plug in one end of the tube and pour in the sand with the tube held in a vertical position.
3. Pack the sand by striking the outside of the tube with an appropriate tool.
4. When the sand is consolidated, drive a tight-fitting tapered plug into the open end of the tube.
5. The tube is now ready for bending.

FIG 4.13 True and deformed bends: (a) true bend; each division represents a 'throw' in making the bend, (b) deformation of bends in unsupported tube

BENDING PROCEDURE (OXYACETYLENE FLAME)

1. Clamp the tube in a vice in such a manner that maximum leverage is achieved.
2. Select the correct size of heating tip to suit the diameter of the tube.
3. Heat the range of the bend to a dull red heat using a neutral to slightly carburising flame.
4. When the tube is evenly heated, commence to pull evenly in a constant horizontal plane.
5. Maintain an even heat on the throat and sides of the tube (but not the back, due to the possibility of the tube splitting at the heel of the bend because of excessive stretching). Do not heat to too high a temperature as this can cause excessive grain growth in the material being heated. Overheating may also cause an undesirable surface to develop. At worst, it can lead to rupture of the metal during bending.
6. Bend the pipe approximately 3° beyond the desired angle, then open it out again to remove any flattening of the throat and heel.
7. Check the accuracy of the bend for tube roundness and angle. A template made from flat sheet metal can be cut to the given throat radius specification. This enables a check to be made during and after bending.

Tube benders (lever type)

A variety of mechanical devices can be used for tube bending. One of the most common is the lever tube bender illustrated in Figure 4.14.

This tool is versatile, accurate, and easy to use and is ideal for bending small bore (15 mm–20 mm) copper and stainless steel tubes. It is suitable for both hard drawn and annealed copper, making smooth, short radius bends up to 180°. Most benders are calibrated in degrees to give a precise angle. They require that the tube is *not* annealed.

Bending springs

Bending springs are used extensively in hot and cold water plumbing installations and particularly with small bore copper tube. The bending spring consists of a length of tough coiled spring steel which may either be inserted into a tube (internal) for tight radius bends or fitted over the tube (external) for larger radius bends. Bending is normally achieved by pulling the section of tube that is supported by the spring over the knee. Figure 4.15 shows an external bending spring. The tube must be annealed and cooled before the spring is placed in position. It should be slightly overbent and then returned to the correct angle to allow the spring to be released from the tube.

FIG 4.15 External bending spring

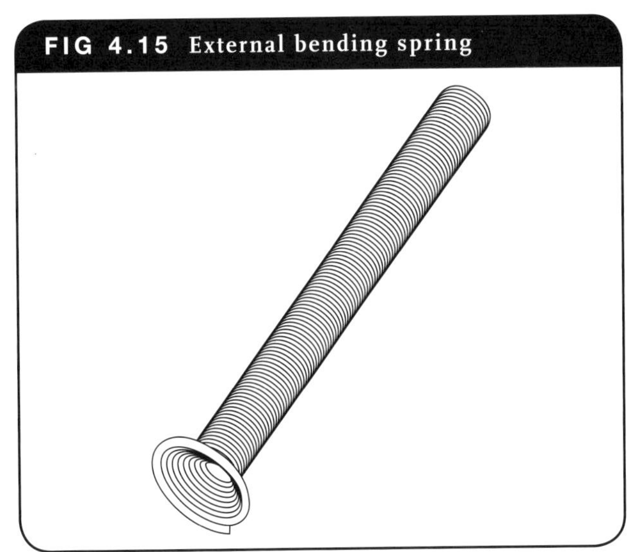

Bending plastic piping

Bending is not recommended with most plastic piping materials, particularly when the line is to operate at or near maximum pressure and temperature, since the process leaves residual stresses in the pipe. The use of factory-made bends is recommended in preference to field bending of most thermoplastic material.

Bending polyethylene piping

Polyethylene pipe can be 'cold bent' to a radius of 8 pipe diameters for pipe sizes up to 50 mm in diameter.

Hydraulic pipe benders

Hydraulic pipe benders use the 'ram and press' principle in which the steel pipe being bent is supported between two members (rollers) while a third member (the ram) provides the bending force.

FIG 4.14 Lever-type tube bender

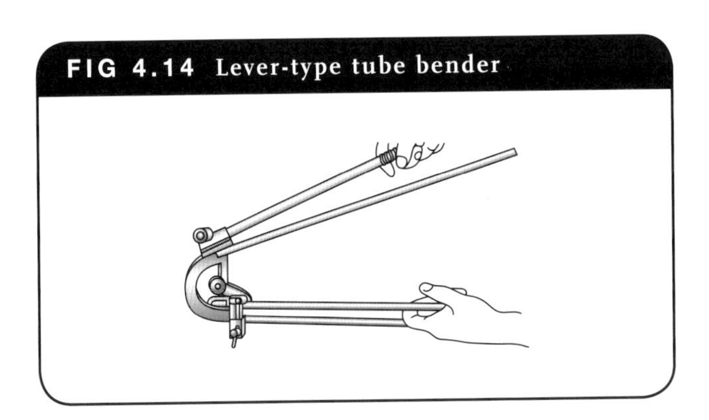

FIG 4.16 Manually operated hydraulic bending machine

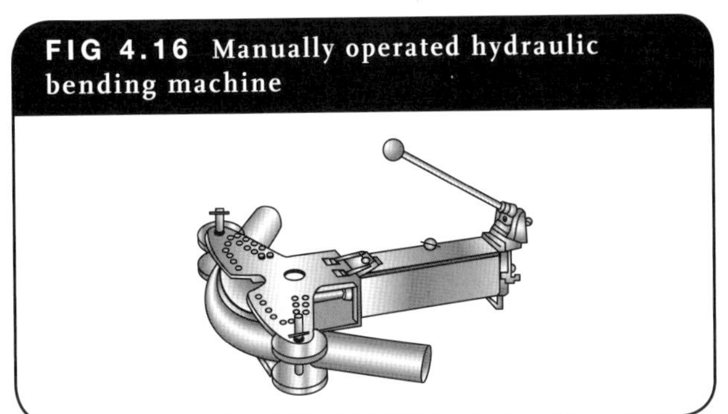

PROCEDURE FOR MANUALLY OPERATED PIPE BENDERS

1 Select a former to suit the pipe being bent and place the pipe in position on the ram and the two rollers. The centre of the former should contact the pipe at the centre of the proposed bend.

2 Close the release valve and operate the hand pump until the ram has travelled sufficient distance to produce the desired angle. Due to the spring reflex of the pipe, it may be necessary to travel a little beyond the desired angle. With a little practice, the operator will soon make allowances to overcome this problem.

3 To remove the bend, open the release valve to permit the ram to return to its normal position.

Figure 4.16 shows a hydraulic manually operated pipe bending machine.

Power-driven benders

Electrically driven hydraulic bending machines are available and may be either free standing or portable. They are used for bending extra heavy and large diameter pipes. The electric motor drives the hydraulic pump which in turn delivers power to the hydraulic cylinder for advancing the bending ram. Figure 4.17 shows an electrically driven hydraulic bending machine.

FIG 4.17 Motor-driven hydraulic bending machine

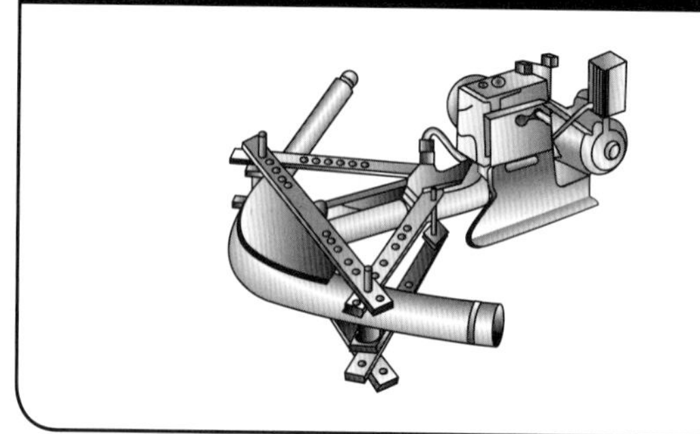

FORMING COPPER TUBE

Forming differs from bending in that more severe deformation is required and appreciable changes in wall thickness occur because stretching of the metal is involved. However, over a great many years seamless copper tube has proved to be a highly suitable material for almost all plumbing services in buildings. The use of this superior metal in sanitary plumbing installations has made necessary the introduction of a range of tools for forming various shapes in copper tube. Seamless copper tube is used in soil, waste and vent applications because it possesses many advantages over other approved plumbing materials. Although it is not manipulated as easily as lead, copper can nevertheless be worked to meet most situations because of its malleability, and its excellent corrosion-resistant properties permit copper services to last almost indefinitely.

Manually formed junctions

Plumbing pipe systems, including waste connections from fixtures, require connection at some point of discharge. These connections usually terminate at a junction. The following is a standard procedure for fabricating junctions in seamless copper tube.

BRANCH PIPE PREPARATION

1 Obtain the angle of entry (using a bevel square or template).

2 Cut the branch pipe to this angle (we will assume the angle is 60° as in Figure 4.18).

3 Remove any burrs by filing with a half-round file.

MAIN PIPE PREPARATION

1 Mark the centre-line on the main pipe.

2 Measure the length of the oblique cut from point 2 above and mark the toe and heel positions on the main pipe.

Allowances for upstand

An allowance for upstand is estimated depending upon the diameter of branch pipe and angle of entry. The angle of entry is an important consideration because the flatter the angle, the more allowance for upstand is required. Table 4.1 gives a guide for upstand allowance.

FIG 4.18 Pipe toe cut to angle

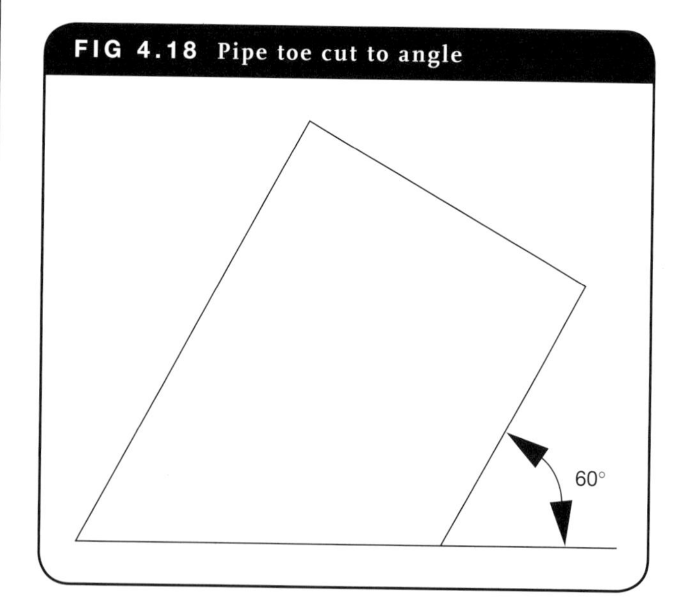

TABLE 4.1 Upstand allowances

Diameter of pipe (mm)	Angle of entry (°)	Upstand allowance (mm)
38	60	12
50	45	15
75	45	20
100	45	25

MARKING OUT FOR UPSTAND

Assume that the branch pipe is 38 mm in diameter and has a 60° angle.

1 Refer to Table 4.1 for upstand allowance.
2 Mark and centre punch drill holes 12 mm (plus half drill diameter) from toe and heel positions as indicated in Figure 4.19.
3 Drill 6 mm holes on centre-line positions.

OPENING OF MAIN PIPE

1 Anneal the opening and cut between the holes with a hacking knife or snips.
2 Commence working the opening with a bent bolt or turn pin and hammer as in Figure 4.20(a).
3 Enlarge the opening using successively larger steel mandrils.
4 Lift up heel and toe on the main pipe.
5 Lay the upstand over at 60° and frequently test for angle of entry, as shown in Figure 4.20(b), until the upstand is parallel to the branch pipe.

FITTING OF BRANCH

1 File the upstand parallel to the top of the main pipe.
2 Slip the branch pipe into the opening of main pipe and dress the branch to a snug fit with the mandril inside the branch, as shown in Figure 4.21.

FIG 4.19 Preparation of branch joint: (a) plan view, (b) side elevation

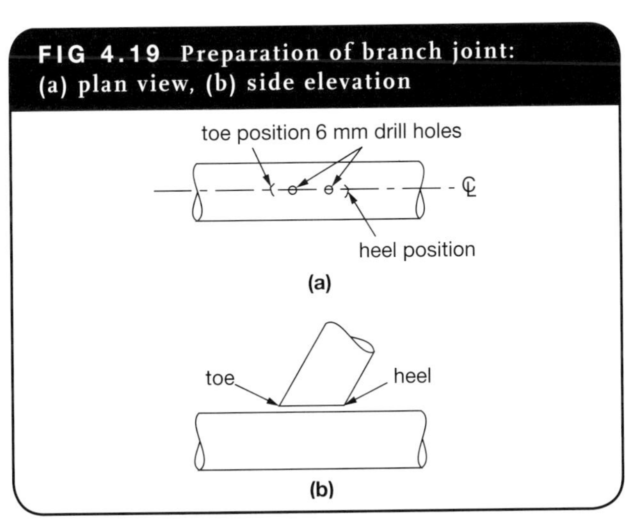

FIG 4.21 Branch pipe in position outside the mandril

Annealing copper tube

Seamless drawn copper tube is not readily worked without first being annealed. It is therefore necessary to heat the area to be worked. If oxyacetylene equipment is used, a neutral to slightly carburising flame is recommended. The flame should be moved continuously over the area to be softened, the inner cone approximately 12 mm away from the surface of the tube. Care should be exercised not to allow the flame to remain stationary and melt the tube. Melting of the tube is likely to damage the copper, resulting in crystallisation of the affected area. Selection of the correct blowpipe tip is essential; as a general guide the sizes for the corresponding tube diameters are shown in Table 4.2.

TABLE 4.2 Correct tip size

Copper tube diameter (mm)	Tip size
15	10–12
20	10–12
25	12–15
38	15–17
50	17–19
75	21–26
100	21–26

FIG 4.20 (a) Work the opening with bent bolt or turn pin, (b) enlarge opening with mandril

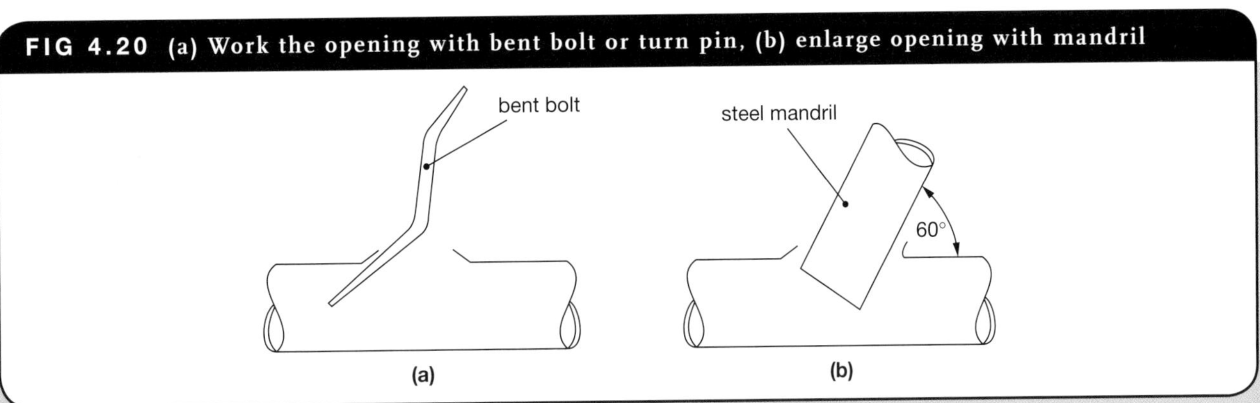

JUNCTION FORMING (USING SPECIAL TOOLS)

The following is only one of the many adopted procedures in fabricating junctions in seamless copper tube, using special forming tools. The method illustrated below is similar to the manual method. It is, however, always advisable to follow the instructions issued by the tool manufacturer. Figure 4.22 illustrates a sequence of operations involved in forming a junction.

Branch joints in small bore copper tube

The following is only one of the many adopted procedures in junction forming kits which are suitable for making square branch tees in small bore copper tube. They are simple to use, ensure a good fit and produce an approved joint. It is essential that this type of joint is a lap fit and not merely butted. It is, however, always advisable to follow the instructions issued by the tool manufacturer.

PROCEDURE FOR FORMING A JUNCTION

1. Mark out the position of the drill holes. Refer to Table 4.1.
2. Drill the holes using the correct drill size.
3. Mark out the area to be annealed.
4. Anneal the tube using the correct size of tip, selected from Table 4.2.
5. Incise the tube between the drilled holes to accommodate the forming tool.
6. A typical copper tube branch-forming tool showing the component parts.
7. Draw the dome-shaped die through the tube by turning the threaded bolt.
8. File the branch upstand parallel with the main pipe.
9. Completed branch opening.

FIG 4.22 Junction forming

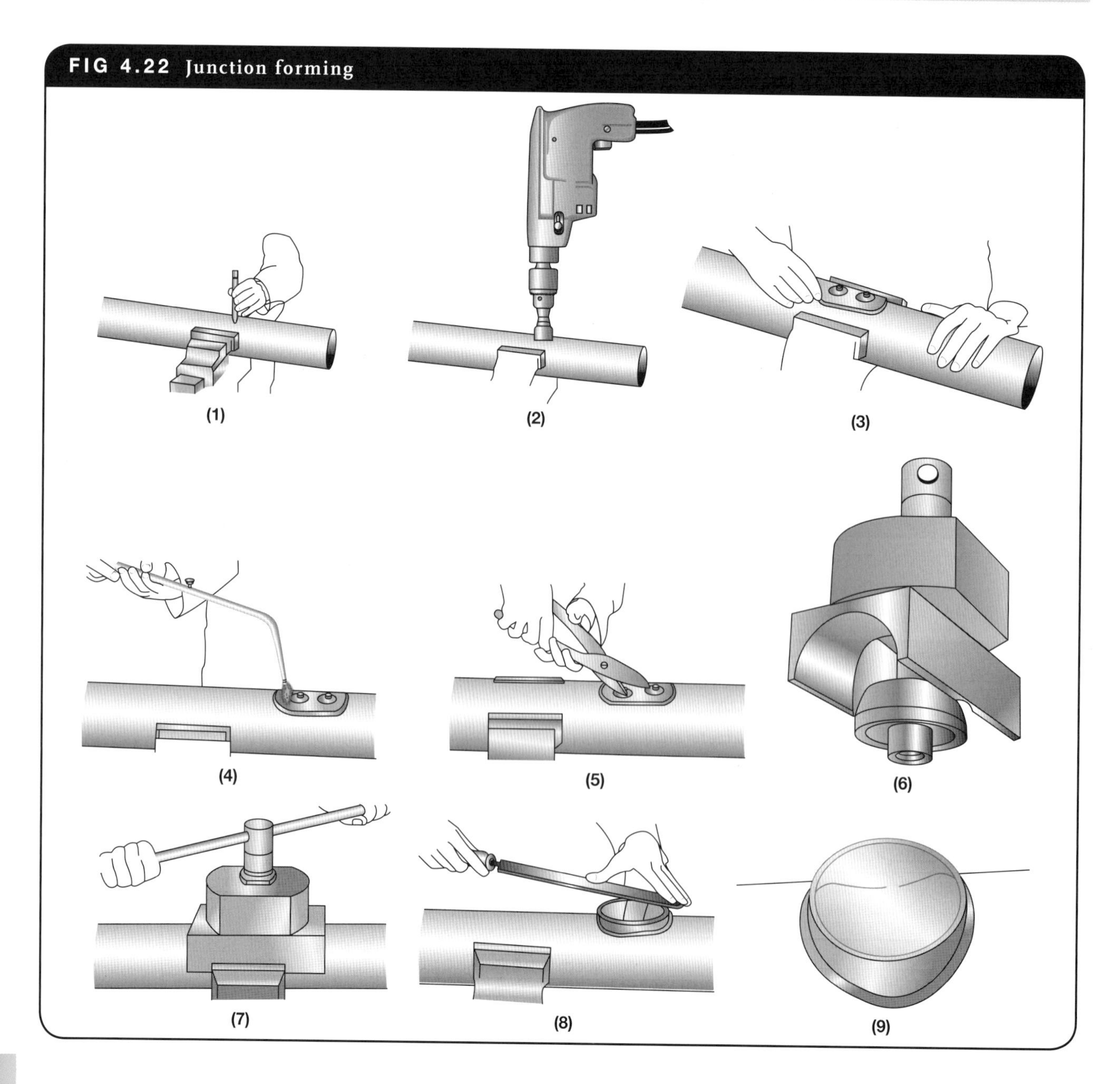

PROCEDURE FOR MAKING THE JOINT

1. Drill a hole in the main tube. The size of the hole depends upon the diameter of the branch tube.
2. Anneal the localised area around the drilled hole.
3. Clamp the main tube in a split block so that the hole is centred over the appropriate opening in the block.
4. Work the drilled hole using a bent bolt so that the annealed copper is forced against the wall of the block.
5. Insert a forming tool into the enlarged hole and twist it clockwise (Figure 4.23(a)) until an upstand is formed.
6. Remove the main tube with its newly formed upstand from the split block.
7. Prepare the branch tube: cut square, deburr, expand and shape to fit over the upstand of the main tube. The completed tee branch, shown in Figure 4.23(b), forms a suitable joint and is ready for joining.

Socket joints

Expanded longitudinal socket joints are used for joining tube sections in both small bore and large diameter copper tube (Figure 4.24(a) and (b)). Tube expanders are designed specifically to form a socket at one end of a copper tube thus saving time and eliminating costly intermediate fittings. The preparation of the joint prior to forming is important. The tube end must be square and free from internal and external burrs. The tube must also be annealed to enable the joint to be formed without splitting the tube.

Hand-operated lever type

Hand-operated two-stage lever action tube expanders (Figure 4.25) are ideal in water plumbing applications because the expander heads are interchangeable, and will therefore suit a range of tube diameters (12–25 mm). This type of tube expander is also suitable for use in confined spaces.

Screw-operated type

The screw-operated type of expander shown in Figure 4.26 is designed for large diameter copper tube, and offers the advantage of being suitable for a range of different tube diameters. The tube must be annealed and placed over the end of the expander jaws. The unit is screw operated and the socket joint is formed by turning the handle clockwise.

FIG 4.23 Branch joints. (a) Forming an upstand, (b) showing the branch tee

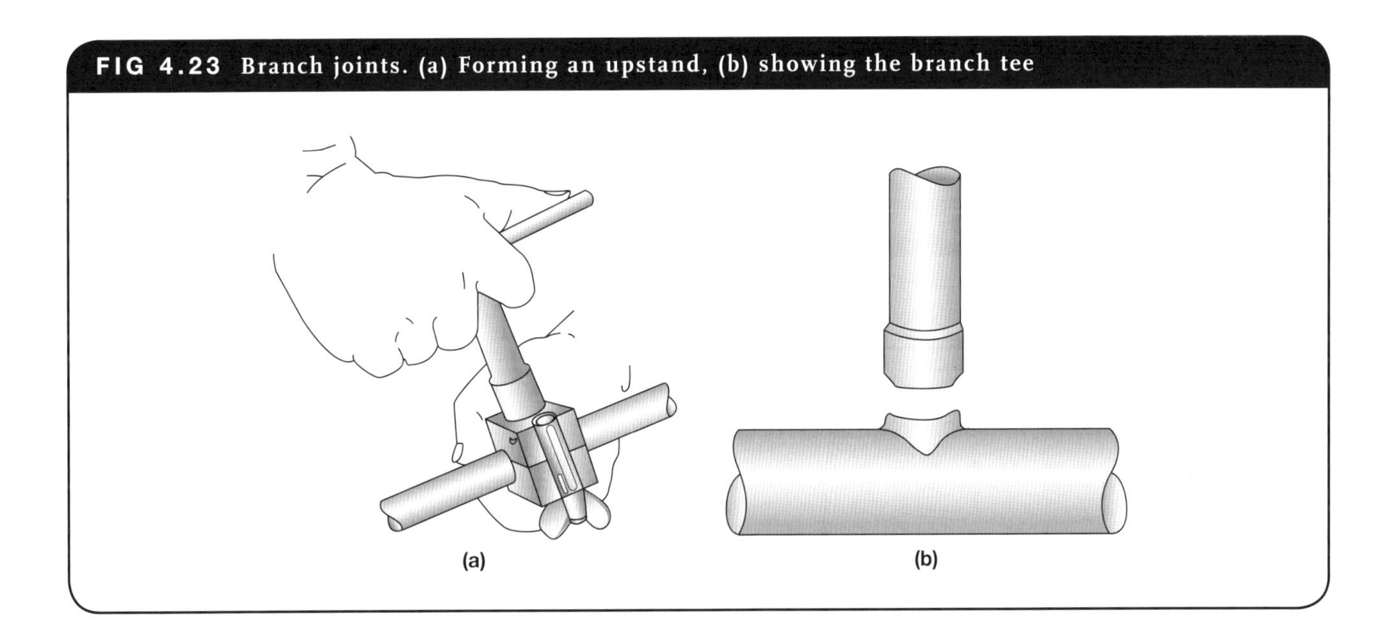

FIG 4.24 Socket joints. (a) Socket joint, (b) reduced socket joint

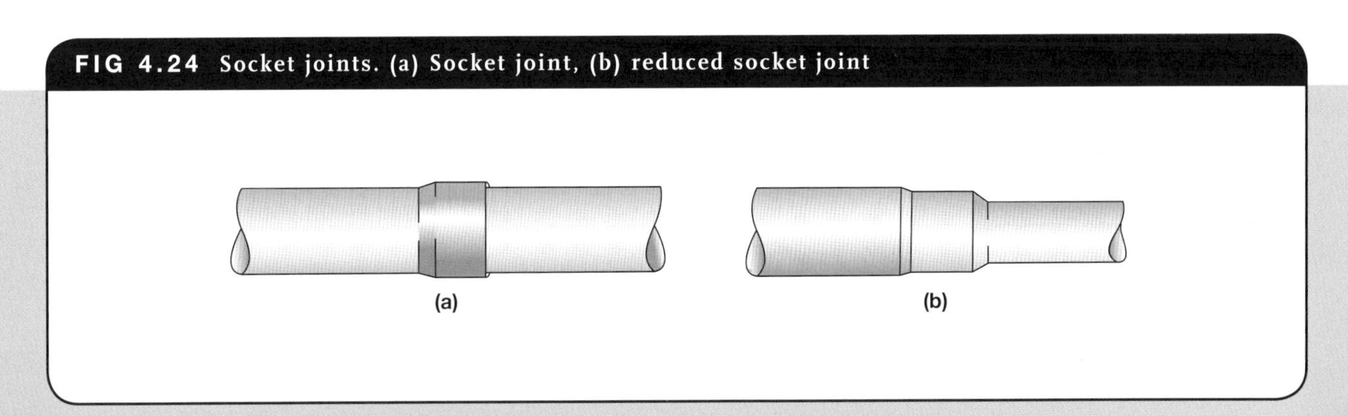

FIG 4.25 Two-stage lever type copper tube expander

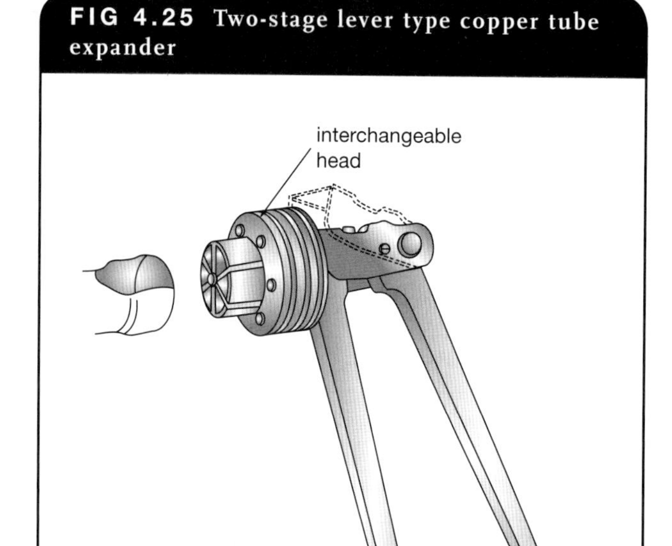

FIG 4.26 Screw type copper tube expander

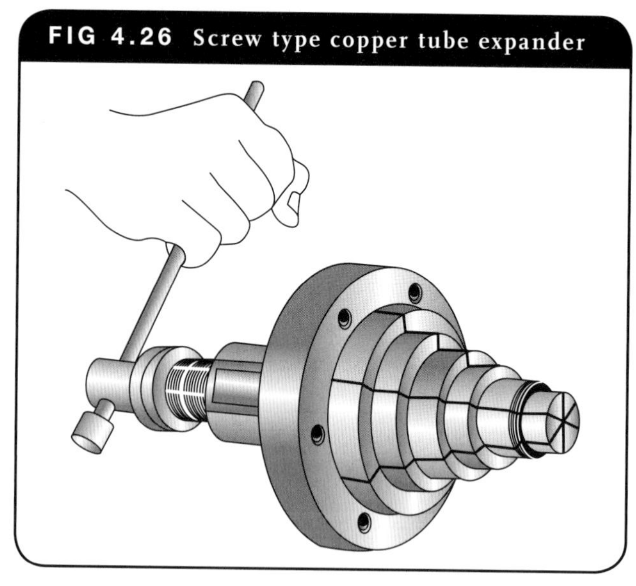

FOR STUDENT RESEARCH

- AS/NZS 3500.0: 2003 Plumbing and Drainage. Glossary of Terms

PLUMBER PROFILE 4.1

SIMON MCGANN

Job Title: Director of Simon and Co. Pty Ltd, New South Wales

Simon first got into the plumbing industry after following his father's advice, who suggested it to him as a career path. He moved to Sydney from the Central Coast for his apprenticeship after his pre-apprenticeship at TAFE. After some years as a tradesman he travelled overseas and upon his return set up his own contracting business back in the Central Coast of NSW, which he still has today. He has managed projects both small and large and was also given the opportunity to work on a drainage project in Samoa.

How did your career progress after your apprenticeship?

After my apprenticeship I stayed working as a tradesman for a commercial plumbing company for a few years, then I went travelling for a while to Europe on holidays for about four months, just travelling around, then I came back and started sub-contracting for my old company for a year and I just started working for myself and I'm still doing that today.

How did you set up your own business: was this easy for you to do?

I found it relatively easy as I had work sub-contracting for my old employer, which got me on my feet and then I just started building contacts from there.

What does a typical day of work look like?

I get up at 4.30 am and I am in the office at around 5 am each morning doing administration, quoting, and I aim to spend approximately three days a week through the day—7 am onward—working on the tools, and the other days quoting and organising jobs for my employees. I get home around 6 pm and spend time with my kids until 7.30 or so, and then go and work in the office again until around 9.30 or 10 in the evening. There is a level of satisfaction in what we do but after a few years of self-employment money is a big motivator to work those hours.

What is the biggest challenge you face as a business owner?

Finding good employees, people who are generally interested and enjoy being plumbers has been the hardest part. We have had a lot of people who don't really want to be there, and don't have a genuine interest in the trade so they are really bad for morale. I like to have a really motivated team around me and it only takes one bad guy to bring all that down. Motivated, positive and enthusiastic people, and capable comes after that.

What has been the biggest job you have taken on?

We did the installation of a sewer scheme in Mooney Mooney on the Hawkesbury River, which was about a $1.5 million job. I had to put some extra staff on that job and find good reliable people who wouldn't destroy our equipment, which was also quite difficult—but again, finding the right people was the biggest problem.

What is the worst job you have done?

The worst one ever was cleaning out a blocked commercial grease trap, I have never smelt anything like it. It wouldn't wash off for over a week—I could smell it and it was painful.

What has been your most memorable experience on the job?

Only last year actually, we had a project for about a month where we went to Samoa and we worked on a sewer scheme over there. That was just fantastic combining some travelling and cultural experience with work. It was the first time I had done something like this and I was there for around five weeks. It came about after the project I said was my biggest one. The head contractor there asked myself and one other guy who worked for us to come and help finish the sewer scheme they were working on and running a bit behind with in Samoa.

When we arrived we had the choice of about 35 local labourers to help us with what we had to do. We had very minimal tools to do it with—a lot less than what we were used to working with: we took a toolbox each with us on the plane and that was all. We had to be a lot more creative than we do at home to get things done without much equipment.

But it was great, we had a great relationship with the local boys we were working with. There were a couple of Australian guys over there, a project manager and an engineer, who were also great. We worked a lot but when we had time off the social life was good there too.

What was the biggest difficulty you faced on that trip?

I think the lack of modern tools like we are used to in Australia. Using sledge hammers instead of jack hammers, that sort of thing. We are used to having a range of different drills for concrete which just wasn't available there. It was back to using your hand tools.

What do you think are the perks of working in the plumbing industry?

I like having a mixture of working indoors and outdoors—I really like working outside. I think the best thing about the plumbing trade is how diversified it is—you can do so many different things. When I was in Perth I used to love doing roofing, I hate it now but I love doing drainage so there is plenty of scope knowing what you are doing can change within the trade, which is what I really recommend to young people about plumbing.

Is there any particular project you would say you are very proud of?

We have done a few projects where I have been, but probably the one in Samoa—when we were finished those people who had never had sewers before suddenly had them, which was amazing. In Australia, I don't know, we have done a lot of new factories and some of the factories we have built are pretty impressive, which are huge projects for a small business.

What do you most like about having your own business?

I guess the freedom to change the scope of what work we do. It's all up to me, so it's nice. Instead of people telling me what to do with my life I can choose.

Do you have any advice for future plumbers?

I think just really apply yourself to your trade while you are an apprentice and be respectful to those people who are trying to teach you. You have such a short time to learn so much so just generally apply yourself to it.

CHAPTER **FIVE**

Fastening, joining and sealing

LEARNING OBJECTIVES

In this chapter you will learn about:

- **5.1 types of anchor fixings**
- **5.2 powder-powered fastening**
- **5.3 soft soldering**
- **5.4 oxyacetylene welding**
- **5.5 welding techniques**
- **5.6 oxyacetylene cutting**
- **5.7 hard soldering techniques**
- **5.8 arc welding**
- **5.9 welding safety**
- **5.10 screwed pipe joints and fittings**
- **5.11 UPVC pressure fittings**
- **5.12 polyethylene pipe applications**
- **5.13 cast iron pipe and fittings**
- **5.14 sealants.**

INTRODUCTION

Within the plumbing industry many forms of anchors or fixing devices are used. They are used to secure plumbing fixtures, pipework, appliances and machines to walls, floors, ceilings and other building structures. The type of anchor or fixing device used for a particular job will depend upon a number of factors:

- the material to which the anchor is being fixed—concrete, brick, block, steel or timber
- the centre-line of the thrust or pull on the fixings—shear, push, vertical pull
- the weight that the anchor is required to support—heavy, medium or light loads.

After you have considered all these factors the most suitable fixing device for each application can be selected. A large variety of anchors are available.

TYPES OF ANCHOR FIXINGS

The types of anchor fixings available are:

- plastic universal anchor
- impact anchor
- concrete anchor
- sheath anchor
- drop-in anchor
- chemical anchor.

Fixing procedure

All these methods require a hole to be drilled into the material to which the anchor will be fixed. The size of the hole will be dictated by the size of the bracket or support being fixed.

When the hole has been drilled (ensuring that the correct size of drill bit has been used), the screw or bolt should be started into the plug before installing the bracket to ensure that the fixing screw is located in the centre of the plug for maximum strength. Having started the fixing screw, remove the screw and install the bracket, then drive the screw firmly into the fixing material. Avoid over-tightening as this can strip the thread that has been formed in the fixing material and severely weaken the fixing.

Plastic universal anchor

The plastic universal anchor is a versatile plastic anchor for a whole range of fastening into various base materials, such as concrete, masonry and dry wall installations (Figure 5.1).

FIG 5.1 Plastic universal anchor

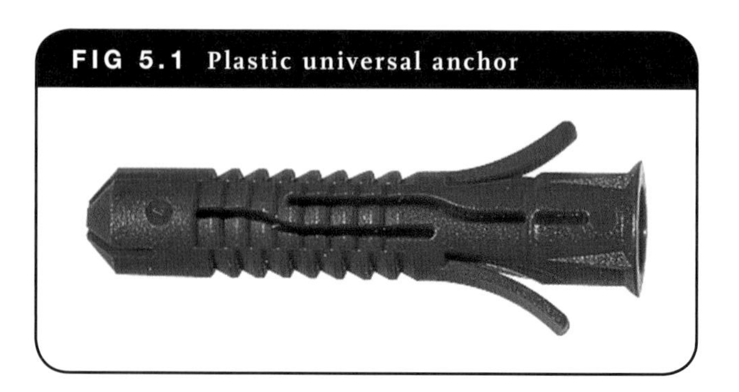

Impact anchor

The impact anchor is a ready-to-use anchor consisting of a plastic sleeve and a pre-assembled screw, which can be hammered home to expand the plastic sleeve (Figure 5.2).

The applications of use are to secure metal flashings, to install plumbing components (e.g. brackets and clips) and dry wall metal partitions fixing into concrete, solid and hollow brick walls.

FIG 5.2 Impact anchor

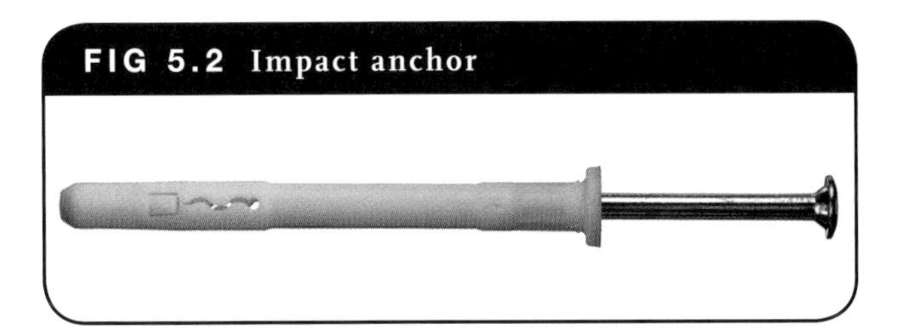

Concrete anchor

The concrete anchor is designed for a time-saving installation with a secure hold in concrete (Figure 5.3). The fastening system can be used for indoor and outdoor applications for securing formwork, pipe supports, door frames and windows. Drill a pilot hole with a masonry drill and drive the anchor into the concrete.

FIG 5.3 Concrete screw anchor

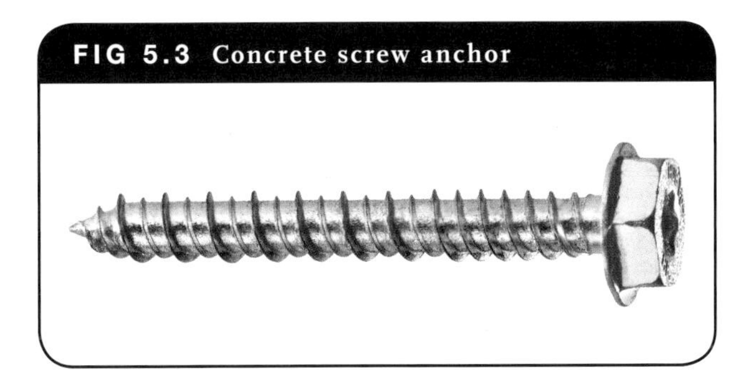

Sheath anchor

The expanding outer case or 'sheath' anchoring system has been developed to suit different situations.

The selection of the correct length of bolt depends upon the thickness of the material to be fastened and the minimum depth of anchor required in the concrete for maximum holding power.

The anchor illustrated (Figure 5.4(a)) is a complete unit ready to install and the problems of alignment of the bolt into the sheath are eliminated.

Method of fixing

Having selected the correct length and diameter, drill a hole equal to the diameter of the fastener. The anchor's performance depends upon the accuracy of the drilled hole; oversized holes reduce the holding power of the bolt. The anchor is driven into the hole (Figure 5.4(b)) until the nut and washer are flush with the fixture. The nut is then tightened.

Drop-in anchor

The drop-in anchor illustrated in Figure 5.5(a) has excellent holding power, does not require precise hole depth, and the anchor size equals the hole size. This anchor is designed for use with machine bolts of any length. The pre-assembled plug in the anchor cannot fall out, and the anchor can be set at any desired length.

These anchors are also used for suspending pipework from threaded rod, which is screwed into the internal anchor thread.

Method of fixing

For the drop-in anchor shown in Figure 5.5(a), the hole is drilled equal to the length of the anchor and the anchor inserted using a special setting tool (Figure 5.5(b)). This tool drives a built-in expander plug until the shoulder of the punch meets the anchor. The bracket is then positioned over the anchor, the bolt screwed in and tightened.

Chemical anchor

Chemical anchors have been designed with injection technology, by using adhesive capsules which are designed for a wide range of materials.

The adhesive is forced into the predrilled cavity where the threaded rod or internally threaded sleeves are placed. The application is for all types of masonry, hollow and solid brick, concrete, aerated light-weight concrete and glass construction.

FIG 5.4 (a) Fixing bolt, (b) method of fixing bolt

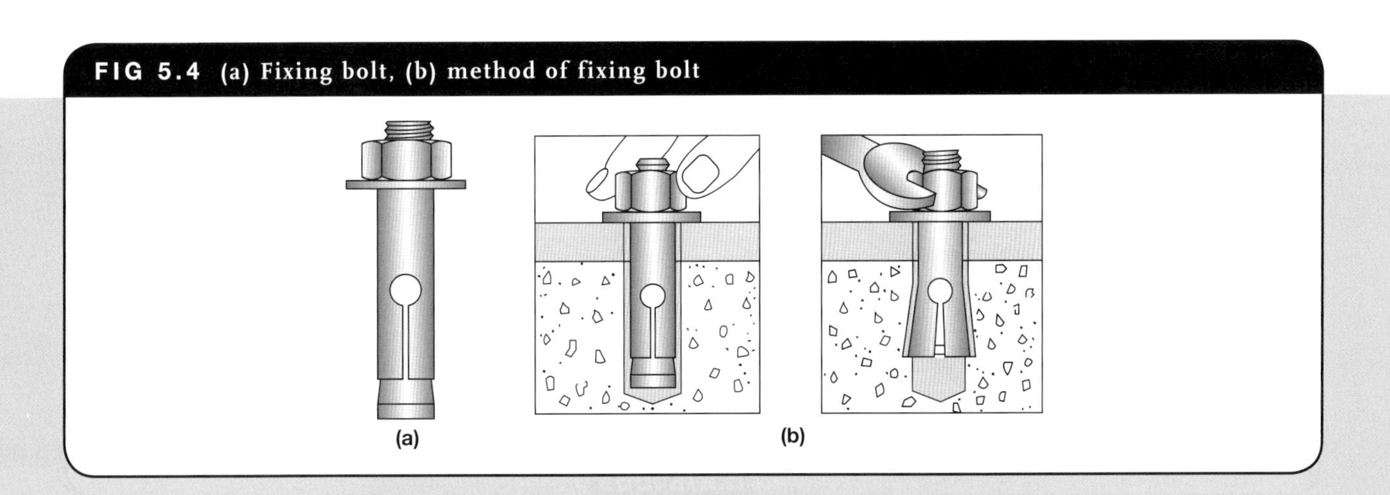

FIG 5.5 (a) Drop-in anchor, (b) setting tool, (c) method of fixing bolt

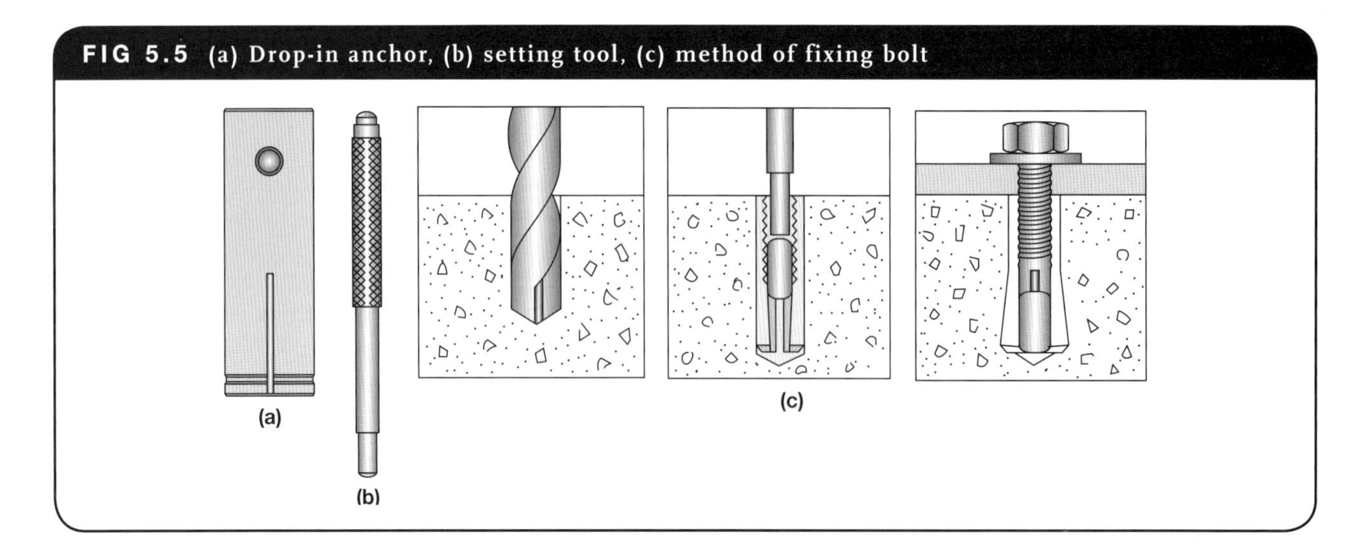

POWDER-POWERED FASTENING

Powder-powered fasteners are a common fastening method in the construction industry. Portability is the main reason for their use, because no external power supply is required for their operation. Also, the need to predrill in concrete and masonry is eliminated. The tools use specially loaded cartridges which are matched with a range of fastening devices for fixing into steel, concrete and other suitable structural materials.The tools are safe when used in accordance with manufacturer's instructions but can become dangerous in the hands of careless operators. It is therefore essential that the following safety precautions are observed.

SAFETY PRECAUTIONS FOR POWDER-POWERED FASTENING

1. Only trained operators are permitted to use the equipment.
2. All equipment must be used in accordance with the manufacturer's operating instructions.
3. Never fire the tools into free space: flying fasteners are extremely dangerous.
4. Never use the tool in an explosive (i.e. oxygen-enriched) atmosphere. Failure to observe this rule could result in a major explosion.
5. Maintain the equipment in good order and repair.
6. Always use the lightest charge within the range recommended for the job as a test shot: too heavy a charge may force the fastener completely through the structural material.
7. Ensure that the correct type of fastener is being used for a particular job.
8. Store cartridges away from excessive heat and corrosive substances.
9. Do not modify the equipment in any way to permit the fixing of a fastener at any angle other than 90° to the surface. Most powder-powered tools are fitted with a device which prevents firing at any angle greater than 5° or 6° to the perpendicular.
10. Always wear safety glasses when using powder-powered tools.

Types of fasteners and their application

A large range of fasteners is available. The selection of the correct type of fastener for each application is essential and the manufacturer's recommendations must be followed at all times.

Basic fasteners can be used on steel, concrete or any other structural material, but not timber or brittle materials. Brittle materials such as cast iron, tiles and tool steel will shatter when the fastener penetrates them. Fixings to various materials are shown in Figure 5.6.

FIG 5.6 Common types of powder-powered fasteners: (a) steel to concrete, (b) wood to concrete, (c) wood to steel, (d) steel to steel, (e) steel to brick, (f) fabricated hard materials to steel, (g) concrete penetration, (h) steel penetration, (i) steel to concrete

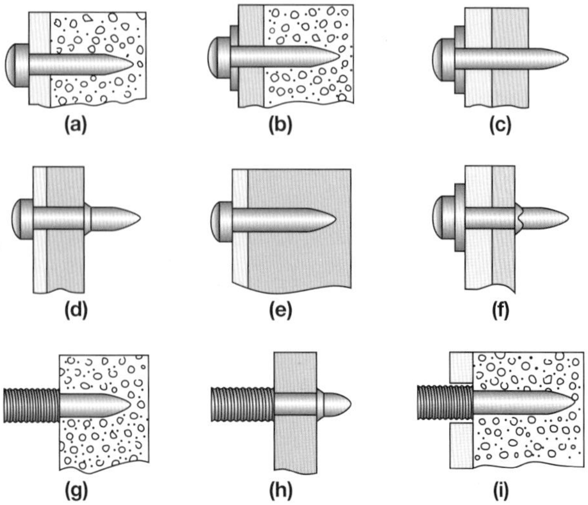

(a) (b) (c)

(d) (e) (f)

(g) (h) (i)

Powder-powered fastening tools

Powder-powered fastening tools make use of the powerful force of a specially loaded cartridge to seat a fastener into the structural material. These tools are the piston type.

Piston type

The piston type of powder-powered tool is a low velocity piston-action fastener which uses energy from a special blank cartridge to drive a piston that in turn sets a fastener into structural material with a single hammer blow. With this action, the danger of the fastener passing completely through the structural material is greatly reduced and, in some cases, eliminated. Figures 5.8 and 5.9 show a diagrammatic layout of the piston type tool and its operation.

When the hammer action fires the charge, the energy released is transferred to the piston, driving it at a low velocity onto the head of the fastener, which in turn is driven in a single blow into the structural material.

There are only three steps required to set a fastener with a powder-powered tool and these are illustrated in Figure 5.7.

Strength of charges

Powder-powered fastening may be used for a variety of fixings on various structural materials. Due to differences in the density of the materials, various strengths of charge are required for different applications. The strength of the charge ranges from 'sub-charge' through 'medium' and 'heavy' to 'extra powerful'; the different charges are usually identified by a colour-coded tip.

FIG 5.7 Method of loading and firing high-velocity type powder-powered fastening tool: (a) insert fastener and powder charge; close tool, (b) hold tool tightly against work; press barrel, (c) pull trigger. Fastener is instantly and properly set

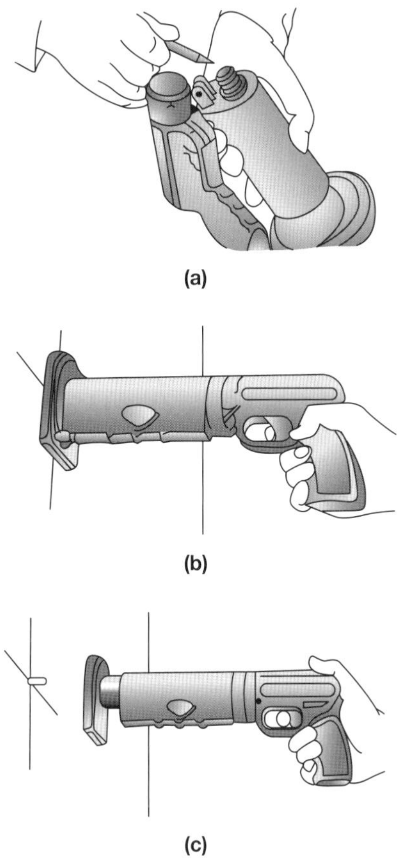

(a)

(b)

(c)

FIG 5.8 Piston type powder-powered tool

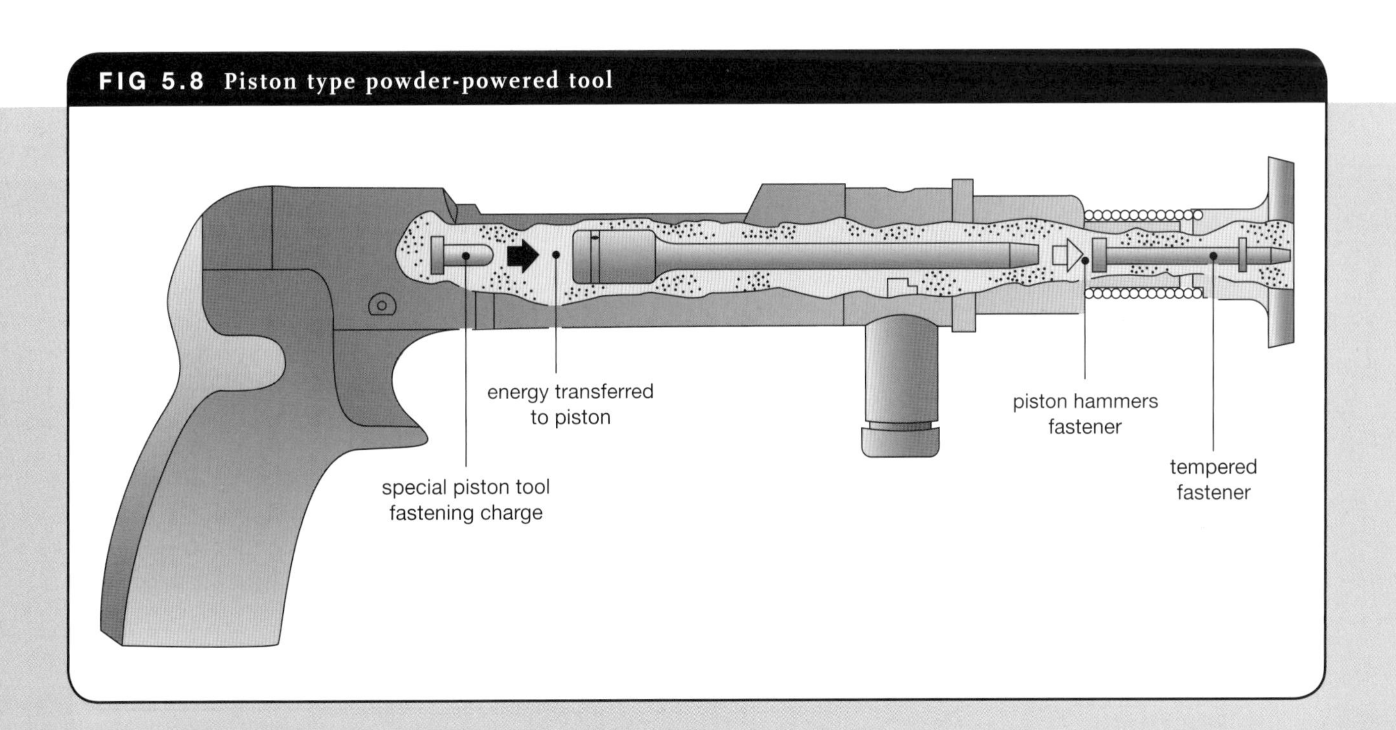

FIG 5.9 Piston type powder-powered tool detail

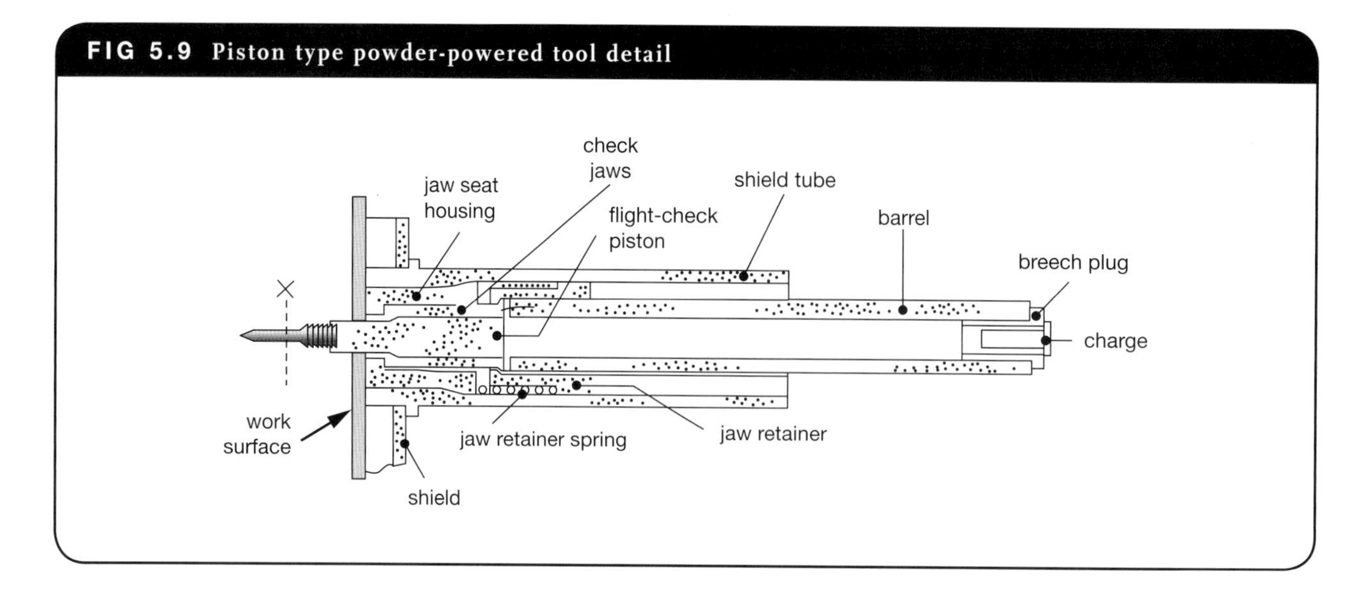

SOFT SOLDERING

The fabrication of sheet metal sections used in the building industry, particularly for roofing, frequently involves mechanically-fastened joints that require sealing. In the past, the most commonly produced coated sheet metal for fabrication work was zinc-coated steel. Zinc-coated steel is easily soldered and soldering therefore became the traditional method for sealing it.

Purpose of the soldering process

Soldering is a method of uniting two metallic surfaces by the application of a non-ferrous 'filler' metal. The filler metal is distributed between the prepared joints by capillary attraction. Metals used for this purpose usually consist of alloys of tin and lead and are known as 'soft solders'. These alloys have a melting point below that of the base metal and they can therefore be applied in the molten state without risk of melting the metals to be joined.

If a properly-soldered joint is examined under high magnification, it is possible to see that the solder is keyed into the surface cavities of the material and the soldered strip will have a crystalline appearance. The process relies on 'wetting' of the joint surfaces by molten solder. The dissolving action of the tin in the solder occurs at a relatively low temperature, with the result that the surface of the metal being joined is intimately intermingled with tin from the solder. The penetrating action of the solder into the surface of the material is sometimes referred to as 'intermolecular penetration'. Most soft-soldering operations are carried out with the aid of a soldering iron.

Joint design

Tin–lead alloy solder in a joint is primarily a low mechanical strength sealer capable only of holding joint members in place. Where any significant strength is required, the joint should be mechanically fastened (riveted) prior to soldering. Joint design should permit easy application of solder and heat, and provide a ready escape route for the flux and flux vapours. Joint clearances are important to allow proper sweating. The necessary drawing in of solder by capillary action can not occur with loose seams.

Solder

Solder is an alloy, available in stick form, and is used to form a continuous metallic joint between similar or dissimilar metals. Soft solders usually fuse at temperatures below 350 °C and can therefore be distinguished from hard solders. Selecting the correct solder alloy is important for any soldering application. Different solders are available, but because this section deals with sealing of zinc-coated sheet steel, only solder acceptable for joining this material will be discussed.

A good quality 50:50 solder should be used. This consists of one part lead and one part tin (by weight), having a melting point of 183 °C. Solder sticks should be clean and the use of badly-oxidised sticks should be avoided.

Soldering iron

The common method of applying and transferring solder to the materials to be joined is by the conventional 'pot' type soldering iron, shown in Figure 5.10. Soldering irons can also be connected to liquid petroleum gas (LPG) with a specially-designed hand-piece and hose attachments. The working part of the soldering iron is the head or bit, which is the same for the conventional pot soldering iron as well as the soldering iron designed for LPG. The head is made from a solid block of copper, generally of square section, and tapered at one end to form a narrow point.

The head has three essential functions to perform and must possess the following characteristics:

- the capacity for storing heat
- the ability to rapidly transfer heat to the job
- the ability to be readily 'tinned', or coated in solder.

The head must therefore be of adequate size for its task, particularly for outdoor work where cooling may be rapid. For lengthy jobs it is good practice to alternate two soldering irons so that heat continuity is maintained.

The use of LPG gas (propane) as an alternative heat source will also achieve this objective.

FIG 5.10 A soldering iron

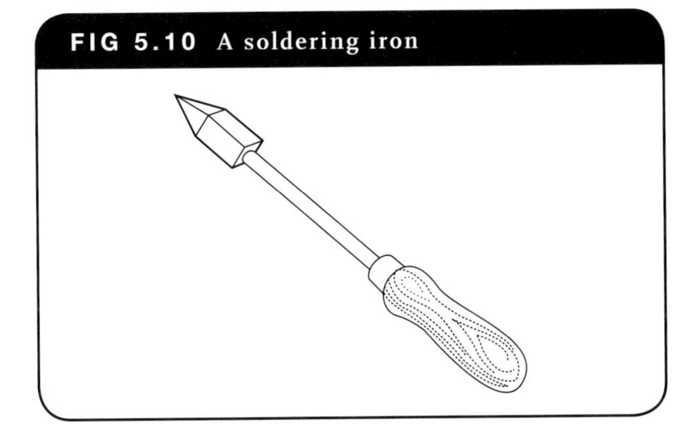

Preparation of soldering

To make a successful joint, the solder should unite freely with the surface to be soldered. Metals exposed to the atmosphere will generally oxidise which prevents the solder making contact with the metal. Unless the solder unites freely, or 'tins' the metal, the joint can not be properly executed. The oxide coating on the metal must therefore be removed and precautions taken to prevent its re-forming.

Before soldering is attempted, the surface must be thoroughly cleaned. Mechanical abrasives such as sandpaper or steel wool will produce a sufficiently clean surface on most metals. Abrasion is seldom necessary on new metals, but if the metal surface shows evidence of corrosion or small depressions (which tend to harbour dirt) the surface must be thoroughly cleaned.

Fluxing

The purpose of fluxing is to chemically clean the surface to be soldered by removing the surface oxide layer, and to prevent the formation of oxide while soldering is in progress. Fluxing is not a substitute for surface preparation. The correct flux and fluxing technique should:

- provide effective liquid cover, wet the surfaces to be soldered and create an oxygen-free atmosphere conducive to the alloying of the solder to the workpiece surface
- remove impurities from the joint surfaces
- aid the solder flow and be displaced by molten solder during the soldering operation.

Types of flux

Corrosive

These fluxes are extremely active, being both quick and effective in use. Provided that there is no trace of residue left after cleaning, they are highly suitable for general plumbing work.

WARNING: Corrosive fluxes are extremely dangerous and due care should be exercised.

Non-corrosive

These fluxes, as the term applies, are relatively inactive. Tallow is one such flux and is used in joining lead sheet and pipe.

Common corrosive fluxes

Hydrochloric acid (HCl)

When used as a flux, this acid is often referred to as 'raw spirits of salts'. Hydrochloric acid is extremely active and may be diluted by the addition of clean water. It is useful in soldering galvanised sheet steel, especially dirty sections, in which case it assumes the dual roles of pickling agent and flux.

Zinc chloride ($ZnCl_2$)

This flux, commonly called 'killed spirits' because it has been reacted with an excess of zinc, is suitable for general work that has a clean surface. Zinc chloride can be used as an effective flux on mild steel, brass, copper and tinplate.

Mixture of HCl and $ZnCl_2$

A mixture of equal parts of hydrochloric acid and zinc chloride can be used for soldering stainless steel. The surface of the metal will need to be roughened because polished metal does not allow the solder to penetrate its surface.

Proprietary liquid fluxes

These can be in the form of flux or soldering solution and are suitable for zinc-coated steel. These fluxes wet the zinc surface instantly, are easier to control than a spirit-based flux and produce better joint penetration.

Ammonium chloride (NH_4Cl)

This flux is extensively used in its block form for the purpose of tinning soldering irons. It is referred to as 'sal ammoniac' in the trade, and is the main ingredient of the dip pot for cleaning the heated tip of the soldering iron head.

Fluxes suitable for various metals

The choice of the correct flux depends on the condition of the surface being soldered and also on which metals are to be joined.

Table 5.1 indicates suitable fluxes for various metals.

TABLE 5.1 Types of flux

Metal	Flux
Galvanised sheet steel	1. Hydrochloric acid 2. Proprietary fluxes
Copper	Zinc chloride
Brass	Zinc chloride
Lead	Tallow
Stainless steel	Mixture (equal parts) of hydrochloric acid and zinc chloride
Terneplate	Zinc chloride
Tinning copper head	1. Ammonium chloride 2. Zinc chloride
Tinplate	Zinc chloride
Wrought iron	Zinc chloride
Zinc	Dilute hydrochloric acid

Tinning a soldering iron

Before a satisfactory soldered joint can be executed, the soldering head must be properly tinned (coated with solder). Every time a soldering head is overheated it should be retinned. To prepare the soldering head for tinning, it must first be heated to a cherry red heat. If a vice is available, the soldering iron is secured between the jaws of the vice and the soldering head quickly filed to remove oxides and other foreign material (Figure 5.11). Use the file to square off the point of the iron. This minimises heat loss from the iron.

While the head is still hot, solder is applied and the head rubbed on a block of sal ammoniac (Figure 5.12). The soldering iron is now ready for use.

FIG 5.12 Tinning the copper bit

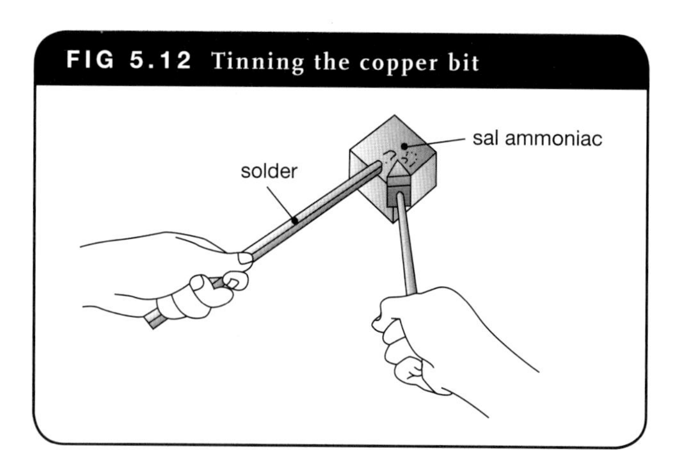

FIG 5.11 Filling the copper bit

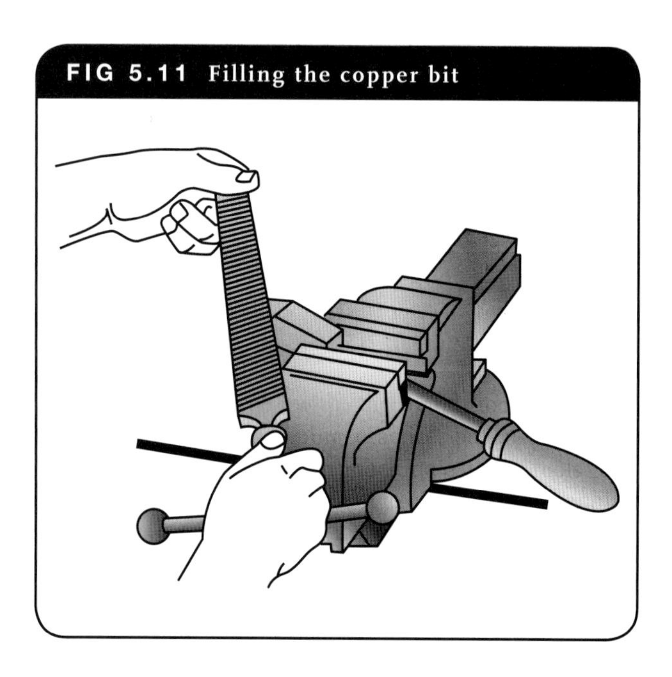

FIG 5.13 Reconditioning a soldering iron

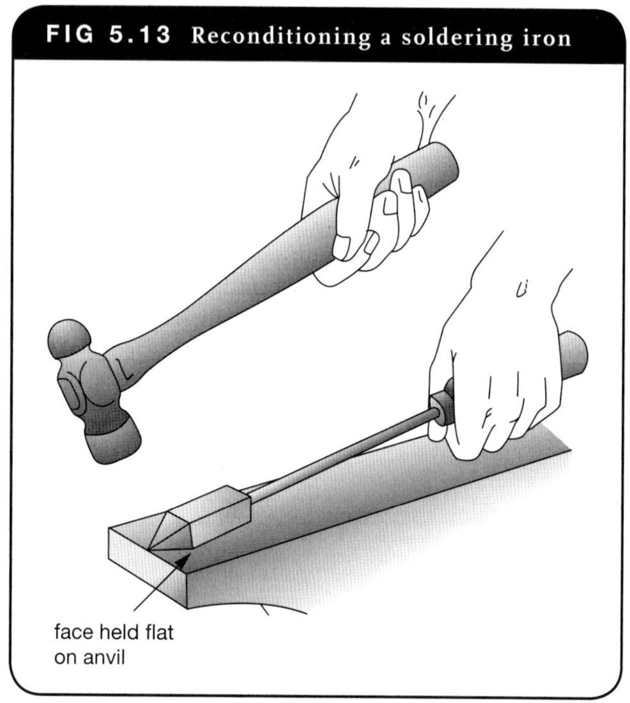

RECONDITIONING A SOLDERING IRON

When a soldering iron has become blunt or out of shape it may be reshaped by forging. Use the following procedure:

1. Heat the head to dull red heat (under normal workshop light conditions) and then file until all solder, burnt tin and oxide scale is removed.
2. Reheat the soldering head to red heat and commence reshaping by hammering the head on a steel block (Figure 5.13).
3. Continue to shape the head by striking it on alternate faces. Ensure that the soldering head is continuously heated during the forging operation so that it remains in a soft state, thereby making it possible to shape the head with a minimum striking effort. When the desired shape has been obtained file the surfaces to remove hammer indentations.
4. Reheat the soldering head and tin it with solder, using sal ammoniac as the flux.

The soldering iron is now ready for reuse.

USEFUL HINTS

- Avoid soldering in draughty conditions.
- Use the correct flux, heat and solder.
- Use a clean, well-tinned soldering head.
- Make sure the surfaces of the materials to be joined are clean.
- The soldering iron should be drawn along the seam towards the operator.
- When 'sweat' soldering, work the soldering head across the seam.
- Use a soldering iron with a high rate of heat transfer.
- Solder the joint in one run wherever possible.
- Ensure a proper joint clearance when 'sweat' soldering.
- Remove excess flux with a damp cloth immediately after soldering is complete.

Associated heating problems

If an iron is too hot it will:

- burn the tin component of the solder, rendering it coarse, less fluid and gritty—this will result in a weak joint
- cause discoloration on the metal surface, especially when soldering copper or brass materials.

If an iron is too cold it will:

- not permit proper surface fusion of the solder with the metal to be joined
- produce a scratchy finish and, because of lack of penetration, resulting in a weak joint.

PRINCIPLES OF FUSION WELDING (OXYACETYLENE WELDING)

The terms used in the following sections on the principles and techniques of fusion welding can be found in the Glossary.

Accurate flame adjustment is essential for successful oxyacetylene welding.

There are three types of flame (Figure 5.14):

1 neutral
2 oxidising (excess oxygen)
3 carburising (excess acetylene).

Neutral flame

The neutral flame is used for most welding requirements. Set the pressures at the regulators to suit the size of the tip fitted to the blowpipe (see Table 5.2). Open the blowpipe acetylene valve; light the issuing gas. Increase the gas flow by turning the valve in an anti-clockwise direction until the flame no longer smokes. Then open the oxygen valve slowly. The flame will go through a colour change from yellow to blue and will have a long feathery inner cone of a carburising flame. Continue to open the oxygen valve until the secondary or feathery cone just disappears.

Oxidising flame

This flame is obtained by setting a neutral flame, as described above, then reducing the flow of acetylene. This shortens the inner cone and the amount of excess oxygen can be judged by the shortness of the cone compared with the length of the neutral cone.

Carburising flame

This type of flame has an excess of acetylene, which is obtained by opening the acetylene valve further than required for the neutral setting. The luminous inner cone becomes ragged and, as the acetylene increases, the flame changes in appearance. Through welding goggles three parts will be easily recognisable: an intense but feathery-edged inner cone which is surrounded by a luminous secondary cone and a bluish outer envelope which forms a third zone. The amount of excess acetylene is usually judged by the ratio of the length of the secondary luminous cone to the length of the neutral cone.

- a neutral flame: used for steel, stainless steel, cast iron, copper, aluminium, etc.
- an oxidising flame: necessary for welding brass
- a carburising or reducing flame (excess acetylene).

FIG 5.14 Flame adjustment: (a) carburising, (b) neutral, (c) oxidising

TABLE 5.2 Regulator pressures and gas consumption

	Regulator pressures (kPa)		Gas consumption (L/min)	
Tip size (mm)	Oxygen	Acetylene	Oxygen	Acetylene
8	50	50	1	1
10	50	50	2	2
12	50	50	4	4
15	50	50	7	7
20	50	50	11	11
25	50	50	29	29

Note: When hoses longer than 3.5 m are used, the pressure at the regulators should be increased accordingly. However, at no time should the acetylene pressure exceed 100 kPa as acetylene becomes unstable and consequently dangerous above this pressure.

FIG 5.15 Welding equipment

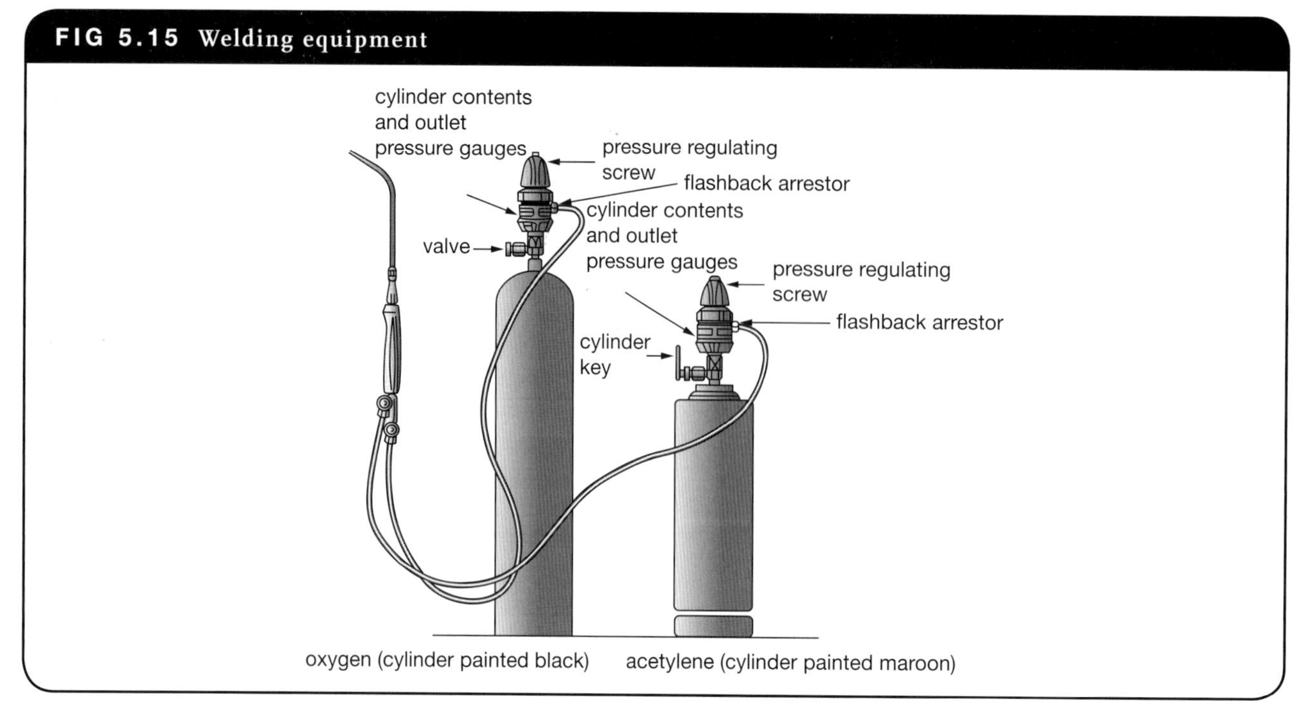

High pressure oxygen-acetylene welding kit

To successfully carry out any welding or cutting operation, the following equipment, maintained in good safe working condition, is required:

- gas cylinders
- gas regulators
- gas hoses
- blowpipe, including mixing chamber
- blowpipe tip
- cutting attachment
- tip cleaners
- flint lighter
- welding goggles.

Gas cylinders

Two cylinders are required, one of oxygen and one of acetylene (Figure 5.15). They are identified by colour—black for oxygen and maroon for acetylene. In addition, the acetylene cylinder is fitted with a left-hand thread. This is a safety requirement. All fuel gas cylinders are fitted with left-hand threads to avoid cross-connection with oxygen cylinders. (See Table 5.3 for cylinder sizes.)

Gas regulators

The regulators supply gas from the cylinder to the blowpipe at a constant pressure to enable trouble-free operation (Figure 5.16). The regulators deliver constant pressure, regardless of variations in cylinder or pipeline pressure.

Gas hoses or tubing

Reinforced rubber hoses deliver the gas from the regulator to the blowpipe. The hoses are coloured maroon for acetylene

FIG 5.16 Regulators for (a) oxygen and (b) acetylene

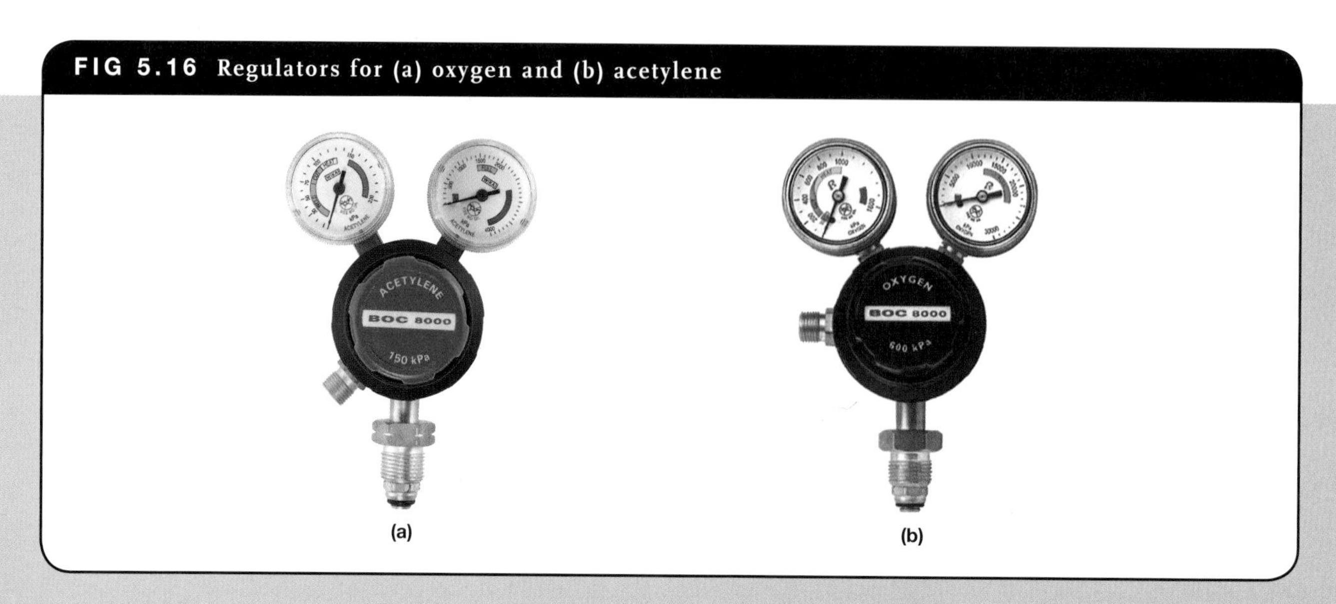

TABLE 5.3 Gas cylinder sizes

Oxygen		Acetylene	
Size	Contents (m^3)	Size	Contents (m^3)
G	7.6	G	7.0
*GM	6.4	*GM	5.7
E	3.8	*GS	5.0
D	1.5	E	3.2
C	0.4	*EM	2.8
		*ES	2.4
		D	1.8
		DS	1.0

Note: Asterisks(*) indicate that sizes may vary in some states.

and black for oxygen. The connection on the maroon (acetylene) hose has a left-hand thread to avoid accidental cross-connection (Figure 5.15).

Blowpipe

The blowpipe, or 'hand piece' (Figure 5.17(a)), consists of four main sections:

- control valves. These control the flow of gas to the tip from the hoses. They turn in the conventional way—turn anti-clockwise to open, clockwise to close.
- handle. This is moulded to fit comfortably into the right hand, allowing the left hand easy access to the control valves.
- mixing chamber. This is attached to the handle by a threaded outer section and seals or seats with the aid of O-rings. The function of the chamber is to mix the two gases without allowing either gas to flow back into the hoses if a pressure drop occurs.
- welding tip. The welding tip screws into the small end of the mixing chamber to receive the mixed gases. The mixture issues from the orifice at the end. The tips are rated by size (see Table 5.4).

TABLE 5.4 Tip size in relation to tip number

Tip number	Orifice size (mm)
8	0.8
10	1.0
12	1.2
15	1.5
20	2.0
26	2.6

Cutting attachment

The cutting attachment (Figure 5.17(b)) is screwed onto the blowpipe after the mixing chamber has been removed. This converts the blowpipe into a cutting blowpipe. It is the most common type in use in the plumbing industry. The cutting blowpipe (Figure 5.17(c)) is used for cutting, flame cleaning, powder cutting and flame hardening.

Tip cleaners

Tip cleaners are used to maintain the nozzle orifice in a clean condition. It is essential that the smallest tip cleaner be used to prevent oversizing of the tip orifice.

FIG 5.17 Blowpipe attachments: (a) blowpipe, (b) blowpipe fitted with cutting attachments, (c) cutting blowpipe

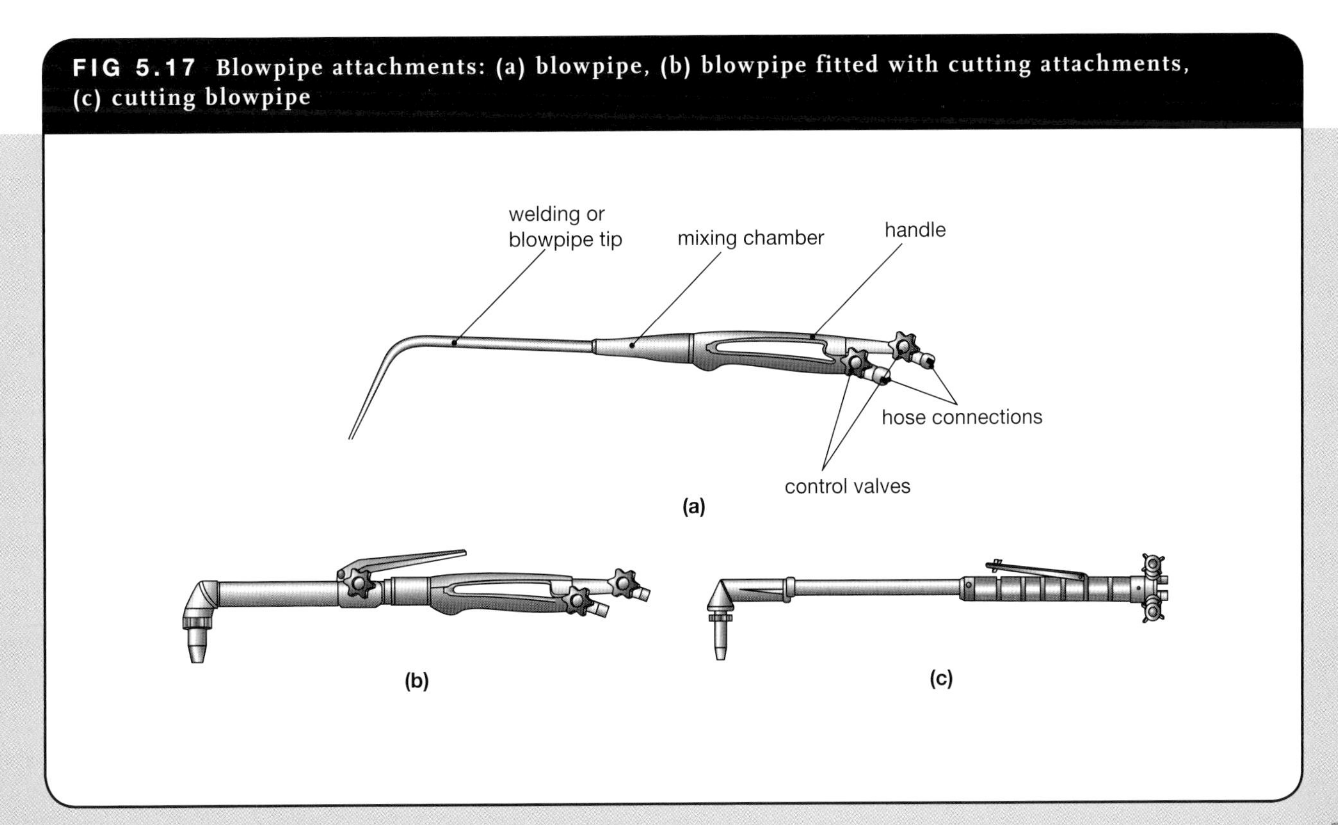

Flint lighter

To ignite the mixed gases, a flint lighter should be used to avoid burns to the hands.

Welding goggles

Light is necessary for clear vision of the work in hand, but welding or cutting operations cast too much light of high intensity. Unless this light is reduced, the operator is dazzled and can not see the work properly. The primary use of the welding filter is therefore to reduce this light to a satisfactory level for eye comfort. Equally important is the ability of the filter to reduce exposure of the eye to ultraviolet light that may cause irritation.

Setting up

Ensure that your equipment, hands and gloves are clean and free from oil and grease. Place the oxygen and fuel gas cylinders in a clear position where oil is not likely to drip on them and where they are not likely to be knocked by moving or falling objects. Before connecting regulators to the cylinders, blow out any foreign matter which might harm seats or clog orifices. Inspect all glands, couplings and seats to ensure that they are not scored and that they are clean. Attach the oxygen and fuel gas regulators to the respective cylinders and tighten them sufficiently to prevent leaks. Use spanners designed for the particular nuts.

Ensure the oxygen and fuel gas regulator adjusting knobs are released. Oxygen and fuel gas cylinder valves should always be opened very slowly, so that the high pressure gauge hand on the regulator moves up gradually. A rush of high pressure gas may strain the gauge mechanism. Oxygen may cause seat ignition, while acetylene may cause decomposition.

Never stand in front of regulator pressure gauge faces when opening cylinder valves. Always stand to one side. Never open cylinder valves more than one and a half turns.

Attach the black oxygen tubing to the regulator outlet. Screw in the oxygen regulating adjusting knob slightly in a clockwise direction and allow a small quantity of gas to blow through to remove dust from the regulator and French chalk from the tubing. Release the adjusting knob and attach the other end of the tubing to the blowpipe oxygen inlet valve, which is marked ‘O’.

Repeat the above procedure with the fuel gas regulator and tubing. The blowpipe fuel inlet valve is marked ‘F’.

Purging before lighting up

After connecting the welding or cutting apparatus to the oxygen and fuel gas cylinders or when starting to reuse the apparatus after an interval of half an hour or more, each gas should be allowed to flow through its respective tube separately for a few seconds to purge the tubes of any mixture of gases. Do not perform this operation in a confined space.

Lighting up

Open the blowpipe acetylene valve and light the issuing gas with a flint lighter or pilot light. Then open the valve further until the acetylene flame no longer produces soot. If the flame burns away from the tip or blows off as soon as lit, slightly close the blowpipe acetylene valve.

Open the blowpipe oxygen valve and the flame will resolve itself into two visible sections, that nearest the tip being white and ragged, and the further section bluish and feathery. Open the oxygen valve until the inner cone becomes clearly defined. Check and readjust the working pressure if necessary before making the final flame adjustment.

PROCEDURE FOR LIGHTING UP

When using a cutting blowpipe, neutral preheating flames are the most suitable. When using the cutting attachment, the procedure should be:

1. Open the blowpipe acetylene valve and ignite at the nozzle.
2. Open the blowpipe oxygen control valve fully, then open the heating oxygen control valve on the cutting attachment until the neutral flame is obtained when the cutting oxygen level is pressed.

It is recommended that the following procedure be adopted to check for leaks:

1. Shut off the blowpipe oxygen valve.
2. Open the oxygen cylinder valve and set the regulator to show approximately 50 kPa on the delivery gauge.
3. Close the oxygen cylinder valve.
4. Watch the cylinder pressure gauge. If pressure decreases, there is a leak in the oxygen system. If no pressure drop is experienced, there are no leaks up to the blowpipe valve.
5. Conduct similar tests on the fuel gas system.

To locate leaks, use a clean paint brush, water and a cake of soft soap or liquid soap to make a soap–water mixture, then brush this mixture on to the screwed joints.

After you have checked for leaks:

1. Close the blowpipe oxygen valve. Select a suitable size welding tip and screw it onto the blowpipe.
2. Open the blowpipe oxygen valve and screw in the oxygen regulator adjusting knob until the correct pressure shows on the delivery gauge. Always follow the equipment manufacturer’s pressure recommendations.
3. Close the blowpipe oxygen valve.
4. Repeat this operation with the fuel gas regulator adjusting knob and blowpipe fuel gas valve.

For large-flow equipment it is recommended that the pressures be set approximately on the regulator then finally adjusted by briefly opening each blowpipe valve. Improper pressures are not only wasteful but prevent you doing your best work and can be dangerous; flashbacks can be caused by improper pressures.

Use the manufacturer's pressure chart as a guide to correct pressure for your job. The maximum safe working pressure for acetylene is 50 kPa.

PROCEDURE FOR CLOSING DOWN

1. Close the blowpipe fuel gas valve.
2. Close the blowpipe oxygen valve.

The above is satisfactory for temporary halts. However, when closing down fully, use the following procedure:

1. Close both cylinder valves.
2. Open the blowpipe oxygen valve and allow the gas to drain out.
3. When both gauges on the regulator have fallen to zero, close the blowpipe oxygen valve.
4. Release the oxygen regulator adjusting knob.
5. Repeat this procedure with the fuel gas valves and regulator.

SPECIAL PRECAUTIONS

- Before attaching any air–fuel gas equipment to an oxygen-fuel gas welding blowpipe, disconnect the oxygen tubing from the blowpipe.
- Never use oxygen or fuel gas regulator pressures higher than the job requires. To do so is unnecessary, wasteful and dangerous.
- The pressure in a cylinder should always be appreciably higher than the pressure required for the particular tip in use, otherwise the flame becomes unstable.
- Never attempt to stabilise the flame by screwing in a regulator T-screw. Disconnect the cylinder and couple on a fresh supply.
- Whenever the flow of either oxygen or fuel gas is stopped by an obstruction, immediately close the valves on both cylinders and leave them closed until the obstruction is removed.

Backfires and flashbacks

Backfire

A backfire is a momentary extinguishment or burning-back of the flame into the blowpipe tip. It is caused by touching the tip against the work, by particles entering the tip and obstructing the gas flow, or by overheating the tip. The trouble will sometimes clear itself immediately. If the work is hot enough, the blowpipe will relight automatically. However, if this fails to happen, close the blowpipe valves at once. Before relighting, check your equipment.

Flashback

A flashback is the burning-back of the flame into the blowpipe or the ignition of an explosive mixture in one of the gas lines. Flashbacks can burn right back into the tubing.

The danger in case of a flashback can be practically eliminated by recognising immediately what it is and knowing what to do about it. In the case of a flashback into the blowpipe, a flame burns at the mixer with a shrill hissing sound. Close the blowpipe oxygen valve at once. Then close the fuel gas valve. If the flashback enters the rubber tube, close the cylinder valves immediately.

The causes of flashbacks vary. Among them are wrong pressures, distorted or loose tips or mixer seats, kinked tubing, clogged tip or blowpipe orifices, and overheated tips or blowpipes.

The occurrence of flashbacks is always a sign that something is very wrong—either with your equipment or your handling of it. Before attempting to relight your blowpipe, check for trouble at the points suggested above. First of all check your pressures. Are they the recommended pressures? Then remove the tubing from the blowpipe and inspect for damage. If the flame has burned back into the tubing, replace the affected length of tubing. In setting up, be sure the tubing does not lie where it may be stepped on or run over. Avoid using long lengths of tubing.

Flashback arrestors

To ensure total safety and protection from the causes and effects of flashbacks, flashback arrestors are required be fitted to each gas line as the risk of reverse flow of gas exists with both oxygen and acetylene (Figures 5.15 and 5.18).

The requirements comply with Australian Standard AS 4839-2001—Safe use of portable and mobile oxy,

FIG 5.18 Flashback arrestors

fuel gas systems for welding, cutting, heating and allied processes and Australian Standard AS 4603-1999—Flash back arrestors–Safety devices for use with fuel gasses and oxygen or compressed air.

OXYACETYLENE WELDING TECHNIQUES

Good quality, well-maintained equipment is essential for correct welding, but the equipment must also be used in conjunction with good preparation. Preparation is necessary to permit correct manipulation of the blowpipe and filler rod, achieving good rod penetration and fusion of the weld deposit on the metals being joined together.

The portability of the equipment allows the surface on which the weld is to be deposited to be in a horizontal or vertical position or at any angle in between. Forehand, backhand or vertical techniques may be used.

Joint preparation

Joint preparation varies with the type and thickness of material and must be considered in conjunction with the type of joint, the angle of the rod to the metal surface and the angle of the tip to the metal surface (Tables 5.5 and 5.6).

In all fusion welding, the technique involves the forming of a molten puddle into which the filler rod is deposited. The movement of the blowpipe will vary in accordance with the type of weld being carried out.

Forehand welding

Forehand welding is used on steel for flanged edge welds, for unbevelled plates up to 3 mm and for bevelled plates up to 5 mm thick. It is also the method usually adopted for cast iron and non-ferrous metals. The procedure used in forehand welding is to weld with the blowpipe flame pointing in the same direction in which the weld progresses. The blowpipe is moved from side to side to establish the width and build-up of the weld (Figure 5.19(a)).

Backhand welding

Backhand welding is recommended only for steel plates exceeding 5 mm thickness. Plate edges from 5 mm to 8 mm thickness should be bevelled to 30° to give an included angle of 60° for the welding 'V'. The procedure used in backhand welding is to weld with the blowpipe flame pointing in the opposite direction to that in which the weld progresses. The blowpipe is moved in a circular motion throughout the entire weld length (Figure 5.19(b)).

Vertical welding

Vertical welding is used on unbevelled steel plate of up to 5 mm thickness (or up to 16 mm thickness where two operators are employed). The weld is started at the bottom and proceeds vertically to the end of the beam. The manipulation of the blowpipe and rod is shown in Figure 5.19(c).

TABLE 5.5 Angle and distance

Metal	Angle of rod with metal surface (°)	Angle of blowpipe with metal surface (°)	Distance of cone from metal surface (mm)	Type of joint
Mild steel	30–40	60–70	1.5–3.0	forehand
Stainless	60–70	70–80	close	forehand
Copper	40–50	50–70	1.5–3.0	forehand
Brass	40–50	40–50	3.0–5.0	forehand
Mild steel	40–50	40–50	1.5–3.0	backhand
Mild steel thickness (mm)				
1.5	30	25	1.5–3.0	vertical
2.5	30	35	1.5–3.0	vertical
3.0	30	50	1.5–3.0	vertical
5.0	30	50	1.5–3.0	vertical

TABLE 5.6 Joint preparation

Metal	Thickness (mm)	Type of preparation	Penetration gap (mm)	Type of weld
Mild steel	1.5–3.0	Nil	0.5	forehand/backhand
Mild steel	0.3–5.0	bevel 90°	1.5	forehand/backhand
Mild steel	5–10	bevel 60°	1.5	forehand/backhand
Mild steel	1.5–3.0	nil	1.5	vertical
Mild steel	3.0	nil	3.0	vertical
Mild steel	5.0	bevel 80°	3.0	vertical

FIG 5.19 (a) Forehand welding, (b) backhand welding, (c) vertical welding

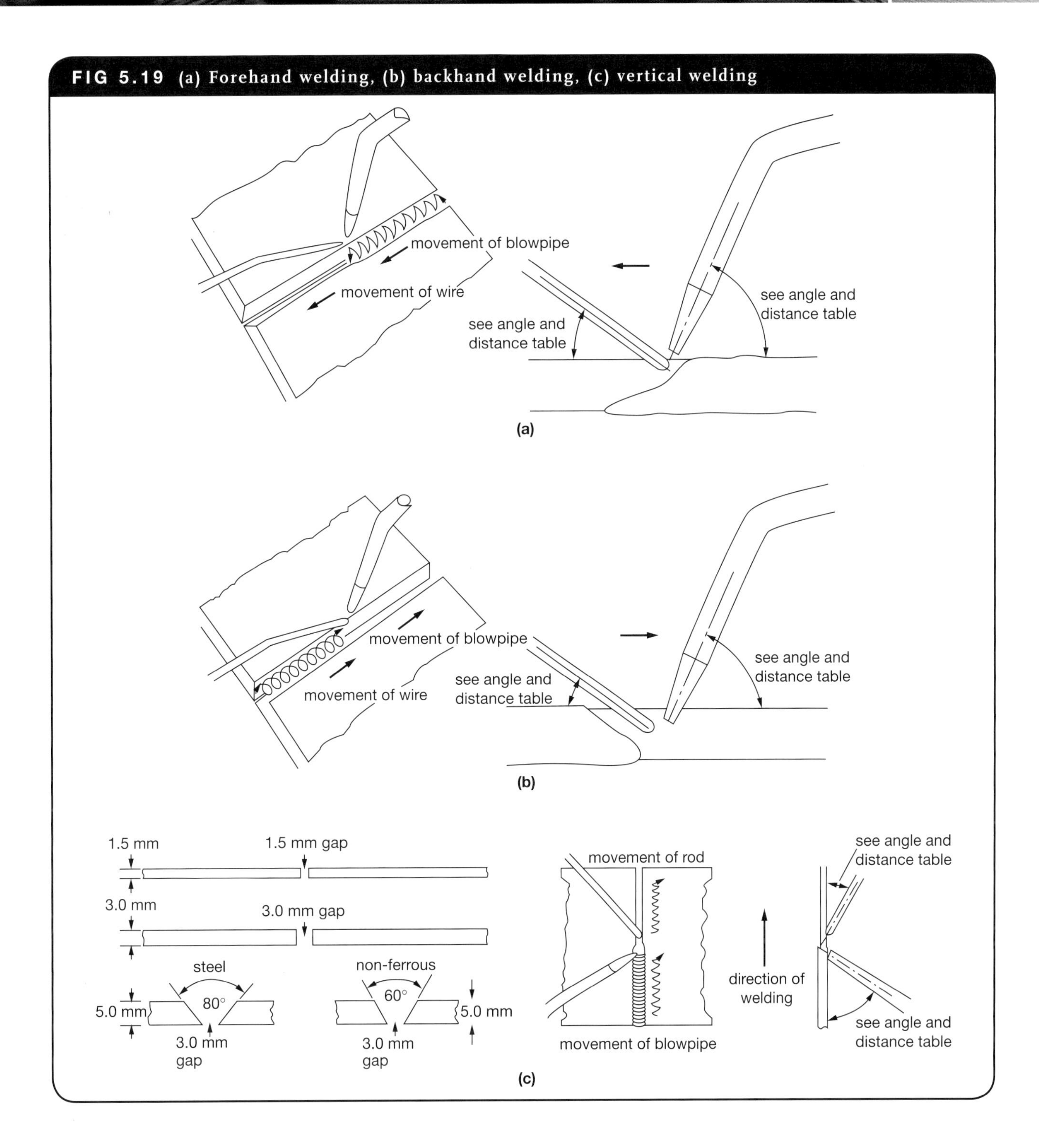

OXYACETYLENE CUTTING

The oxyacetylene cutting of iron and steel is primarily achieved by the oxidation or combustion of the material. Iron and steel attract oxygen, particularly at high temperatures, so that when a jet of oxygen is directed on preheated steel, combustion of the steel takes place with the formation of iron oxide. The additional heat generated during this combustion stage is used to maintain the cutting or ignition temperature of the steel, thus ensuring the continuation of the cut.

The equipment needed for oxyacetylene cutting is the same as shown in Figure 5.17(a). The tip, however, is exchanged for a cutting nozzle or the welding blowpipe is exchanged for a cutting blowpipe (Figures 5.17(b) and (c)). The nozzle style is the same whatever method is chosen. The cutting nozzle consists of preheat jets and a cutting or oxidising jet (Figure 5.20).

The nozzle has two seats at its upper (or gas entry) end; the correct alignment of these seats into the cutting head is important. This prevents mixing of the cutting oxygen and the premixed preheating gases. To install the nozzle correctly, screw it into the head with the locking nut loose—this ensures that the cutting oxygen seat beds in first. Then tighten the locking nut to provide a seat for the mixed gases.

FIG 5.20 Nozzle for oxyacetylene cutting jet

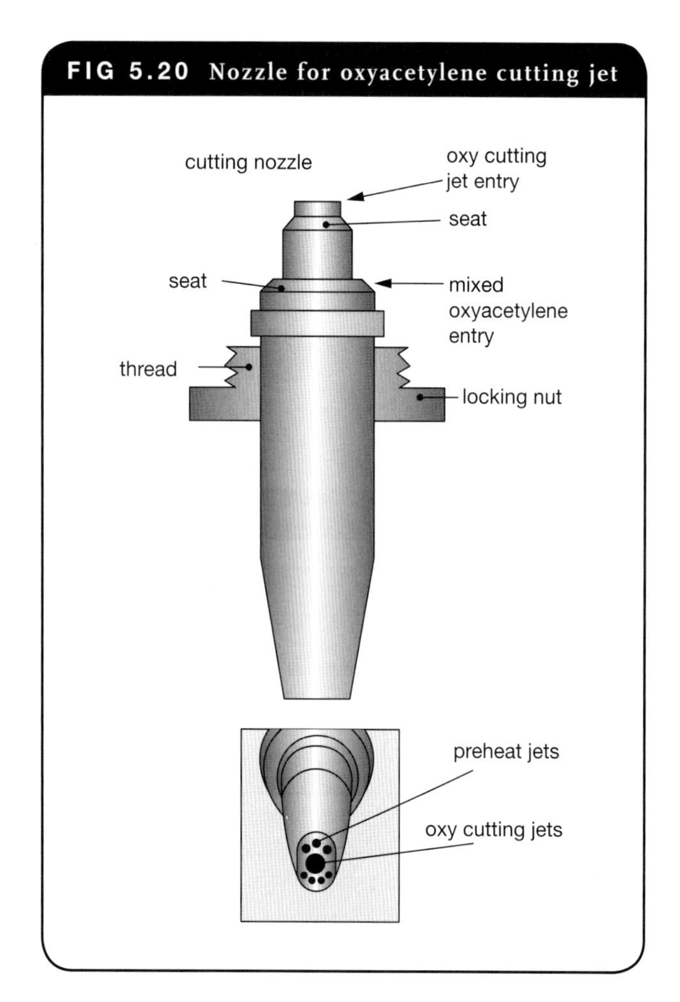

FIG 5.21 Blowpipe handle fitted with cutting attachment, showing controls

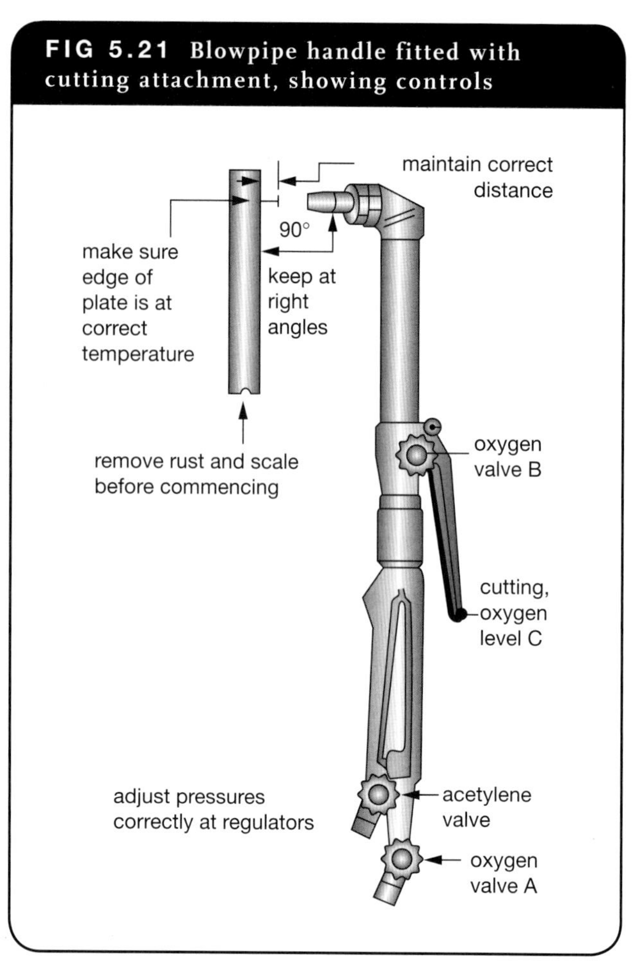

TABLE 5.7 Operating data for oxyacetylene cutting

Plate thickness (mm)	Nozzle size	Cleaning drill size		Pressure (kPa)		Cutting speed (mm/min)	Consumption (L/min)	
		Cutting	Preheat	Oxygen	Acetylene		Acetylene	Oxygen
6	8	8	5 x 6	200	100	450	3.5	17.5
12	12	12	6 x 7	200	100	380	4	38
20	12	12	6 x 7	250	100	340	4.5	42
25	15	15	6 x 8	220	100	320	6	56
40	15	15	6 x 8	350	100	270	7	75
50	15	15	6 x 8	400	100	240	7.5	85
75	15	15	6 x 8	450	100	180	8	95
100	20	20	6 x 10	400	100	150	9	134
125	20	20	6 x 10	450	100	150	10	155
150	20	24	6 x 11	450	100	130	11	211

Lighting the blowpipe

After selecting the correct size of cutting nozzle, open the cylinder valves and set the correct oxygen and acetylene pressures (see Table 5.7) with the gas passing through the nozzle. When these are set, shut all the valves on the cutting blowpipe.

To light the blowpipe, open the acetylene valve slowly (Figure 5.21), ignite the gas with a flint lighter and slowly open the oxygen valve A to its full extent. Now adjust the preheat flame to a neutral flame with oxygen valve B. Check the operation of the cutting stream by pressing the cutting oxygen lever C. Note the preheat flame. If it changes from a neutral to an oxidising flame, the cutting tip should be removed and re-installed correctly. If no change occurs, the cutting blowpipe is ready for use. To close down the cutting blowpipe turn off the acetylene valve; then turn off the oxygen valves A and B, and release the pressure in the cutting section by pressing the cutting lever.

CUTTING PROCEDURE

1. Ensure that all dirt and scale are removed before starting the cut.
2. To commence cutting, hold the inner cone of the preheating flames so that they almost touch the edge of the metal
3. When the metal becomes a bright red colour (under normal workshop lighting conditions), slowly press the oxygen cutting lever and commence to move the cutting blowpipe at the speed recommended. A shower of sparks falling from the underside of the sheet indicates that the cut is penetrating the sheet. Care should be taken at this point to ensure the cutting blowpipe nozzle is held at right angles to the metal being cut.

To commence a cut other than at an edge requires a slightly different technique.

1. Preheat the area in the same manner as for the edge cut, but as the cutting lever is pressed lift the cutting nozzle approximately 15 mm from the surface to be cut; this action helps to prevent the molten metal spraying over the operator.
2. When the penetration is made, lower the nozzle and commence cutting at the speed indicated (Figure 5.22).

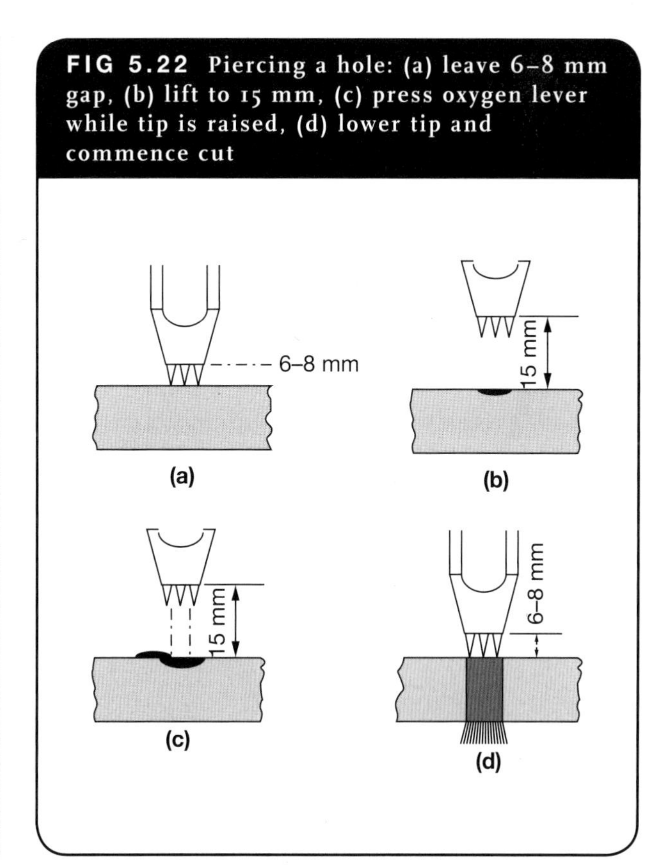

FIG 5.22 Piercing a hole: (a) leave 6–8 mm gap, (b) lift to 15 mm, (c) press oxygen lever while tip is raised, (d) lower tip and commence cut

TABLE 5.8 Common faults in cutting and their causes

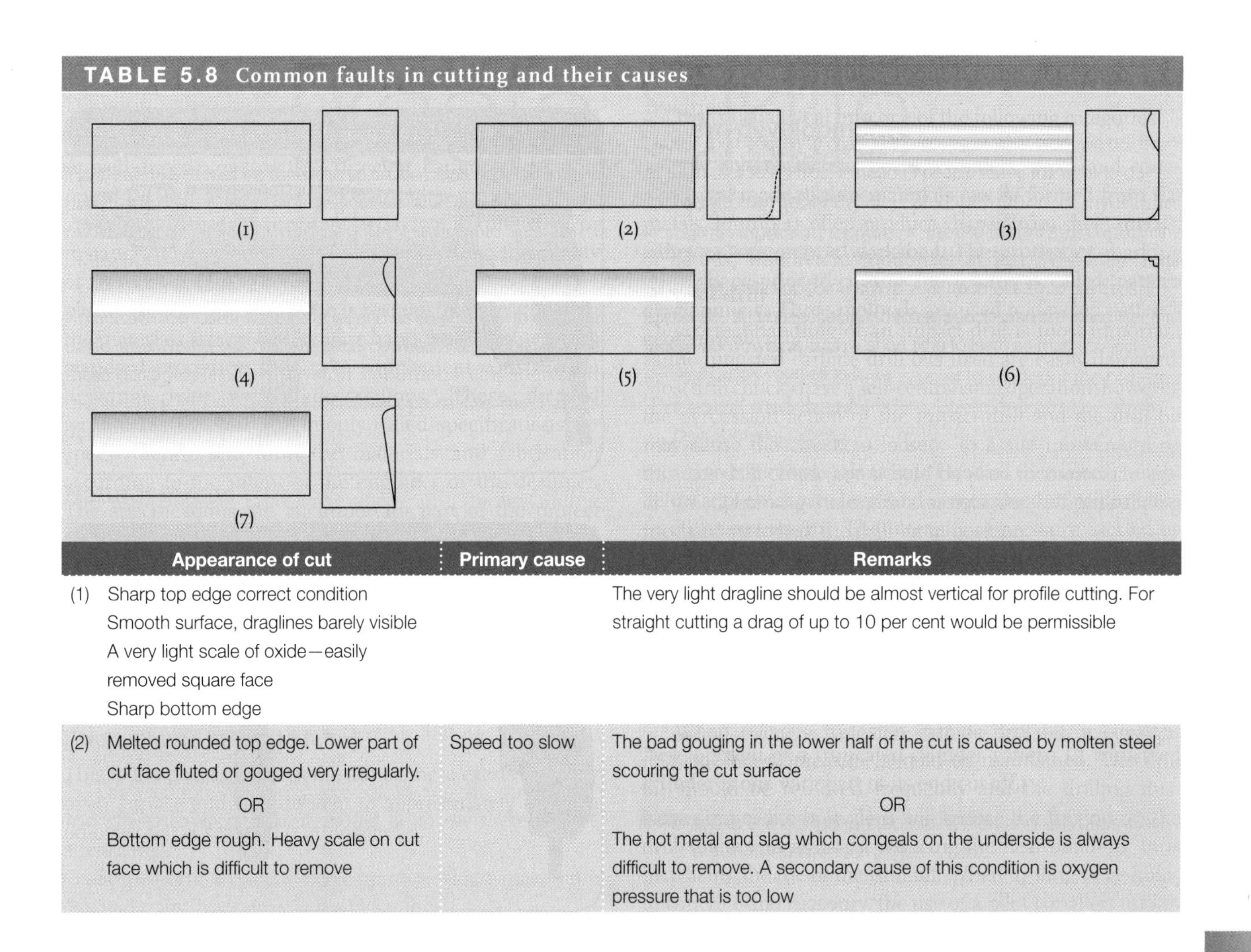

Appearance of cut	Primary cause	Remarks
(1) Sharp top edge correct condition Smooth surface, draglines barely visible A very light scale of oxide—easily removed square face Sharp bottom edge		The very light dragline should be almost vertical for profile cutting. For straight cutting a drag of up to 10 per cent would be permissible
(2) Melted rounded top edge. Lower part of cut face fluted or gouged very irregularly. OR Bottom edge rough. Heavy scale on cut face which is difficult to remove	Speed too slow	The bad gouging in the lower half of the cut is caused by molten steel scouring the cut surface OR The hot metal and slag which congeals on the underside is always difficult to remove. A secondary cause of this condition is oxygen pressure that is too low

Appearance of cut	Primary cause	Remarks
(3) Top edge not so sharp and may be beaded. Undercutting at top of face cut Draglines have excessive backward drag Slightly rounded bottom edge	Speed too fast	This excessive backward drag of the cut line would result in the cut not being completely severed at the end. The occasional gouging or fluting along the cut indicates the oxygen pressure is too low for the speed, but possibly not too low for a normal speed, in other words, if the speed was dropped and this oxygen pressure maintained a perfectly good cut would result
(4) Excessive melting and rounding of top edge Undercut at top of cut face with lower part square and sharp bottom corner	Nozzle too high above work	The melting of the top edge is due to "heat spread" each side of the cut and the undercutting is caused by the oxygen stream being above the work—so that it spreads or tends to 'bell out' as it traverses down the kerf
(5) Top edge slightly rounded and heavily beaded. Cut face usually square with fairly sharp bottom corner	Nozzle too low	Having the nozzle too low does not usually spoil the cut face badly, but it does badly burn the top corner. Very often it retards the oxidation reaction and makes it appear that the cut was done too slowly
(6) Regular bead along fop edge Kerf wider at top edge, with undercutting of face just below	Pressure or oxygen too high	This is probably the most common fault in cutting. The turbulence of the oxygen stream due to the high pressure at which it leaves the nozzle causes the rounding of the top part of the cut face. On thinner material it may cause a taper cut which sometimes leads to the assumption that the cutter is incorrectly mounted in relation to the plate.
(7) Rounded top edge with melted metal falling into kerf Cut face generally smooth, but tapered from top to bottom Excessive tightly adhering slag	Preheat flame too large.	This is the easiest and most obvious condition to correct which, provided other conditions are correct, usually gives a fairly clean, although excessively oxidised cut face, but with a very heavy rounding at the top edge

HARD SOLDERING

There are two major joining processes within the hard soldering classification: silver brazing (commonly referred to as 'silver soldering') and braze welding (often called 'bronze welding'). Although somewhat similar to soft soldering, brazing differs in temperature range, filler material and technique.

One of the main advantages of hard soldering is that dissimilar metals can be joined together. In addition, the joining temperatures necessary are lower than those required by other welding processes.

Silver brazing

Silver brazing, within the plumbing trade, is generally associated with the joining of copper, brass and stainless steel. The process is effected with the aid of any of the available industrially-produced gases, particularly oxyacetylene.

Silver brazing is a process where a joint is made by the application of an alloy filler material drawn by capillary attraction into a fine space between the surfaces to be joined when the heat is applied. The importance of this capillary attraction can not be overemphasised and is influenced by the joint preparation and the fit between the mating surfaces. If the joints do not have closely fitted surfaces the full potential of the process will not be realised.

There is no need for weld reinforcement or build-up and very little molten alloy is used in the process. The surfaces to be joined must therefore be carefully fitted and the joint clearances should be between 0.1 mm and 0.2 mm to produce the most successful joint.

Joint structure and strength

In hard soldering processes skin fusion usually occurs, inasmuch as the molten filler metal will alloy to a slight depth into the parent material. The molten hard solder will unite with the surfaces of the heated joint. The strength of the joint is determined by the surface area of the 'wetted' portion.

Thus lap and sleeve or socket-type joints are stronger than butt joints. The strength of a brazed joint will depend upon several factors. The most important are:

- the amount of alloying between the brazing alloy and the metal being joined
- the temperature of the brazing operation
- the composition of the brazing alloy
- the clearance between sections being brazed.

Silver-brazed joints are much stronger than soft-soldered joints due to the bond being deeper and the solder itself being stronger.

WARNING: **Case should be taken when heating for silver brazing, as excess heat will change the structure of the copper. This can weaken the metal that surrounds the joint.**

Selection of the brazing alloy

The selection of the correct brazing filler material for a given application will be influenced mainly by the composition of the materials to be joined and the temperature at which the

TABLE 5.9 Brazing alloys

Grade	Silver content (%)	Melting range solid to liquid (°C)	Recommended brazing temperature (°C)	Use
Silver brazing alloy Colour code (Yellow tip)	2	640–785	695	Flux-free brazing of copper Use with flux on copper alloys
Silver brazing alloy Colour code (Silver tip)	5	640–805	705	Flux-free brazing of copper Use with flux on copper alloys
Silver brazing alloy Colour code (Brown tip)	15	640–815	705	Flux-free brazing of copper Use with flux on copper alloys

final assembly will function. A number of brazing alloys are in common use and fall into the following categories:

- silver-copper series
- silver-copper-zinc series
- silver-copper-zinc-cadmium series
- silver-copper-zinc-tin series
- silver-copper-phosphorus series.

Silver brazing filler rods vary in silver content between 2 per cent and 60 per cent. Generally, the higher proportion of silver in the filler alloy used, the better flowing and more durable the joint achieved.

The brazing alloys most suitable for joining copper–copper sections are the silver-copper-phosphorus series (commonly known as 'phos-copper') shown in Table 5.9. Fluxing is unnecessary with this type of alloy for brazing copper because the phosphorus is self-fluxing and acts as a deoxidising agent.

Silver brazing alloy 2% (yellow tip)

This alloy can be used on copper without a flux, since its phosphorus content makes it self-fluxing. Although its joint strength is more than required for most applications, its low ductility makes it undesirable where frequent straining or bending is encountered, such as repeated expansion and contraction. Because of its low silver content this alloy provides maximum brazing economy.

Silver brazing alloy 5% (silver tip)

This alloy is also self-fluxing on copper. Although providing slightly greater ductility than the 2% silver content brazing alloy, it should nevertheless be used only on applications where its ductility can be tolerated.

Silver brazing alloy 15% (brown tip)

This alloy is also self-fluxing if used on copper materials. It is recommended for applications requiring greater ductility such as those involving rigid connections and subjected to temperature changes.

Fluxing

The importance of joint preparation, particularly cleaning the joint area by fluxing, must be understood if satisfactory joints are to be produced. Oxides on the metal faces prevent 'wetting' of their surfaces by the molten brazing alloy.

Fluxes perform similar functions to those already stated in the section dealing with soft soldering:

- to dissolve and sweep away surface films and oxides during brazing or soldering processes
- to prevent oxides from reforming during brazing or soldering
- to promote fluid flow and increase adhesion of the filler metal.

All fluxes are formulated for specific applications and so far it has not been possible to produce a universal flux which will serve its purpose at all temperatures and on all metals. The tendency is for new fluxes to be developed to perform special duties. Proper selection of a suitable flux is equally as important as the selection of the correct silver brazing alloy.

It has been mentioned above that fluxing is not necessary when joining copper to copper. A silver brazing flux must, however, be used if joining metals in the following combinations:

- copper to brass
- brass to brass
- copper to stainless steel
- brass to stainless steel
- stainless steel to stainless steel.

Fluxes used in silver brazing operations may be of the fluoborate type for lower temperature work (below 750 °C) and the borate type for high temperature brazing (above 750 °C).

SILVER BRAZING COPPER TUBE

Procedure for silver brazing copper tube

1. Remove all burrs from the end of the tube.
2. Use either steel wool or emery tape to clean the area of tube to be joined.
3. Ensure that the tube ends provide a proper sliding fit.
4. Heat the joint evenly by moving the heat source around the circumference of the joint area. As the fitting heats up, it will expand away from the tube to leave clearance for the alloy.
5. Apply silver brazing alloy at the dull red colour indication and allow it to flow around the joint until a slight fillet forms around the shoulder of the fitting.

With proper technique the alloy will be drawn into the joint before the fillet has formed. Work should be carried out whenever possible in the 'downhill' position, so that gravity as well as capillary attraction may help to fill the joint.

Figure 5.23 illustrates techniques which should prove helpful in brazing copper tube in various positions.

In designing tube assemblies, particularly in copper, the plumbing industry has realised that brazing contributes substantially to compactness of design. The streamline threadless jointing used in conjunction with brazing eliminates bulky and sometimes unattractive screwed joints.

Braze welding

The term 'braze welding' is applied to the joining of metals both similar and dissimilar, ferrous and non-ferrous, by means of an oxyacetylene blowpipe in conjunction with a filler metal of the copper-zinc series. The high temperatures necessary for fusion welding often cause complications by way of expansion and distortion. These drawbacks can be overcome by using the braze welding process.

The relatively low welding temperature of braze welding makes it an excellent method for joining metals with a high melting point, such as galvanised iron pipe. In addition, it has the advantage of preserving the original zinc coating, the corrosion-resistant property of the material. With fusion welding, it is impossible to prevent damage to fabricated galvanised pipe sections. Using the low temperature braze welding process, a homogeneous union is made below the melting point of the parent material that is 'without fusion'.

Properly carried out, braze welding produces a joint equal in strength to the parent metal, and in the case of cast iron, greater than the strength of the parent metal.

Theory of braze welding

The theory of the braze welding process lies in the fact that molten bronze will flow onto a properly cleaned and heated surface of metals of a higher melting point than the bronze filler rod.

The strength of the joint obtained is due to:

- alloying along the surface of the bronze and the parent metal
- intergranular penetration of the bronze into the surface structure of the parent metal.

This action can be compared with the intermolecular penetration and alloying of the tin present in the solder in soft-soldering operations. Although the parent metal is never melted, the characteristics of the bond between the weld and the parent metals give a joint that is equivalent to a fusion weld.

Strength of the braze weld

The strength of any weld depends largely upon the skill and knowledge of the operator. In addition to the experience and technique of the operator, the following factors will affect the strength of the braze-welded joint:

- correct joint design and preparation—this will depend on the thickness and nature of the material surface to be joined
- correct filler rod and flux
- correct size of welding tip—generally one size smaller than that recommended for fusion welding
- correct flame adjustment—the flame should be slightly oxidising
- time–temperature factor. The time–temperature factor is the time for which the parent metal is held at the required temperature to allow surface alloying with the parent metal.
- removal of residual flux and cleaning of joint on completion of weld.

FIG 5.23 Techniques used when silver brazing copper tube in various positions: (a) horizontal, (b) vertical and (c) overhead

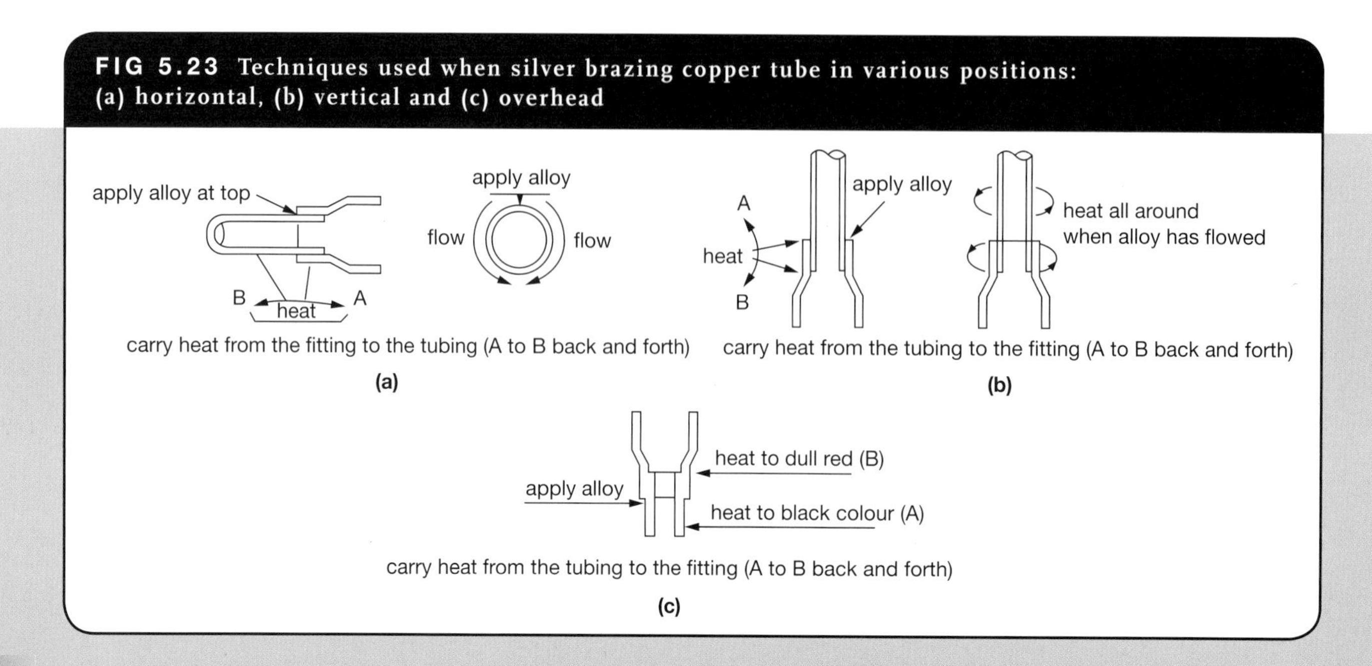

Selection of brazing alloy

The strength of a braze-welded joint depends largely upon the quality and composition of the filler rod. The filler material for ferrous metals is generally a non-ferrous alloy with a melting temperature greater than 500 °C but less than that of the parent metal. Bronze filler rods with melting temperatures of 910 °C or less can be used on various metals. Bronze filler rods can produce high-strength joints on such materials as steels, alloy steels, cast irons, copper and copper alloys. In many cases the strength of the braze-welded joint will exceed that of the parent metal. Table 5.10 shows the recommended filler rods for braze welding and some of their properties.

TABLE 5.10 Filler rods for braze welding

Name	Colour code	Melting point	Nominal composition (%)
Manganese-bronze	Green tip	890 °C	Cu 60 Mn 1 Zn remainder
Nickel-bronze	Maroon tip	910 °C	Cu 47 Ni 10 Mn, Si—trace Zn remainder
Tobin-bronze	White tip	885 °C	Cu 63 Sn, Si—trace Zn remainder

Although a number of different types of bronze filler rods are available, only those applicable to the trade will be discussed here.

Manganese bronze (green tip)

Manganese bronze filler rods are ideal for use on ferrous metals. They possess low fuming characteristics and high strength. When used to braze-weld cast iron, the joint strength exceeds that of the parent metal.

Manganese bronze should not be used for joining copper tubes carrying hot water because the joints would be subject to dezincification. Bronze rods containing manganese produce stronger joints than similar welds made with tobin bronze.

Tobin bronze (white tip)

Tobin bronze filler rods are used on copper and copper alloys where strength and ductility is required and provide joints as strong as the parent metal. Tobin bronze filler rods minimise the possibility of dezincification when used to join copper when the joints are in contact with water.

Fluxing

Selecting the correct flux is as important as using the correct filler rod if a strong joint is to be obtained. An incorrect choice of flux can result in poor tinning of the parent metal, thus preventing a sound bond.

The correct selection for various applications should be made from recommendations provided by the manufacturer. The pink powder copper and brass flux and white powder bronze flux are both suitable for most braze-welding operations.

White

Principally for braze welding cast iron. Also suitable for braze welding of steel, steel castings, malleable iron and so on.

Pink

Particularly suitable for use with both manganese and tobin bronze rods. Used on braze welding copper, brass and bronze, and brazing of copper, steel and other materials.

To apply the flux, one end of the filler rod is first heated and then dipped into the flux.

Flux-coated filler rods are also available in both manganese and tobin bronze. Manganese bronze flux-coated rods are said to produce up to 30 per cent greater bond strength than the standard manganese filler rod. In addition, flux-coated filler rods provide superior flow control and enable the flux residue to be removed more easily.

PROCEDURE FOR HORIZONTAL WELD

1. Set the pipe up in a suitable support for ease of operation and comfort.
2. Heat up the filler rod and dip into the flux.
3. Tin the heated area and form the first ripple (the blowpipe angle and movement relative to the pipe is indicated in Figure 5.24).
4. Pause briefly to allow cooling of the molten bronze filler metal.
5. Proceed to deposit more filler metal and tin ahead for each successive ripple.
6. Rotate the pipe towards the operator through approximately 90°; when the blowpipe reaches the top of the pipe; repeat steps 3 and 4.
7. Rotate the pipe as in step 6 each time the weld reaches the top of the pipe.
8. Overlap the first ripple made in step 3.
9. On completion of the weld, the job should cool slowly, after which the flux residue is removed by wirebrushing and the joint given a protective coating with an approved paint.

FIG 5.24 Blowpipe angle and movement relative to the pipe

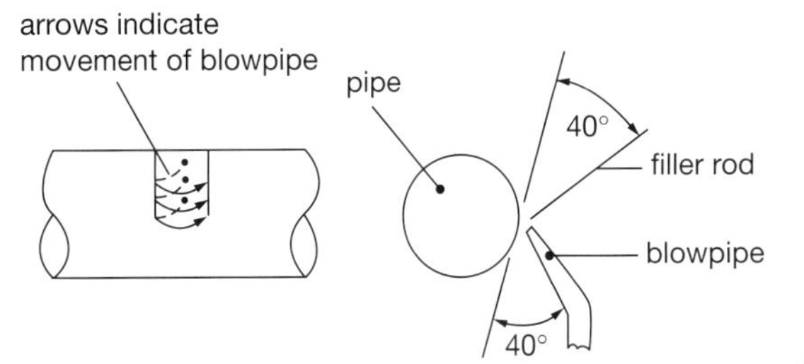

METALLIC ARC WELDING

The 'metallic arc' process is the most widely used welding process. The metallic arc may be applied manually, semi-automatically or fully automatically. Within the plumbing industry, MMA (manual metallic arc) is used exclusively for fabricating brackets and supports, and in the field of mechanical service plumbing.

The welding of mild steel and wrought iron, the materials most commonly used, is readily achieved by the MMA process and electric arc welding has many advantages. The resulting joint will be as strong as the parent material and the welding process may be performed either on-site or in the workshop. There is no wastage of material or subsequent reduction in strength due to the drilling of holes for bolts or rivets. The joints may be made pressure-tight, watertight and weathertight.

In striking the arc in MMA, first touch the metal of the electrode on the work. This causes a short circuit and a heavy current flow. This high current flow produces intense local heating in overcoming relatively high resistance at the point of contact; this causes both the tip of the electrode and the surface of the work immediately beneath it to become incandescent.

The electrode is then drawn away a short distance from the work. The incandescence produces free electrons in the air between the electrode tip and the work. It is these free electrons which ionise the air and allow it to become a conductor of electricity and therefore to permit the electric current to pass through it. The current will continue to flow so long as the necessary power is available and the arc is maintained at the correct length.

When the metal is transferred from the electrode to the work piece, most of it is in the form of globules and the remainder is in the form of metallic vapour (Figure 5.25). No metal is transferred if the electrodes are of equal diameter. One electrode must be larger than the other and metal will always be transferred from the smaller to the larger.

As in all welding techniques, operators should make themselves as comfortable as the work conditions will permit. Relaxation helps to achieve good quality work; a strained posture should be avoided. The electrode holder should be held tight enough to ensure total control of the electrode, but straining should be avoided since total concentration is required at all times during the welding operation. Where possible, the pass made should be equal to the full length of the electrode; this again will require body movement, emphasising the need for comfort and concentration.

The arc length maintained should be as short as practicable, consistent with a steady and continuous flow of metal. This length will be approximately equal to the diameter of the electrodes. If the electrode is held too close to the work piece, 'sticking' or 'freezing' of the electrode to the work can occur. This results in many stopping and starting points in the run and these constitute weak points within the weld. Similarly, if the electrode is held too far away from the work piece, brittle and burnt metal may be deposited due to the relatively long time that the metal has spent in passing through the arc. From this exposure to the atmosphere, oxides and nitrates may be formed and trapped in the weld metal causing weakness in the weld.

In addition, the excessively long arc causes a loss of control and the ability to follow the correct procedure; this can allow molten metal to fall upon relatively cold parent metal, resulting in practically no penetration, the deposit merely adhering to the parent metal instead of fusing into and with it.

Preparation of the work

As in any welding operation, clean working is of the utmost importance. Rust, scale and grease should be removed, as the flux on the rod contains only sufficient ingredients to remove the usual mill scale found on mild steel. As the rust and scale found on metals consist mainly of

FIG 5.25 Metallic arc welding: a horizontal downhand weld

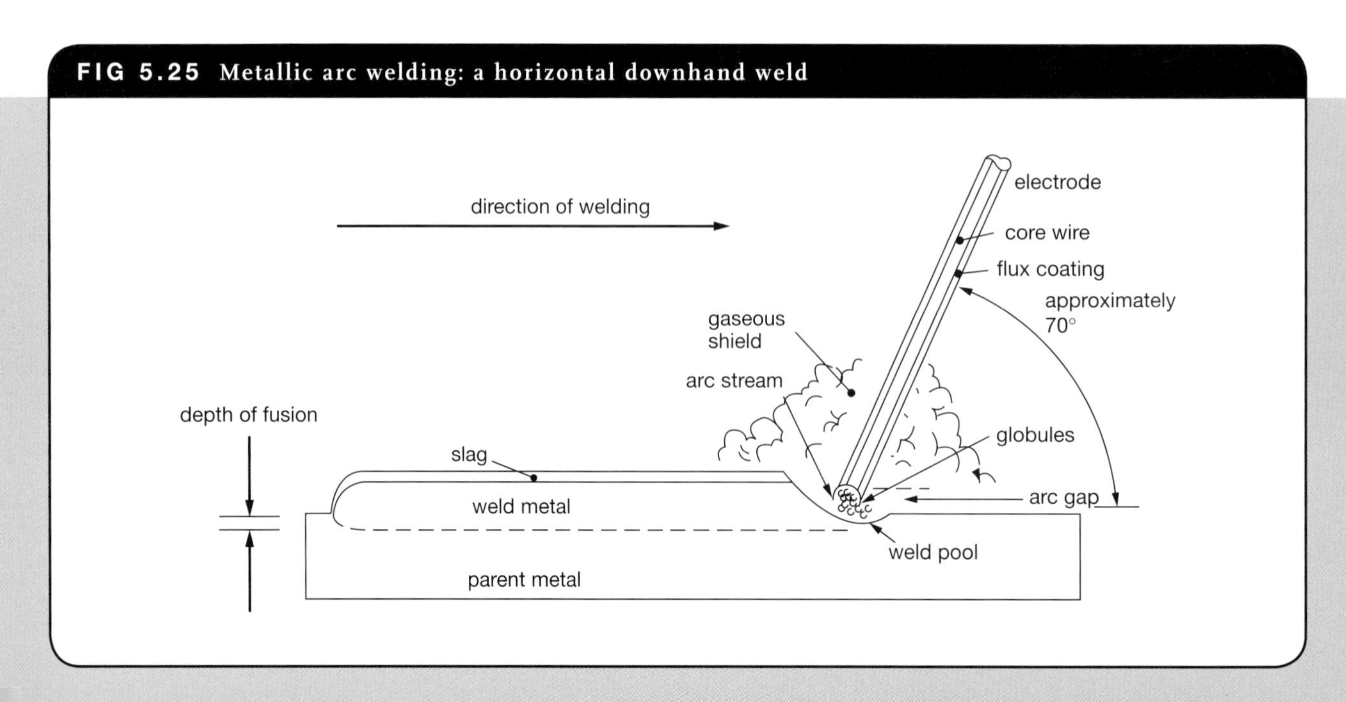

oxides, the incorporation of these into the joint can cause structural weakness. Butt-weld preparation depends upon the thickness of the metal and on whether both sides or only one side is accessible. In a single 'V'-butt, welds can not be reinforced on the back of the 'V' and the joint should be bevelled to an edge (Figure 5.26).

When joint preparation is completed, the joint should, if possible, be set up in the horizontal position. This provides for the most reliable welding and is the most economical in electrode consumption and labour costs.

Welding technique

Firstly, clean off any rust scale, grease or paint to ensure a clean surface for the earthing clamp. When welding light-gauge materials, it is advisable to clamp the earth lead to the job to ensure that a satisfactory circuit results.

The selection of the welding electrode will depend largely upon the material being welded and the position in which the weld will be made. For example, on thin material a smaller or thinner gauge is required, otherwise holes

FIG 5.26 Preparation of the work for metallic are welding: (a) closed butt: for plates up to 2.0 mm no spacing is needed; (b) open butt: for plates from 2.0 mm to 3.0 mm a space of about 1.5 mm is needed; (c) single 'V' butt: plates from 4.5 mm to 12.7 mm require chamfering of the edge of the plate so that the included angle is about 70° along the line of the joint; (d) double 'V'-butt: used for thick plates in excess of 12.7 mm where both sides are accessible; The angles should be 70° included

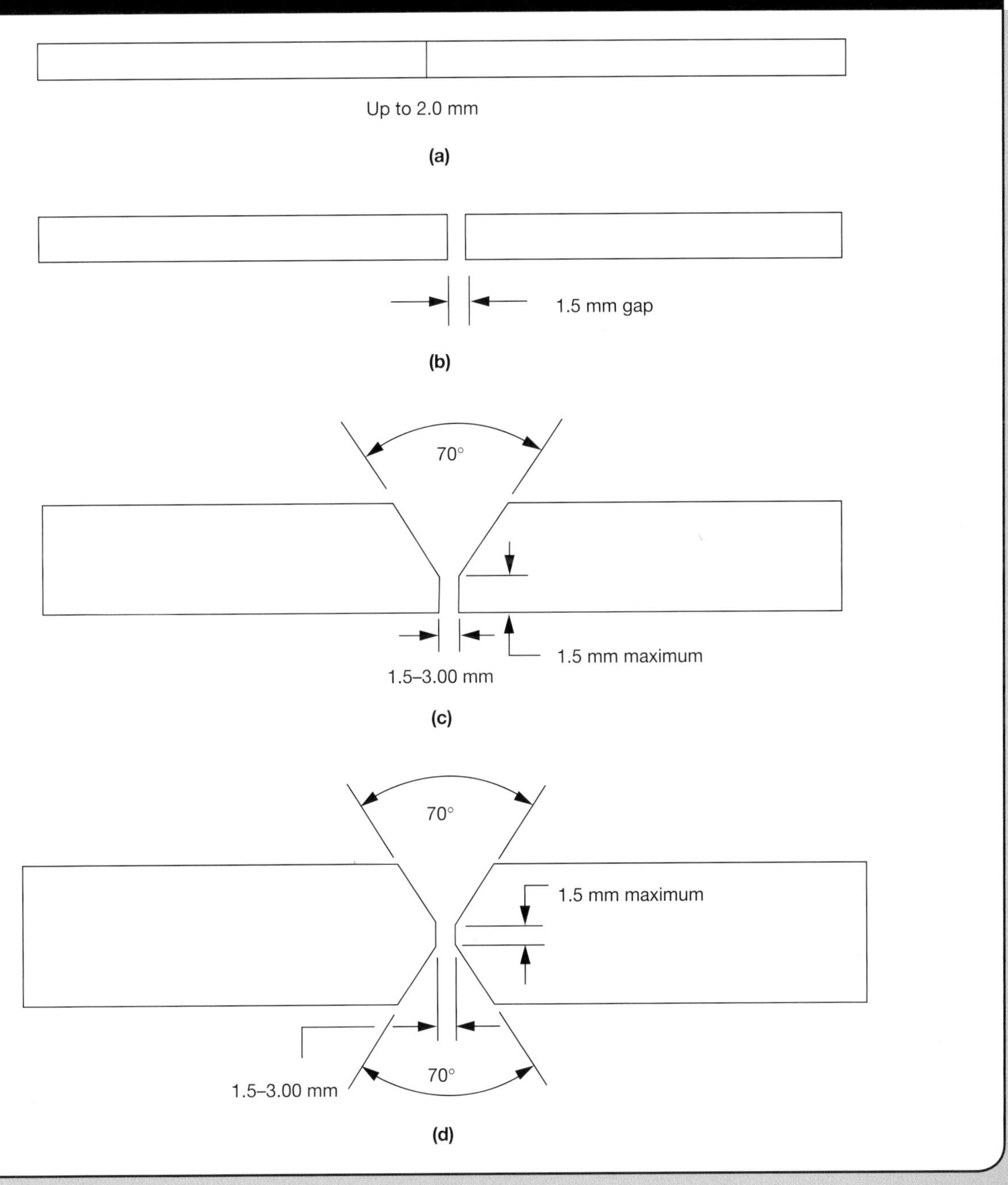

will burn through the material. The electrode should allow for full root penetration. The recommended amperages are generally shown on the electrode packet. Table 5.11 lists those for mild-steel welding. The effects of incorrect amperages are shown in the section on defects due to faulty welding techniques.

TABLE 5.11 Suggested amperages for mild steel horizontal downhand

Size (mm)	Current range (A)
1.75	40–60
2.5	55–80
3.25	95–125
4.0	135–170
5.0	185–230

The position of the body is important in welding, as a good comfortable position allows the welder to concentrate fully on accuracy and technique. Wearing a leather apron and leather gauntlet will reduce anxiety, as the possibility of setting clothes alight is greatly reduced.

The work should be positioned so that the weld is carried out across the body. The electrode holder lead should be clear of any obstruction to enable free movement along the length of the weld as the electrode is consumed. The lead will generally be easier to handle if it is placed over the shoulder to reduce the weight on the welder's hand.

When striking (starting) the arc the tendency to lift the electrode away from the parent metal, thus breaking the arc, must be controlled. Two methods can be employed to strike the arc: the 'scratch' (or 'stroking') technique and the 'stab' technique (Figure 5.27). In the first, the electrode is carried down in a stroking movement; when the electrode makes contact it is lifted to achieve the correct length of arc. In the 'stab' technique, the electrode is 'stabbed' onto the metal; once contact is made the electrode is then lifted up to achieve the correct arc length.

The length of the arc is critical and the correct length is easily achieved. A long arc produces more heat, but a very long arc has a tendency to splutter or crackle and the weld metal is deposited in large irregular blobs; the weld bead is flattened and the spatter increases. A short arc is essential to achieve a high-quality weld. However, if it is too short there is a danger that the tip of the electrode will be frozen into the weld. If this happens, a quick twist back over the weld will frequently detach it.

Having successfully struck and maintained an arc, the speed of travel must now be considered. This requires moving the tip of the electrode towards the molten pool at the same rate at which it is melting. At the same time the electrode has to move along the weld to form a bead. The angle of the electrode into the weld pool is approximately 20° from the vertical. The rate of travel must be adjusted so that a well-formed bead is produced. If the travel is too fast it will produce a narrow bead and may even produce a string of beads along the weld (Figure 5.28). If the speed is too

FIG 5.27 Striking an arc

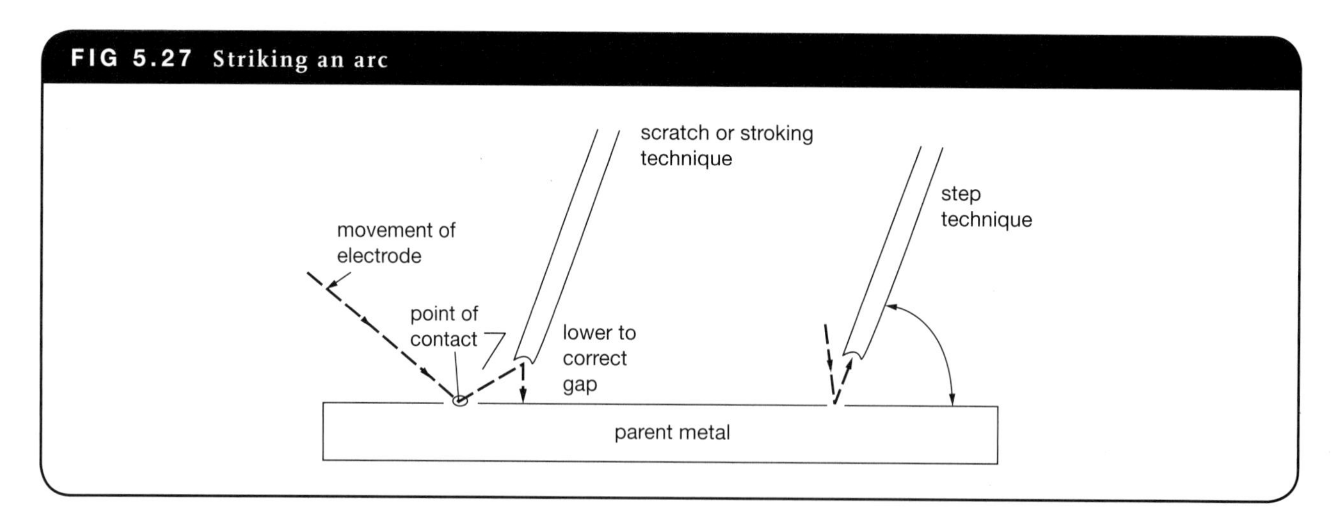

FIG 5.28 Speed too fast: thin beads or individual beads

FIG 5.29 Speed too slow: pause too long, large arc gap

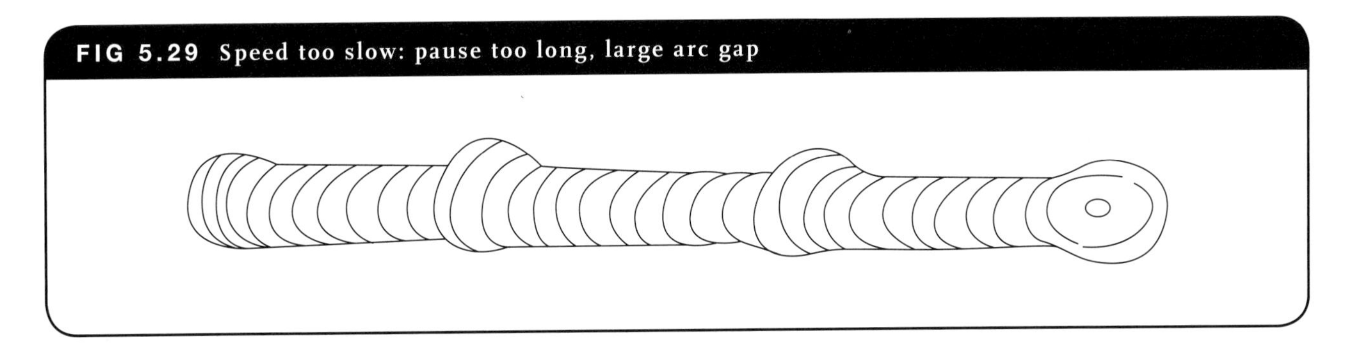

FIG 5.30 Correct weld proportions: (a) correct speed, (b) 2 to 3 times the gauge of the electrode

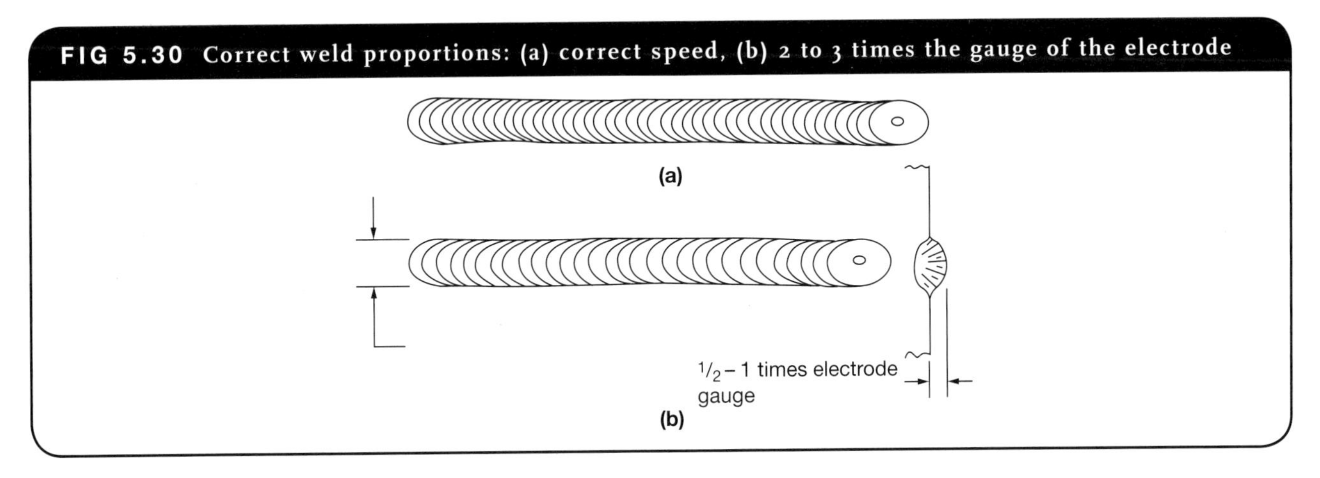

slow the weld metal will pile up causing a large irregular bead (Figure 5.29).

The correct form of a weld and weld proportions are shown in Figure 5.30.

Electrodes

Plumbers using MMA machines for pipe work or general construction welding need to understand the coding of electrodes that are available to them so they can select the most suitable electrode.

All packets of electrodes carry code letters printed on them (for example, E6012GP). The 'E' represents electrode. The first two numbers (either 60 or 40) represent the tensile strength of the metal when deposited as a weld. The last two numbers (11, 12 or 13) denote the covering of the electrode or flux. To a welder it denotes the position of the weld on which it may be used (for example, horizontal/vertical, overhead/horizontal, vertical or flat). 'GP' at the end denotes general purpose; 'P' denotes deep penetration.

The plumber is only interested in a minimum number of classifications, usually E6011 IP, E6012P, E6012GP and E6013GP.

1. E6011P denotes a deep penetration electrode with a high cellulose covering giving a quick freezing slag.
2. E6012P denotes a deep penetration with a viscous (slow moving) slag.
3. E6012GP denotes a general purpose electrode with a viscous slag.
4. E6013GP denotes a general purpose electrode with a fluid slag.

Usage

An electrode with a fluid slag (E6013GP) would not be suitable for a weld position where either the slag can run in front of the weld pool or in a position where the hot slag can drip or fall on to the welder.

These electrodes are best suited for a 'vertical up' weld where it is desirable for the slag to run back from the weld.

For welding a pipe secured in the horizontal position, a suitable electrode would be an E6011P for the initial deep penetration root run followed by the E6012GP for the fill and capping weld.

Defects due to faulty welding

Undercut

This reduction in cross-section weakens the joint and creates a slag trap.

TABLE 5.12 Undercut

Cause	Remedy
High amperage. Arc too long	Reduce to a shorter arc. Keep shorter arc
Angle of electrode too steeply inclined to joint face	Electrode should not be inclined less than 45° to face
Joint preparation does not allow correct electrode angle	Allow more room in joint for manipulation of electrode
Electrode too large for joint	Use smaller gauge electrode
Insufficient depositing time at edge of weave	Pause for a moment at edge of weave to allow build-up. (Weaving is more likely to produce undercut than a straight run. Therefore, where possible use straight runs)

Slag inclusions

Non-metallic particles trapped in the weld metal are called 'slag inclusions'. They may seriously reduce the strength of the welded joint. If slag is present in a weld, chip, grind or flame gauge until removed, then reweld.

TABLE 5.13 Slag inclusions

Cause	Remedy
May be trapped in undercut from the previous run	If 'bed undercut' present, clean slag out and cover with run from small-gauge electrode
Joint preparation too restricted	Allow for adequate penetration and room for cleaning out slag
Irregular deposits allow slag to be trapped	If very bad, chip or grind out irregularities
Lack of penetration with slag being trapped beneath weld bead	Use smaller electrode with sufficient amperage to give adequate penetration. Use suitable tools to remove all slag from corners etc.
Rust or mill scale, preventing full fusion	Clean joint before welding
Wrong electrode for position in which welding is done	Use electrodes designed for the position in which the welding is done, otherwise proper control of the slag is difficult

Incomplete penetration

Incomplete penetration is a gap left by failure of the weld metal to fill the root.

TABLE 5.14 Incomplete penetration

Cause	Remedy
Amperages too low	Increase current
Electrode too large for joint	Use smaller electrode
Insufficient gap	Allow wider gap
Angle of electrode	If too inclined, does not give penetration Keep nearer to right angle to weld axis
Incorrect sequence	Use correct build-up sequence

Lack of fusion

Lack of fusion occurs where portions of the weld do not fuse to the surface of the metal or edge of the joint.

TABLE 5.15 Lack of fusion

Cause	Remedy
Small electrodes used on heavy, cold plate	Use larger electrodes (preheating may be desirable)
Amperage too low	Increase current
Wrong electrode angle	Adjust the angle so that the arc is directed more onto the parent metal
Speed of travel	If too fast, does not allow time for proper fusion
Scale or dirt on joint surface	Clean surface before welding

Note: In overcoming these faults, it is often an advantage if the fob can be positioned to allow welding to be carried out in the 'backhand' position.

GUIDELINES FOR WELDING SAFETY

Personal safety is most important in any welding operation, either oxyacetylene or metallic arc. For safe working, safe habits must be acquired and used in all operations. The following points will be useful in developing safe work.

- Be neat and clean about your work.
- Maintain all equipment in good condition.
- Wear goggles with the correct shade of lens when using a blowpipe to protect your eyes against sparks and injurious rays.
- Wear suitable clothing (that is, aprons, shoes and gloves made from a nonflammable material).
- Watch for sparks in sleeves, cuffs and open pockets.
- Never use oxygen to dust clothes or work.
- Use a flint lighter or pilot light to light the blowpipe.
- Keep sparks or hot metal away from cylinder and tubing.

GENERAL SAFETY REQUIREMENTS WHEN USING OXYACETYLENE WELDING EQUIPMENT:

- Keep all cylinders, whether empty or full, away from excessive heat.
- Keep oil and grease away from cylinders.
- Use only standard cylinder keys on cylinder valves.
- Never open a cylinder more than one-and-a-half turns.
- Never test for leaks with matches or flame.
- Do not use frayed or damaged hoses.
- Use only the black hose for oxygen and the maroon hose for acetylene.
- Use only tubing specifically designed for use with oxygen and acetylene.

GENERAL SAFETY REQUIREMENTS WHEN USING ARC WELDING EQUIPMENT:

- Use only hand shields and helmets that have been approved by Standards Australia.
- Electric cables should be kept clear of walkways and passageways to avoid a tripping hazard or the possibility of damage.
- Always repair or replace faulty electric cables.
- Turn off power to the machine before connecting leads; ensure that all leads are tight before operating the machine.
- Work in an adequately ventilated area.
- Never strike an electrode on a gas cylinder.
- Ensure the correct eye and face protection is used; the ultraviolet and infra-red rays emitted during welding can be very harmful to the eyes and skin.

- Screens should be erected around the operator to protect other people working in the vicinity.
- When chipping flux from a joint bead, clear safety glasses must be worn to prevent hot scale from damaging the eyes.

To achieve safe working, three types of protection are needed:

- a shield for the head that is fitted with an eye filter
- gloves for the hands
- protective clothing for the body.

SCREWED PIPE JOINTS AND FITTINGS

Success in producing a good joint on an installation carrying water or any other fluid depends largely on the skill, ability and product knowledge of the plumber. The variety of pipes and fittings available today requires the student to become familiar with the various system components. Methods of joining pipes vary greatly. In order to meet the piping needs, almost any combination of pipes and joints can be made to order.

Fittings

The need for fittings becomes obvious as soon as the layout design for a plumbing service is studied. Fittings are required each time a pipe: (a) makes a sharp change in direction, (b) branches off into two or more separate runs or (c) changes size.

Fittings are available for almost every conceivable piping situation, whatever the pipe material; each fitting performs a specialised task. The shape and purpose of some fittings are described to some extent by the common name used for the fittings. Some of the most frequently used fittings are: bends, bushes, caps, crosses, elbows, tees, nipples, plugs, sockets and unions.

These fittings are generally available with internal (female) and external (male) threads.

Bends and elbows

Bends and elbows are used where a pipe has to change direction (Figure 5.31). These fittings are available in most pipe materials, are threaded either internally (both ends) or internally and externally at either end. They are also available in the form of compression fittings and are made in either 'equal diameter' or 'reduced' configuration.

Bushes

The bush fitting is used to connect pipes of different diameters (Figure 5.32). The external thread screws into a fitting while the internal thread is bushed down or reduced to accommodate the smaller pipe. Bush fittings are available in galvanised steel, brass and plastic materials.

FIG 5.32 Bush

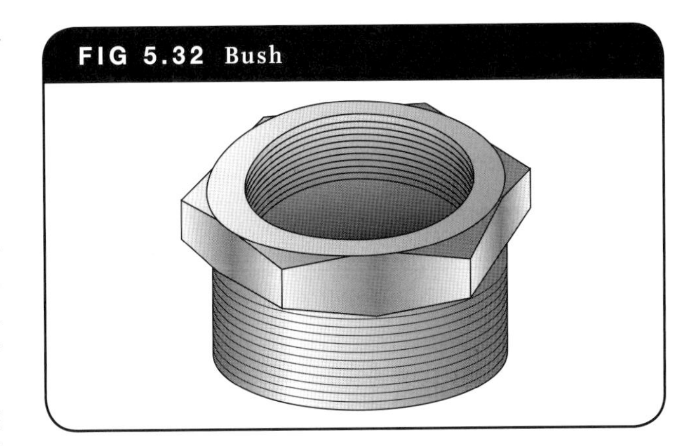

Caps

These fittings are used to close off a pipe end (Figure 5.33). Because end caps are threaded internally they are designed to connect onto an external pipe thread or fitting.

FIG 5.31 (a) Bend F and F, (b) bend M and F, (c) elbow F and F, (d) elbow M and F

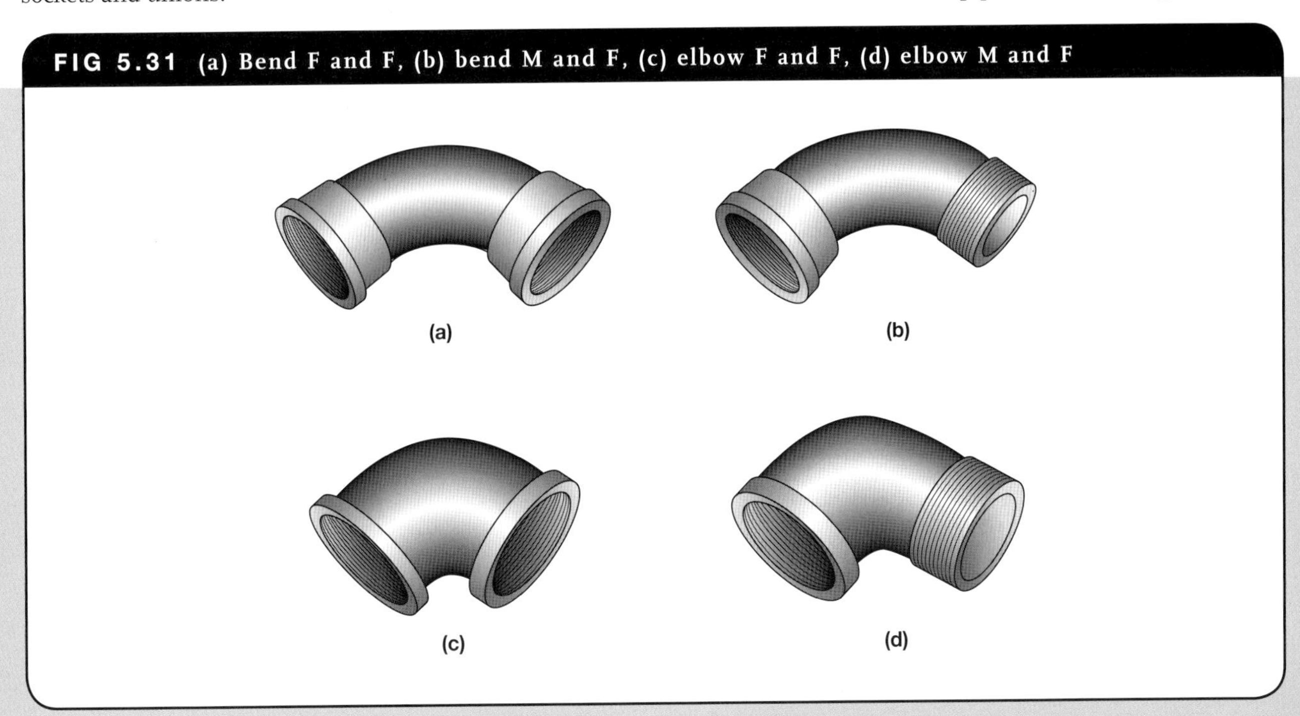

FIG 5.33 Cap

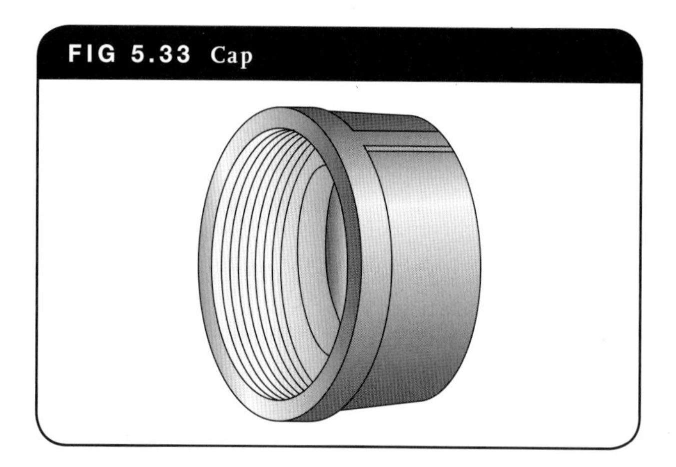

Crosses and tees

Cross fittings are used in situations where four pipes connect at right angles to each other (Figure 5.34(a)). They are threaded internally and are generally used for galvanised steel pipes. Cross fittings give a neat and flush appearance, especially where branch runs occur on exposed walls.

Tee fittings serve a similar function, but are screwed to provide for three pipes to connect at right angles to each other (Figure 5.34(b)).

FIG 5.34 (a) Cross, (b) tee

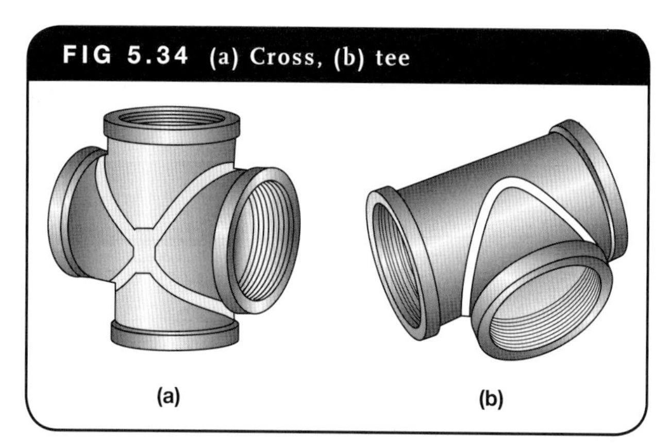

Nipples

Nipple fittings are generally used in conjunction with valves and other internally-threaded fittings. They are produced as a short pipe piece with external threads and may be of either 'plane' or 'shoulder' type. The hexagon shoulder nipple (Figure 5.35) has the advantage of being gripped and screwed easily with a spanner.

FIG 5.35 Hexagon nipple

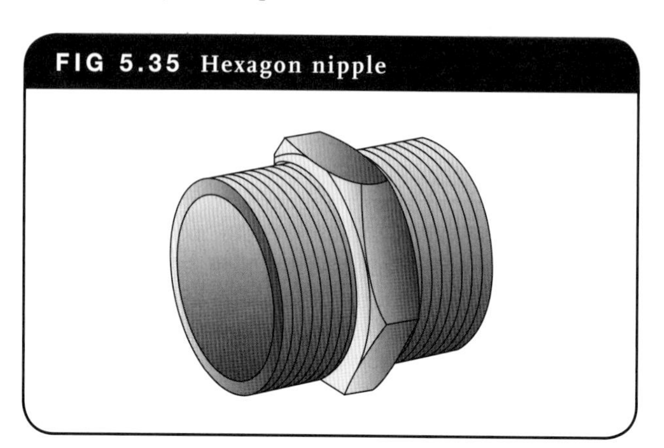

Plugs

Plug fittings perform a similar function to the end cap. Plugs are threaded externally and are therefore used to seal off the ends of fittings, either temporarily or permanently. They are designed with a solid square shoulder to accommodate a spanner.

FIG 5.36 Plug

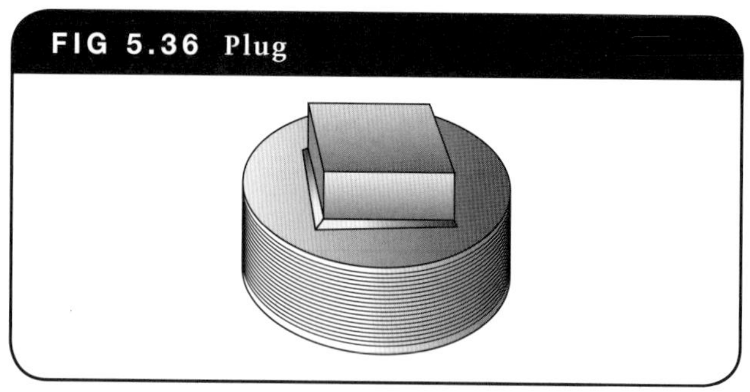

Sockets

Sockets are used for joining pipe lengths together. The fittings resemble short pipe pieces, but are threaded internally. Sockets are available as plain or reducing fittings. The reducing sockets are used to join pipe ends of different diameters.

FIG 5.37 (a) Socket, (b) socket—reducing

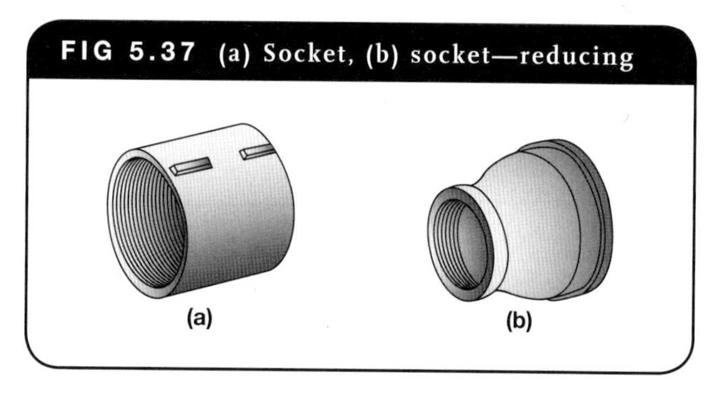

Barrel unions

Barrel unions are used in a piping circuit for much the same reasons as pipe flanges. The union allows a pipe run to connect into and out of its screwed ends. It is designed in two halves to facilitate quick 'making' and 'breaking' of service pipes without interfering with the rest of the piping. Unions are available in materials such as steel, bronze and plastic. They are threaded internally with male and female ends (Figure 5.38(a)), double female ends (Figure 5.38(b)) and screwed to British Standard Pipe (BSP) thread.

FIG 5.38 Barrel unions: (a) screwed M and F ends, (b) screwed F ends

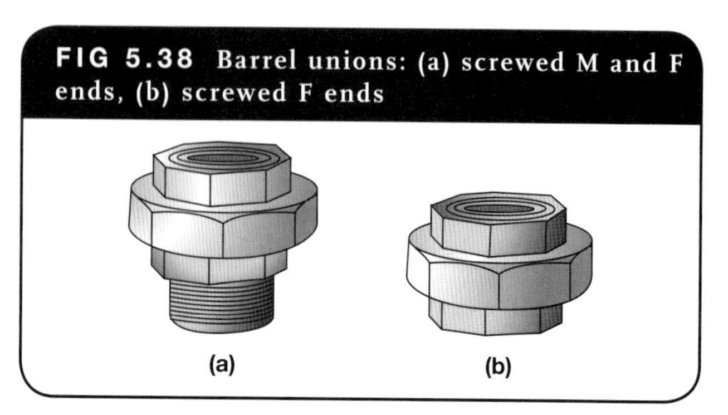

Thread sealants

Generally plumber's hemp, PTFE tape or an approved jointing compound is used on male thread ends to give a positive seal at the joint.

- Hemp is used extensively in the fire protection industry and is available in rope form. Hemp is a reliable thread sealant because when it is in contact with water it swells to make a watertight joint. When used on pipe ends or fittings it must be teased out carefully to a length of approximately three times the circumference of the pipe and wound firmly around the male thread in the same direction in which the thread is screwed (Figure 5.39).
- PTFE tape (Teflon tape) is available in widths approximately 12 to 40mm, with different thicknesses and colours depending on the application. The tape is used extensively throughout the plumbing industry in place of hemp. It is used in a similar way to hemp and any excess should be removed after the joint has been tightened.
- Commercially-prepared jointing compounds are readily available to ensure good, watertight joints. Jointing compounds should not be applied to female threads. As the joint is tightened the compound is forced off the thread and finds its way through the pipeline and into valves when the water is turned on. These paste compounds should be applied strictly in accordance with the manufacturer's instructions.

FIG 5.39 Placing hemp on screw threads (PTFE tape is in the same direction)

Materials for fittings

Galvanised steel

The type of joint most commonly used in conjunction with galvanised steel and brass pipe is a threaded (BSP) joint.

There are two types of threaded joints:

Parallel female and taper male

The most frequently used threaded joint is the parallel female and taper male (Figure 5.40). The taper on the male thread will bind at a particular point along its length. Screwing slightly beyond this point will ensure a watertight joint.

FIG 5.40 Parallel female and tapered male threads

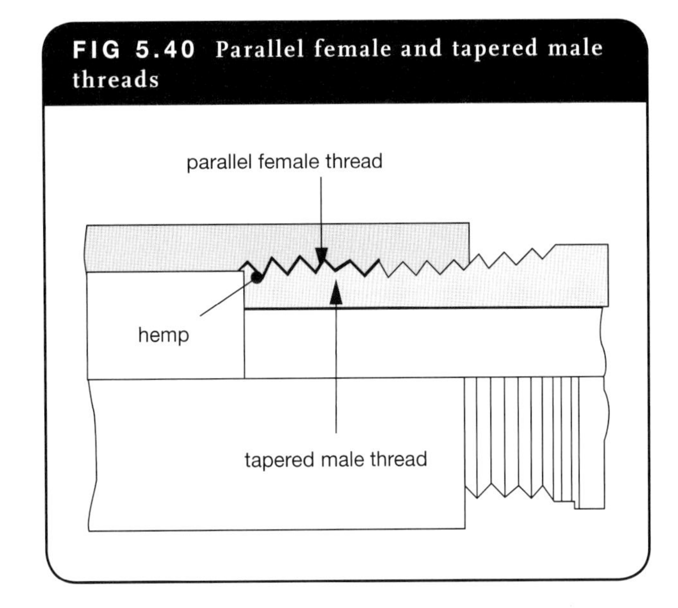

Parallel female and parallel male threads

These are used with longscrews (Figure 5.41). A hemp grummet and locknut provide the seal with this type of joint. However, unions are frequently used to replace this method of joining two fixed pipes together.

The most common types of galvanised steel fittings available are illustrated in Figure 5.42.

FIG 5.41 A pipe joint made with a longscrew

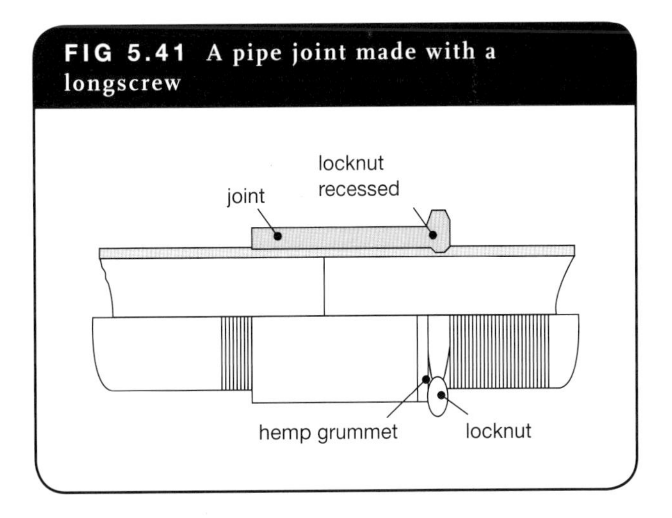

Copper

There are two standard methods of joining copper tube—silver brazing and compression joints. Similar methods are adopted when using stainless steel tube. As the silver brazing process has been covered, only screwed types of fittings and joining techniques will be dealt with in this section. Compression joints used in conjunction with copper tube fall into two categories: manipulative and non-manipulative. Flared compression fittings are suitable for all types of gas joints.

FIG 5.42 Pipe fittings for use with galvanised steel pipe

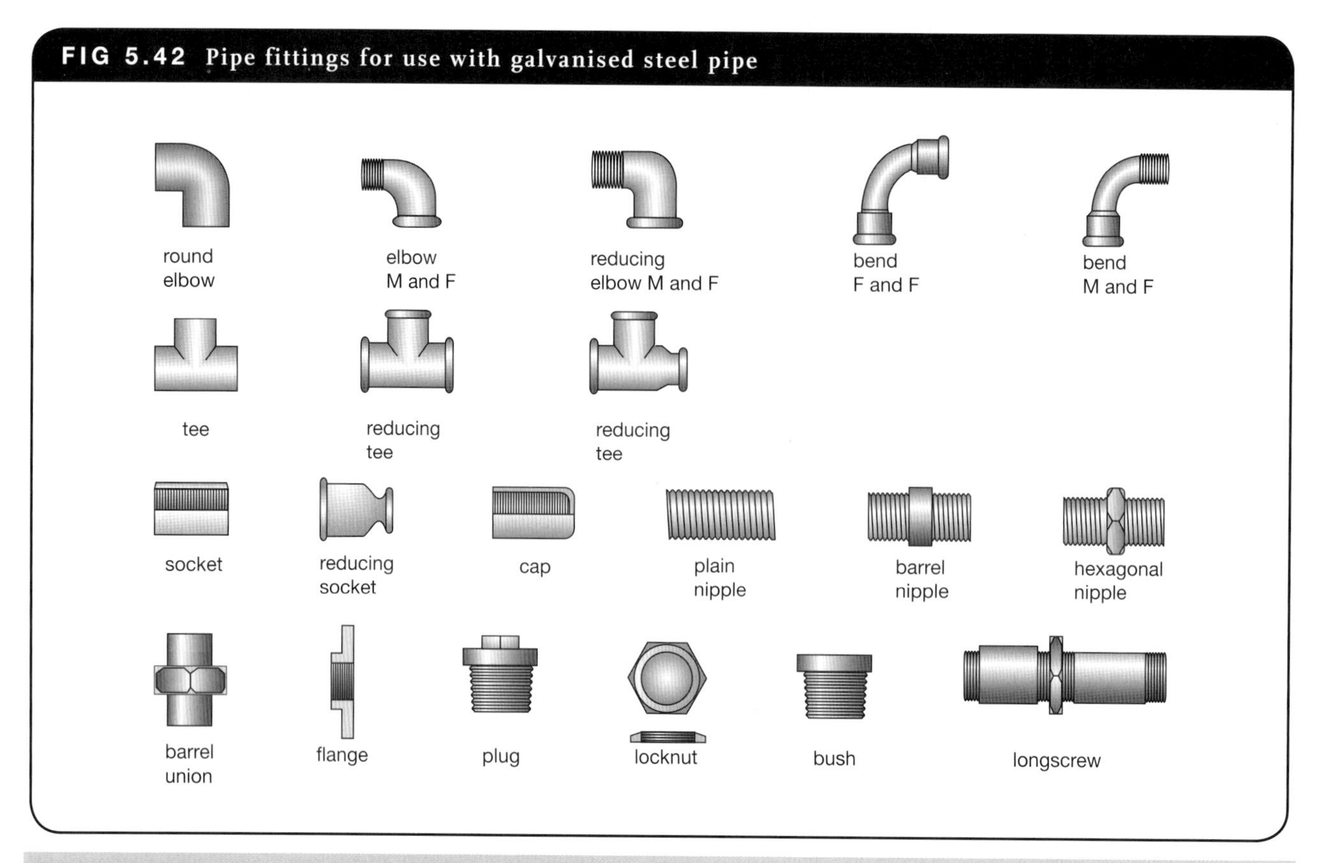

PROCEDURE FOR MANIPULATIVE JOINTS

Figure 5.43 shows an unassembled flared manipulative joint which is made in the following manner:

1 Anneal a square end of the copper tube.
2 Clamp the softened end of the tube into position in a flaring block (Figure 5.44).
3 Drive a coned drift punch into the tube to form a flare to match the inside of the union nut and the outside of the union nipple.
4 Tighten the union nut until the copper seats against the bevel of the nipple.

FIG 5.43 Flared manipulative joint

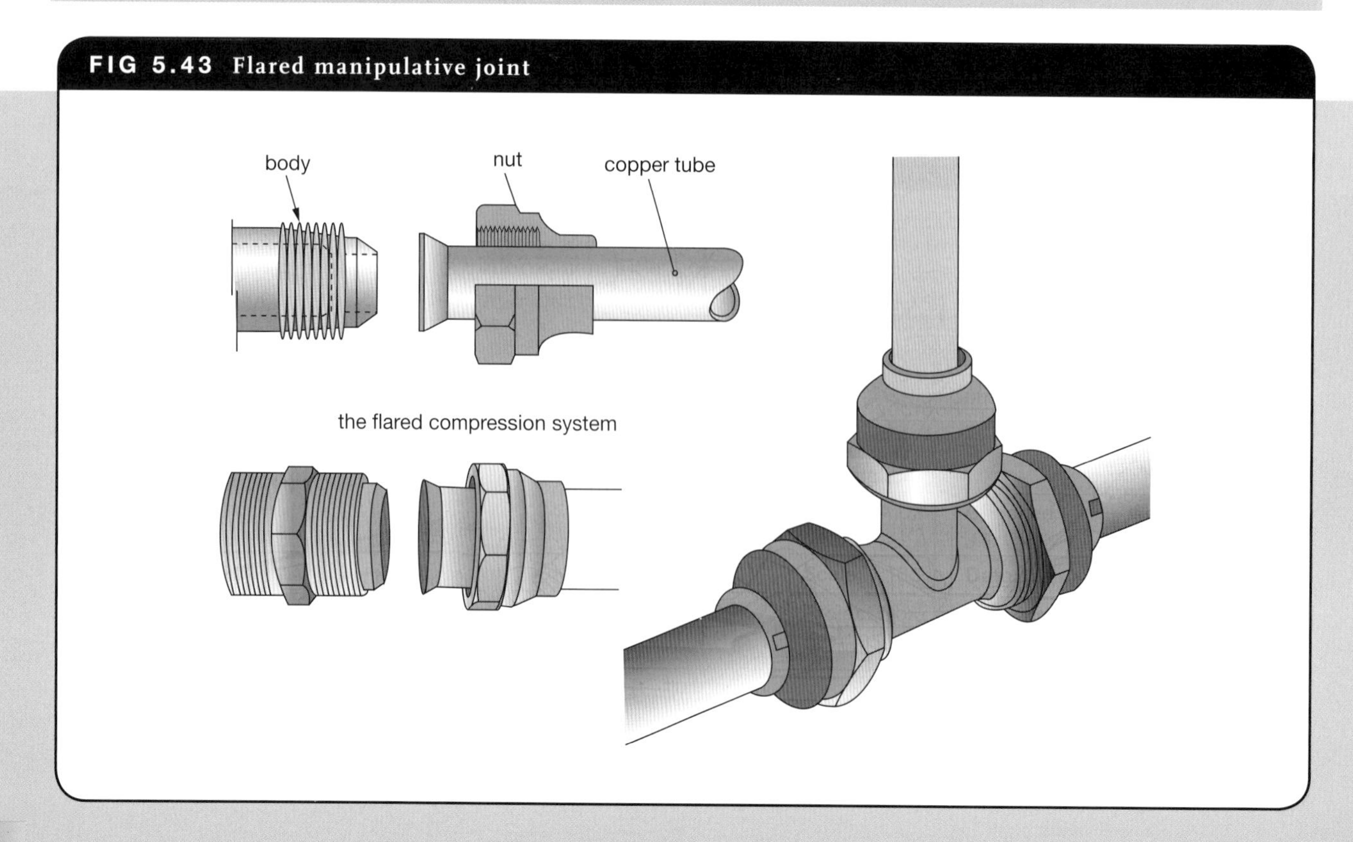

FIG 5.44 Procedure for making a flared manipulative joint: (a) block clamped to pipe, (b) making the flaring

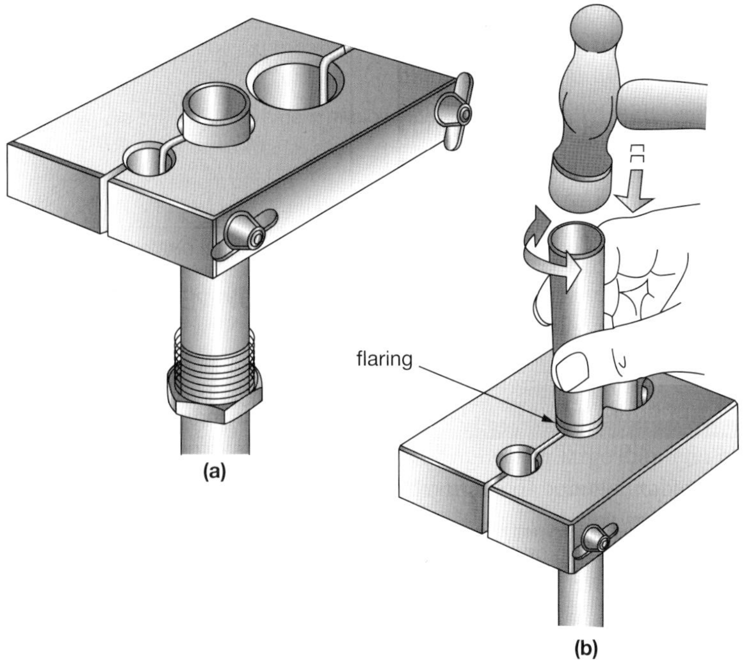

flaring
(a)
(b)

Non-manipulative joints

These fittings use a copper or brass cone or olive and the joint is made possible by tightening a coupling nut against the coned sleeve. This joint should not be used on annealed tube as overtightening of the coupling nut may result in severe indentation of the tube.

Beaded manipulative joints (kingco joint)

These are widely used on water service pipes and are an approved means of connecting a water point to items such as a WC cistern or domestic hot water supply tank. The connection relies for its sealing efficiency upon the compression of a rubber ring which is fitted between an expanded (croxed) section and the end of a copper tube. The rubber ring is compressed round the tube by tightening the kingco nut which screws onto a BSP thread. An example of this joint is shown in Figure 5.45.

Capillary fittings

These are a range of fittings in which the joint is made by the application of a filler metal and heat along the annular space between the outside of the tube and the inside of the socket end of the fitting.

Together with light-gauge copper tube, capillary fittings are suitable for the transmission of hot and cold water, gas, compressed air and other fluids. Capillary fittings are

FIG 5.45 Beaded manipulative joint

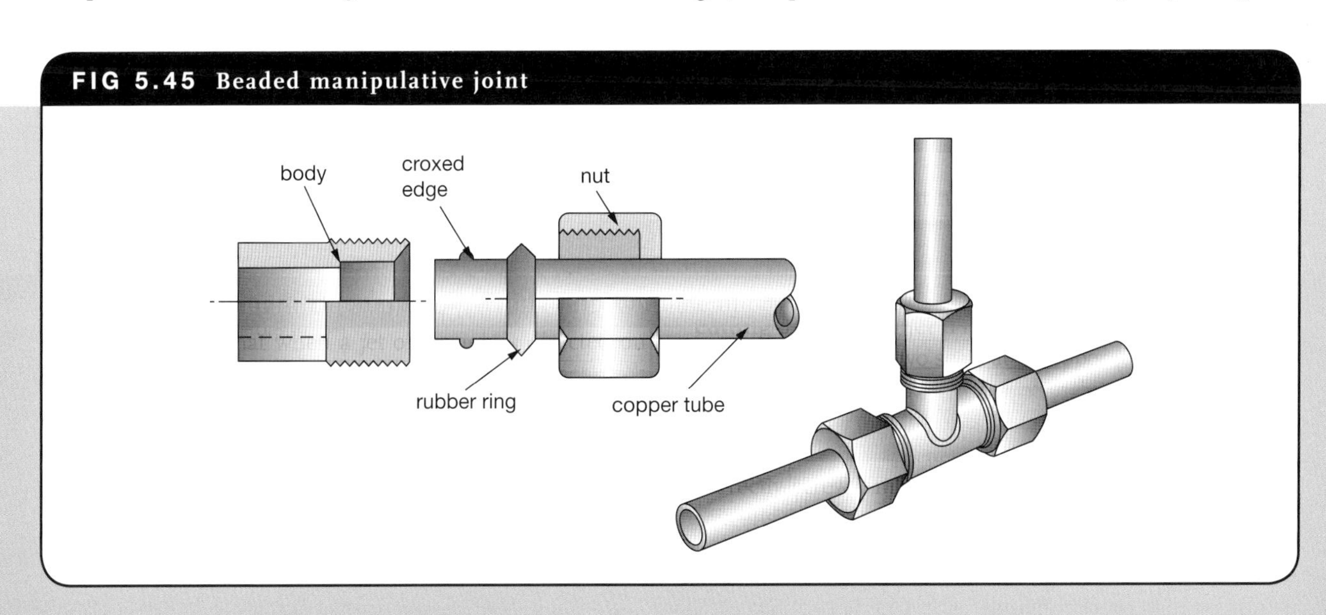

designed to give good flow characteristics, are easily taken apart and are cheaper and less bulky than compression fittings. A range of threaded capillary fittings are shown in Figure 5.46.

FIG 5.46 Threaded capillary fittings

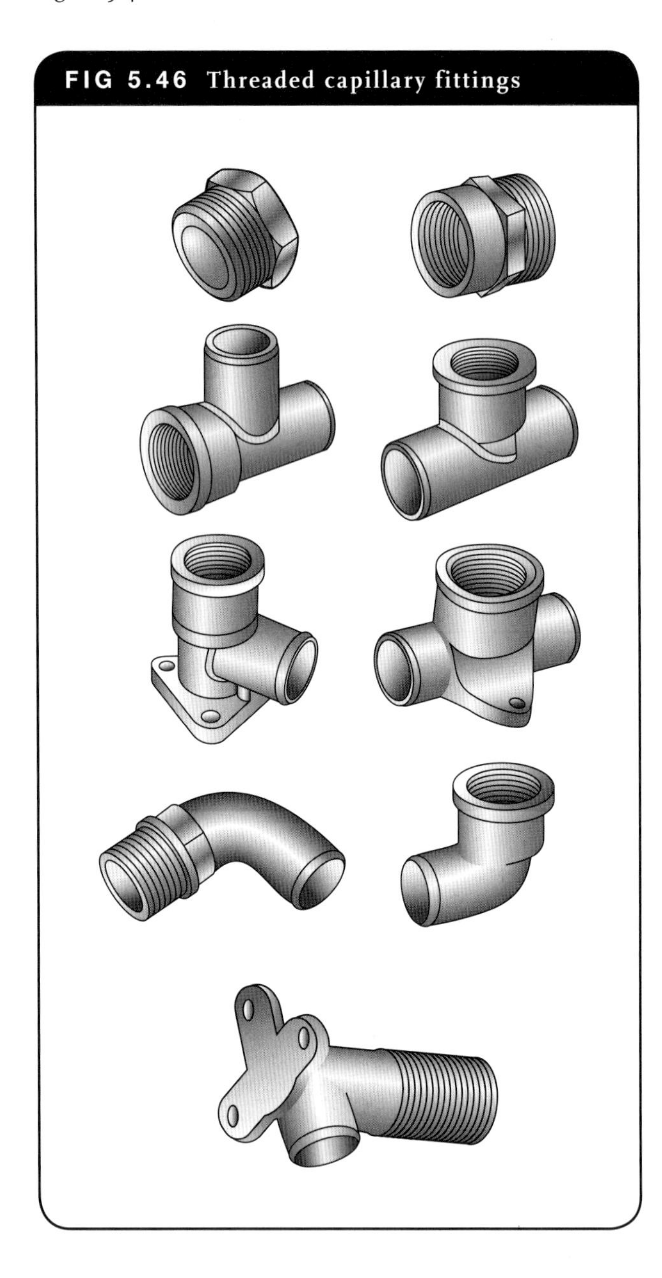

UPVC PRESSURE FITTINGS

The UPVC pressure system offers a range of plastic fittings well suited to domestic, commercial and industrial applications. Most jointing in this system is based on the solvent weld spigot and socket technique, although there are threaded connectors and adaptors to suite mechanical joints.

Solvent-weld UPVC pressure fittings are manufactured and tested in accordance with an Australian Standard and are classified as shown in Table 5.16. Various BSP-threaded fittings are available for take-off connections. The spigot and socket end is connected in the usual way using solvent cement. To connect threaded fittings, the joint is best made by using PTFE sealing tape onto the male end and screwing firmly, avoiding overtightening.

FIG 5.47 UPVC pressure MI coupling

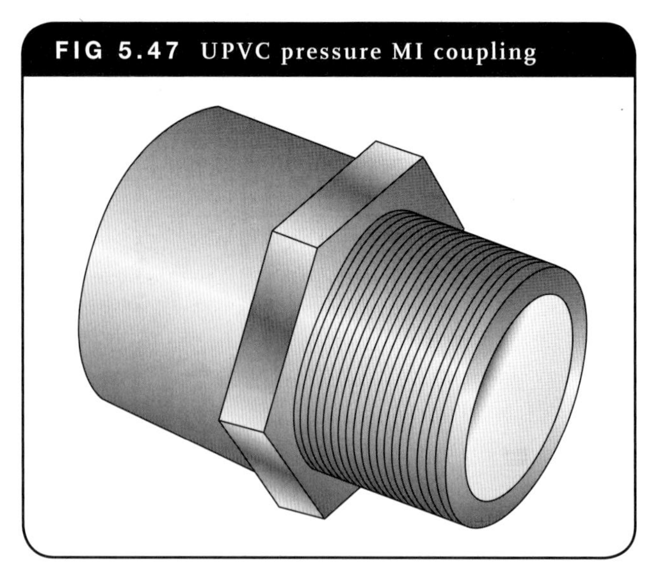

TABLE 5.16 Classification for UPVC pressure fitting

Class	Pressure rating
Class 18	180 m head (1.8 MPa)
Class 12	120 m head (1.2 MPa)
Class 6	60 m head (0.6 MPa)

Note: Pressure ratings apply at 20 °C provided that the material is not exposed to sunlight.

Solvent weld joint

The solvent welding process is not a gluing process, as the solvent cement permeates into the walls of the pipe. The correct solvent cement should be selected for the right application.

PROCEDURE FOR JOINTING STEPS

1. Cut the spigot square using a mitre box and remove the swarf from both edges.
2. Mark the spigot with a pencil at the distance equal to the internal depth of the socket.
3. Using gloves with a clean cloth or a special applicator clean the spigot up to the pencil line and the internal socket with the priming fluid in a well-ventilated area.
4. Apply a thin uniform coat of solvent cement to the socket first and then to the spigot, up to the pencil line.
5. Assemble the joint quickly before the cement dries, pushing the spigot firmly into the socket as far as the pencil mark. Finish with a quarter twisting motion to spread the cement evenly. Hold the joint firmly for 30 seconds without movement.
6. Wipe off the excess solvent cement.
7. The joint requires curing time before pressure testing as recommended by the manufacturers of a minimum of 24 hours.

Application:

Type P is for pressure joints (usually green).
Type N is for non-pressure joints (usually blue).

POLYETHYLENE PIPE

Polyethylene (PE) pipe manufacturing commenced in Australia in the 1950s when small diameter pipes were used for rural, irrigation and industrial applications.

Polyethylene is strong, extremely tough and very durable and is recognised as acceptable in the plumbing industry for water services, drainage and sewer applications.

Applications

There are a range of grades offered in PE products for different applications.

Low-density polyethylene

Low-density polyethylene (LDPE) has high flexibility and retention of properties at low temperatures. The main applications of LDPE are in micro-irrigation, low-pressure drip irrigation, rural irrigation and stock watering applications.

With a working pressure of 300 kPa for a temperature range of 20 °C to 25 °C for pipe IDs 10 mm to 32 mm.

Medium density polyethylene

Medium-density polyethylene (MDPE) or PE80B has been mostly superseded by PE100 grade. The main applications and uses are for rural water reticulation, water distribution for houses, industrial treatment plants, cooling water, and mining slurry and tailing lines.

MDPE is commonly identified and referred to as blue stripe metric pipe. Pipe IDs are from 16 mm to 110 mm.

High-density polyethylene

High-density polyethylene (HDPE) is also known as PE100 polyethylene pipe. Due to its light weight, corrosion resistance and homogenous chemical composition, HDPE pipes are rapidly expanding applications in the mining, coal seam-gas, irrigation, plumbing, civil and electrical market sectors.

High-density polyethylene pipe can carry potable water, wastewater, slurries, chemicals, hazardous wastes and compressed gases.

In the plumbing industry HDPE pipe and fittings are suitable for sanitary systems, trade waste, laboratory waste and storm water discharge in domestic, commercial and industrial applications.

Features of polyethylene

Polyethylene has the following properties:

- easy to install
- durable
- chemical resistance
- high flow capacity
- abrasion resistance
- excellent impact strength
- will not corrode
- due to its smooth inner surface it is resistant to build-up or scale.

Jointing

PE pipe can be connected using these technologies:

- screwed compression coupling
- butt welding
- electro-fusion welding—electro-fusion couplings.

Screwed compression coupling

A special insert housing an O-ring is provided and fits into the central fitting. The joint is achieved by screwing up a coupling nut onto the central fitting.

PROCEDURE FOR MAKING A JOINT

1. Cut the pipe ends square and remove burrs.
2. Place the coupling nut over the pipe end.
3. Fit the insert into the end of the pipe.
4. Push the complete end assembly into the central fitting.
5. Tighten the coupling nut firmly with a wrench, but avoid overtightening. (Figure 5.48 shows the assembled joint.)

FIG 5.48 Compression 'non-manipulative type' fitting used for joining polyethylene pipe

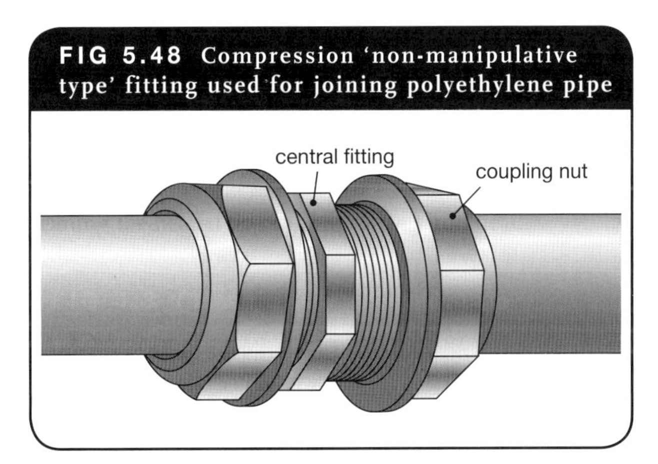

FIG 5.49 Polyethylene screwed fittings

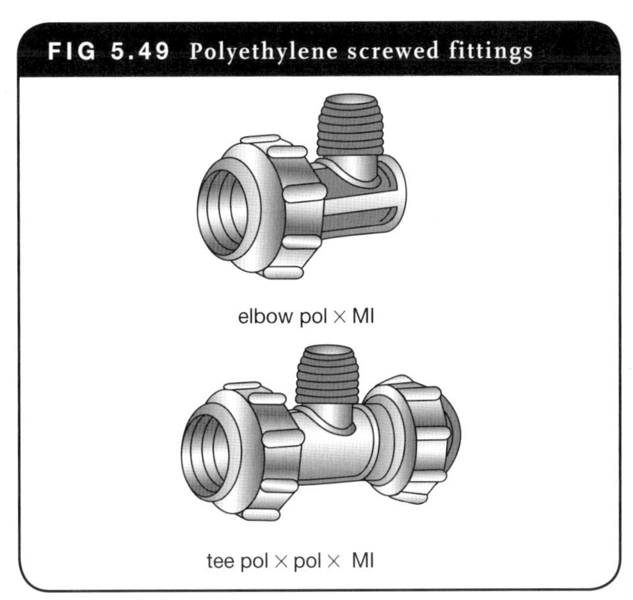

Butt welding

Polyethylene pipe can be joined by butt welding using the machine shown in Figure 5.50, under controlled pressure and temperature conditions. A skilled operator is essential to ensure adequate weld strength is achieved. A cross-section of the weld is shown in Figure 5.51.

FIG 5.50 Butt fusion machine

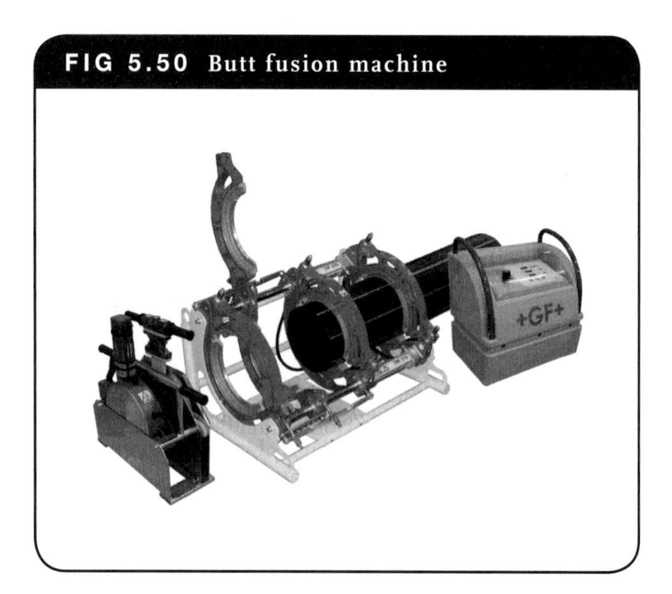

FIG 5.51 Butt weld cross-section

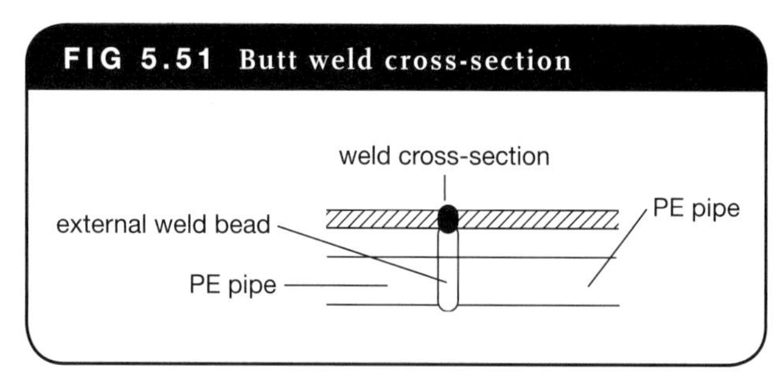

Electro-fusion

Welding polyethylene pipe can be achieved by using electro-fusion coupling (Figure 5.52). Electro-fusion joints consist of resistance wires embedded into the coupling. When the coupling is heated by the passage of a controlled electric current, it causes the surrounding polyethylene material to melt and form a fused joint (Figure 5.53).

FIG 5.52 Automatic electro-fusion unit connected to a fusion coupling

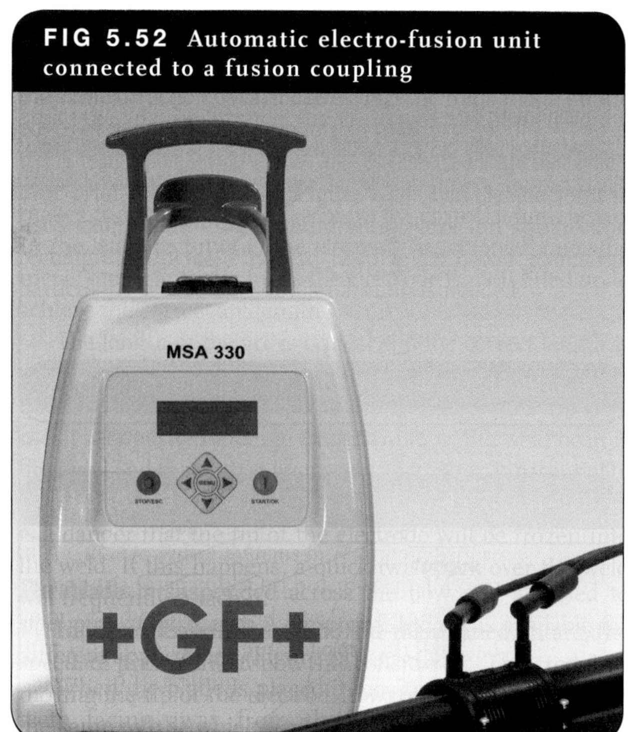

FIG 5.53 Electro-fusion cross-section

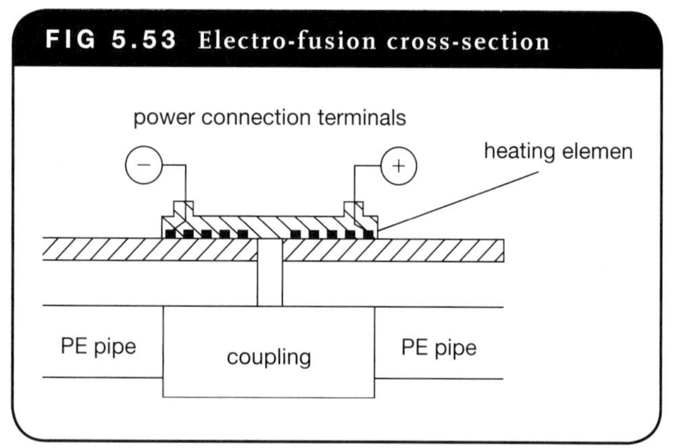

Flanged joints

Flanged joints are generally used on large-diameter pipes and tubes where unions and other screwed fittings are impracticable. Flanged joints have the advantage of allowing sections of piping to be removed or replaced without disturbing any other section of the piping circuit. Flanges are also used to make connections to large valves, pumps and pressure vessels, enabling them to be readily stripped for maintenance.

Flanges are available in a variety of designs and specifications. Working pressures, temperatures, sizes and dimensions are contained in the relevant Standards Australia document.

Flanges are manufactured in plain, threaded, blank, boss and neck designs, and are available in copper alloys, malleable and cast iron and various steels. Threaded flanges are used in conjunction with threaded pipes; the joining procedure is the same as for screwed fittings. Welded flanges are slipped over the plain pipe and then welded. Brazing copper alloy flanges are available for use with copper tubes. Blind flanges have no central holes and are used to seal the end of a pipe or fitting.

FIG 5.54 Flanged joint

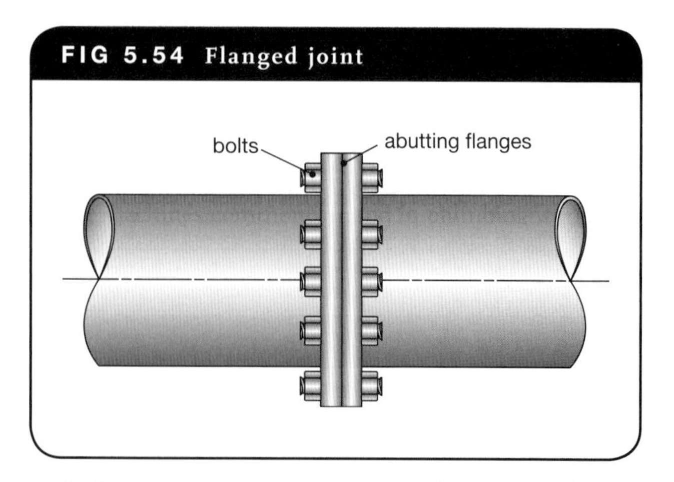

Bolt holes

Bolt holes are drilled off the centre-line (as shown in Figure 5.55) and are spaced equidistant from each other. It is essential that bolt holes are positioned and drilled off the centre-line because flanged valves would not be in an upright position. Bolt holes are marked onto a bolt circle diameter, sometimes known as a 'pitch circle diameter' (PCD), the dimensions of which are derived from Standards Australia (Figures 5.55(a) and (b)).

Bolting materials

Bolts must be made from a material suitable for the specific service conditions applying and should comply with the appropriate Standards Australia document. Materials are selected according to pressure and temperature limitations for flanges. Bolt materials should be compatible with flange materials.

Gasket and O-ring material

Gaskets and O-rings are made from materials suitable to the design conditions of the flanged joint and must be compatible with the fluid the pipe will contain.

Flange identification

Each flange, other than an integral flange, must be identified, as illustrated in Figure 5.56.

Valves

Flanged valves are normally supplied undrilled. If they have to be drilled they must be made according to the relevant Standards table. When valves have to be drilled to a specific template, the position of the holes relative to the centre-line of the spindle must be clearly indicated.

FIG 5.56 Flange identification

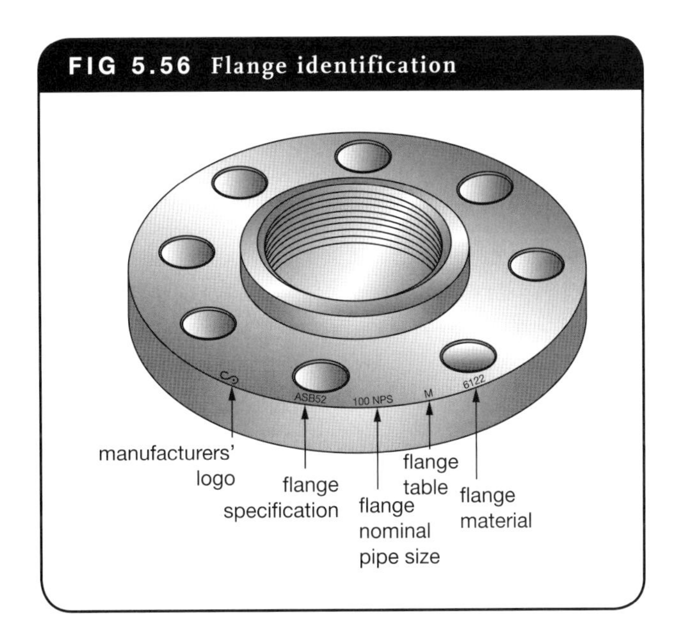

CAST IRON PIPE AND FITTINGS

Flanged cast iron pipes are produced primarily for water mains systems. The flanges are cast integrally with the pipe barrels and the joints are made by bolting—as with other flanged systems. Suitable flanged fittings, such as those shown in Figure 5.57, are available to match the pipe ends. Flanges used in water service pipework all conform to a national Standard.

Handling flanges

Care should be exercised at all times when handling flanges. They should be protected against damage to their faces from such things as score marks and weld spatter, and should be free from cracks and surface flaws. Flange

FIG 5.55 (a) Position of bolt holes, (b) pitch circle diameter (PCD) marked

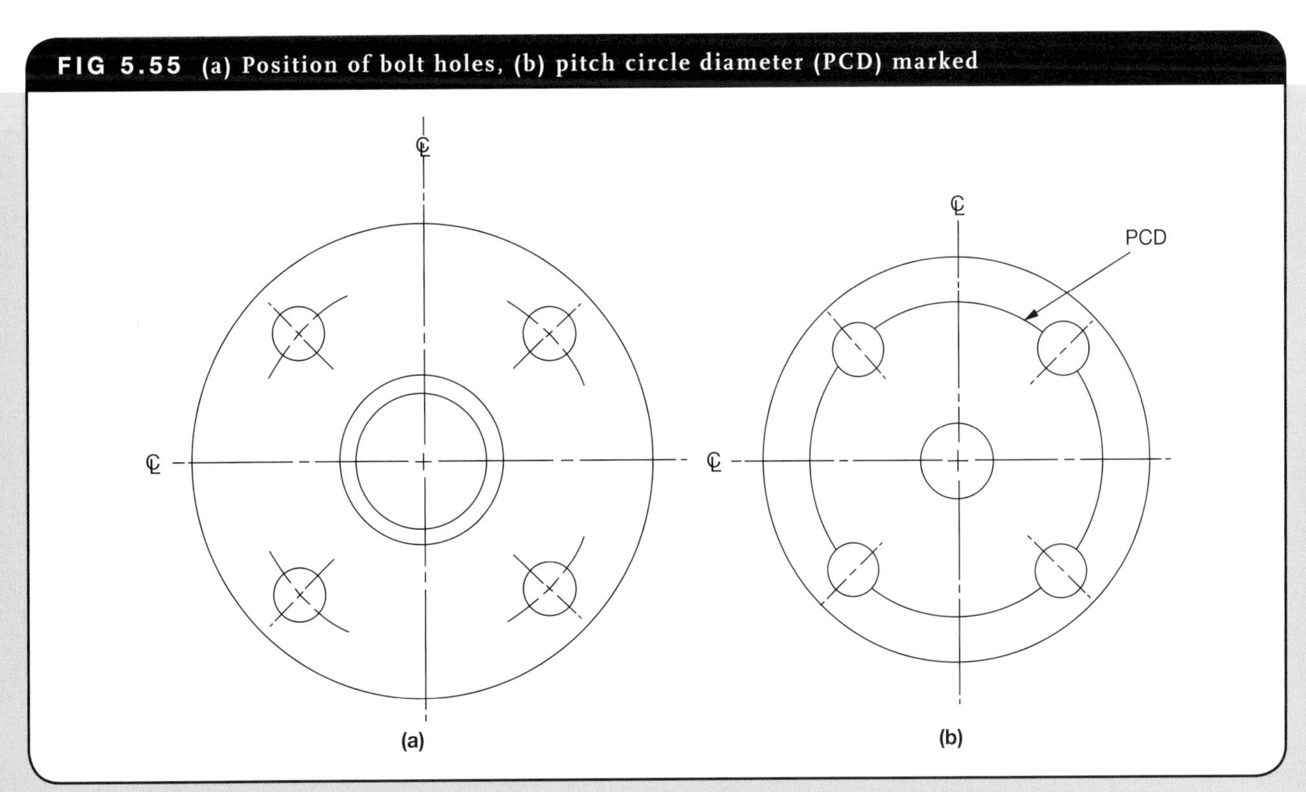

FIG 5.57 Flanged cast iron fittings: (a) bend, (b) 45° branch, (c) tee, (d) cross, (e) concentric taper, (f) eccentric taper

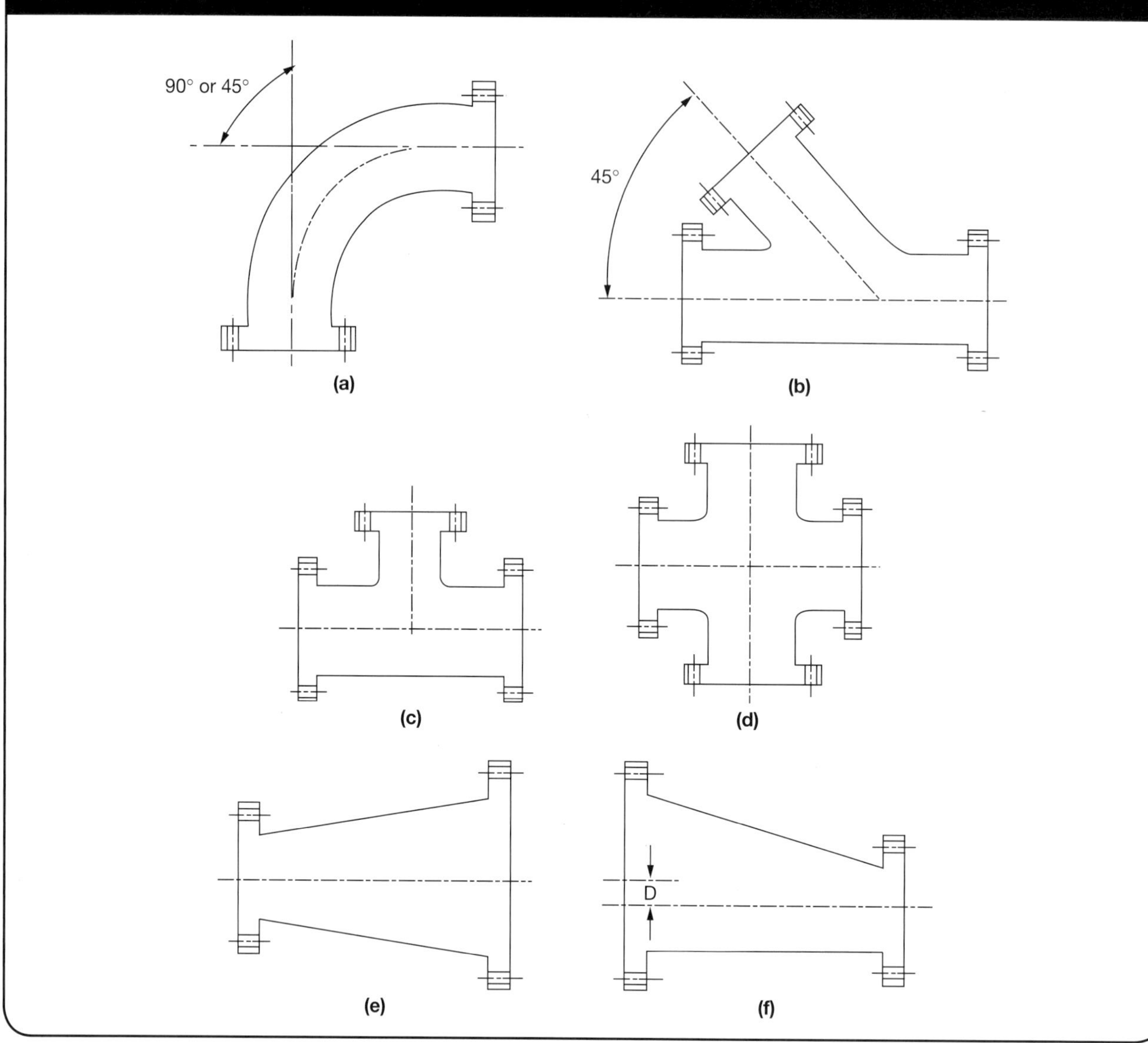

FIG 5.58 Sequence for tightening bolts

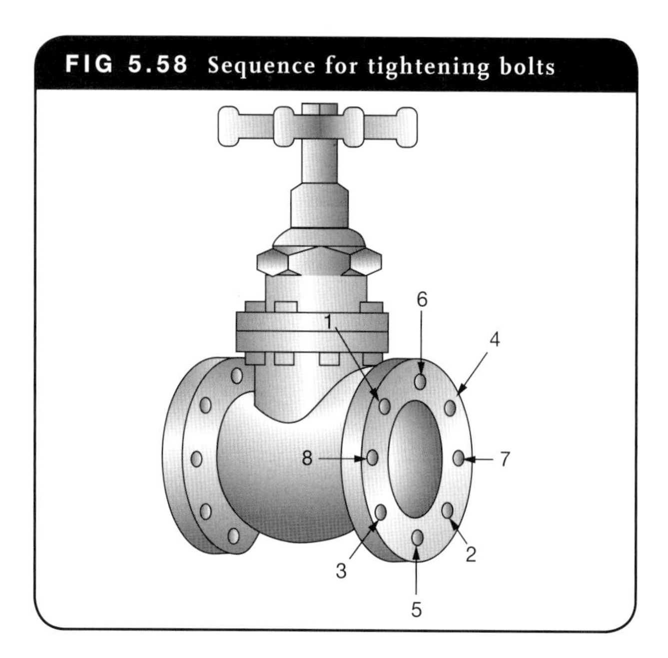

faces and sealing gaskets should be coated with approved compounds. When tightening flanged valve bolts, a 'crossover' method (Figure 5.58) should be used. Uniform pull on bolts reduces stress on the flange and other parts of the valve.

Bolt threads should always be lubricated as lubricant reduces friction between the threads and protects them from rust and corrosion. Also, joints pull up tighter and are pulled apart more easily when lubricant is used.

SEALANTS

Sealants, within the context of this section, refer to the polymeric range and include those sealants made from natural rubber and the various synthetic elastomers. Polymeric sealants are formulated to provide a continuous-surface polymer bonding layer to both opposing surfaces to fill (seal) a gap to set (cure) by polymerisation.

Although, as part of their sealing function, sealants are designed to hold metal sections together, they will always require some form of mechanical aid (fasteners) such as rivets or screws. It is important to realise that sealants are not to be confused with adhesives. Joints are usually subject to some form of movement because the overlapping surfaces expand and contract with temperature changes. Sealants must therefore allow for this movement by curing to a flexible elastic material and by remaining soft.

Approved silicone sealants cure on contact with air to form a permanent, flexible silicone rubber which is very durable and is unaffected by the aggressive weathering elements such as sunlight, moisture and extremes of temperature.

Recommended properties

Sealants used in conjunction with metal roofing sections, flashings, guttering and rainwater accessories should possess these requirements:

- flexibility
- water resistance
- corrosion resistance
- mould resistance
- resistance to ultraviolet radiation
- resistance to temperature extremes
- gap filling (no slumping).

Where doubt exists on the suitability of a particular sealant, the preceding list of properties should be used as a basis for selection.

Joint design

Seams should always be mechanically fastened for strength, whether soft soldered or sealed with a compound sealant. The sealant does not, therefore, require significant adhesive strength but must bond positively to both opposing surfaces as a continuous layer. To ensure complete sealant cure, the width of sealant applied in a lap should not exceed 25 mm when compressed. Fasteners should be spaced approximately 40 mm apart for roofing and cladding applications.

Cartridge-extrudable sealants

Surface preparation

To obtain satisfactory bonding, the workpiece surfaces must be clean, dry and free from other contaminants such as oil and grease. Mineral turpentine is a suitable solvent for removing oily surfaces. Sealants must be applied on the same day the material surface is prepared. When cleaning remedial work—such as stripping old sealants or paint—use a wire brush to ensure a new clean and dry surface.

Sealant application

Place the sealant cartridge into a trigger-operated caulking hand gun. Cut the tapered nozzle close to the bottom end at approximately 45° (Figure 5.59) with a sharp knife to produce a 3 mm diameter bead of sealant. The nozzle is tapered so that a larger orifice and a greater flow of sealant can be obtained if required. The quantity of sealant is controlled by the size of the nozzle opening.

FIG 5.59 Cutting the cartridge nozzle of a compound sealant at 45°

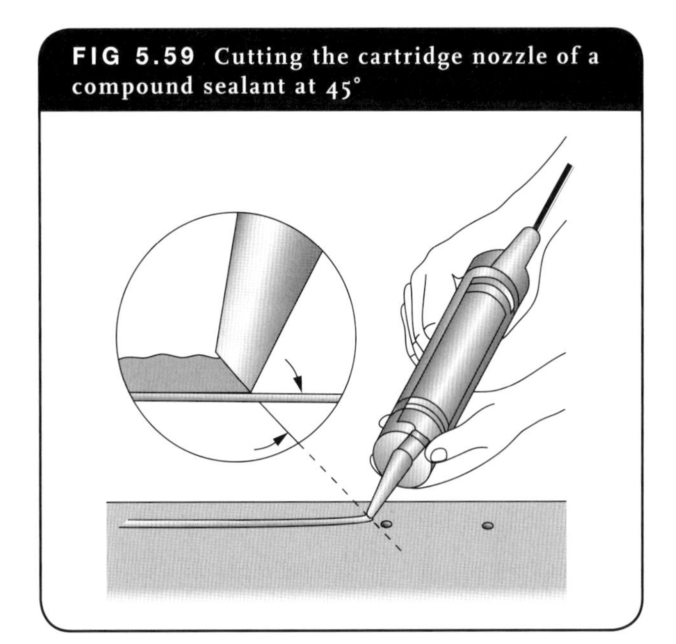

Place the tip of the nozzle on to the workpiece surface and extrude the sealant by applying pressure on the trigger of the hand gun. In placing the sealant, care must be exercised not to entrap air. Always apply the bead of sealant over the holes drilled for the fasteners in a continuous line (Figure 5.60(a)). This technique ensures that under compression the sealant positively seals the fastener, as the sealant is squeezed outward. It is recommended that a single bead of sealant is used for lap joining (Figure 5.60(b)).

APPLYING SEALANT FOR LAP JOINTS

1. Assemble and drill the workpieces.
2. Separate the workpieces and remove the swarf.
3. Clean the joint surfaces as described above.
4. Apply a bead of sealant as described above.
5. Relocate the workpieces, align the holes and fasten with rivets.
6. Seal each fastener externally.
7. Clean excess sealant.

Clean up

For practical or aesthetic reasons uncured sealant can be removed with a clean, dry rag and any excess then removed with mineral turpentine or white spirit. Do not use implements which are likely to damage the surface finish of the material being worked on.

Avoid unnecessary smearing of sealant on surfaces intended for painting.

FIG 5.60 (a) Sealant application in a continuous line (b) Sealant application in a single bead

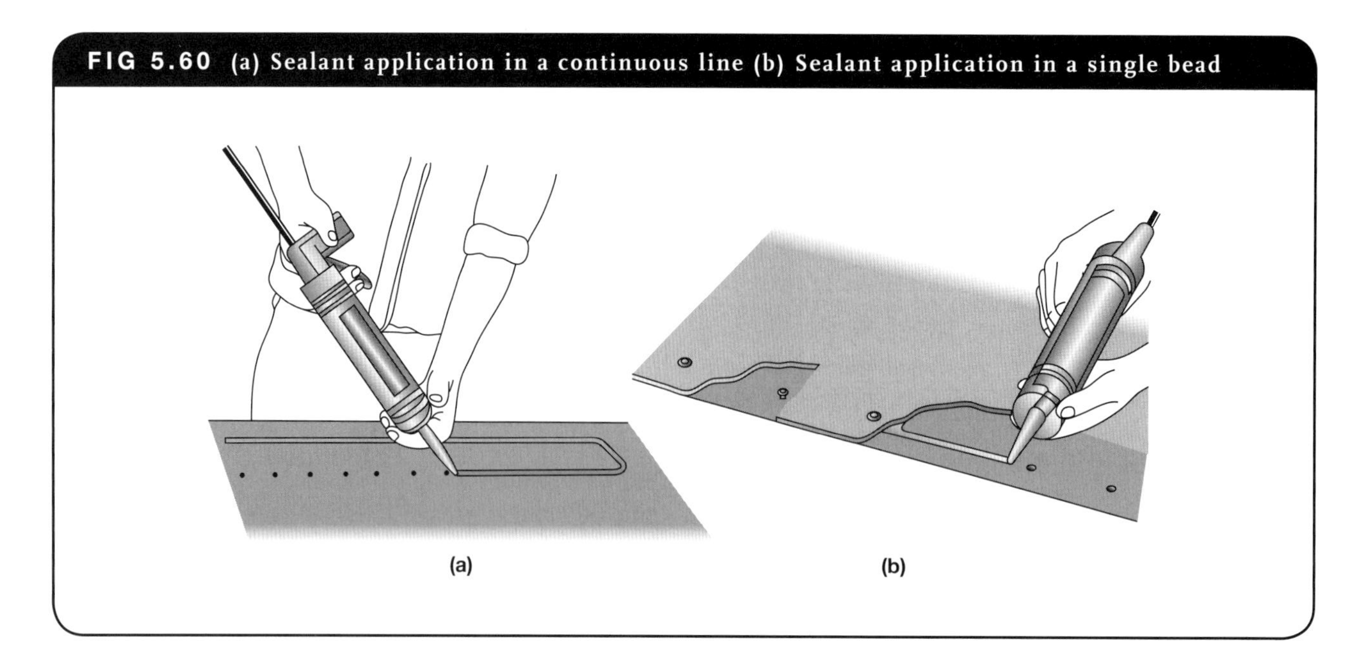

Other sealant types

Apart from the cartridge-extrudable sealant types, the market offers a number of varieties all of which perform predictably in given situations and with certain materials.

Solid sealants in tape form

These are generally semicured butyl rubber possessing positive initial adhesion. The tapes are used in combination with mechanical fasteners.

For the tape to make positive contact with both surfaces and form a seal, it is important that the tape's initial thickness is greater than the gap between the components of the finished joint. Used correctly under suitable conditions, these tapes would be satisfactory for use with exterior finishes. However, some types may be difficult to compress within a joint during cold weather, possibly resulting in a loose joint later.

Preformed gaskets

Gaskets preformed to some of the popular roofing profiles are generally made from cellular material, and are intended for compression between irregular surfaces such as guttering (Figure 5.61). By varying the compression they can perform different functions of sealing against water, dust, noise, draughts and snow. The gaskets are non-adhesive and are often saturated with bitumen or acrylic resin. Fitted correctly they would be suitable for use with exterior finishes in situations where the installation of such bulky seals is indicated.

FIG 5.61 Method of mitre joining a typical gutter using a corner bracket

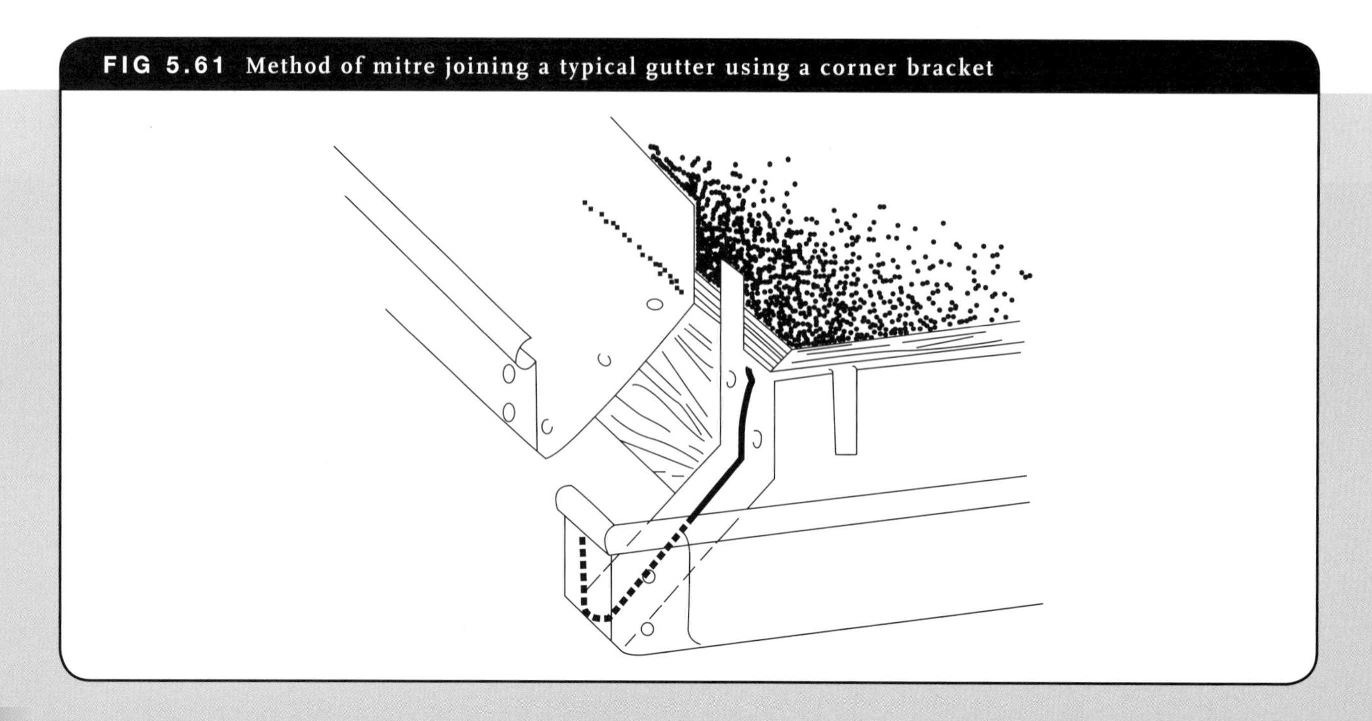

FOR STUDENT RESEARCH

Australian Standards

- AS/NZS 1337: 1992 Eye protectors for industrial applications
- AS/NZS 1554.4: 2010 Structural steel welding
- AS 1579: 2001 Arc welded steel pipes and fittings for water and waste water
- AS 1674.1: 1997 Safety in welding and allied processes (Part 1 Fire precautions)
- AS 1834.1: 1991 Material for soldering—soldering alloys
- AS/NZS 1873: 2003 Powder-actuated (PA) hand-held fastening tools
- AS 2280: 2004 Ductile iron pressure pipes and fittings
- AS 2528: 1982 Bolts, studbolts and nuts for flanges and other high and low temperature applications
- AS/NZS 3500 National Plumbing and Drainage Code, Part 0 (AS/NZS 3500.0) to Part 4.2 (AS/NZS 3500.4.2) inclusive (various editions)
- AS 4087: 2004 Metallic flanges for waterworks purposes
- AS/NZS 4130: 1997 Polyethylene (PE) pipes, pressure applications
- AS 4603: 1999 Flash back arrestors (safety devices for use with fuel gasses and oxygen or compressed air)
- AS/NZS 4680: 2006 Hot-dip galvanized (zinc) coatings on fabricated ferrous articles
- AS 4839: 2001 Safe use of portable and mobile oxy, fuel gas systems for welding, cutting, heating and allied processes

PLUMBER PROFILE 5.1

STEVE ECKERT

Job Title: Director of Eco Building Supplies, South Australia

With 23 years in plumbing, Steve started out working in various roles for different companies, worked at a ski resort in Canada and then tackled setting up his own rainwater solutions products company in the midst of the global financial crisis (GFC) in South Australia, where he also runs plumbing courses.

What jobs have you done over the course of your plumbing career?

I spent seven years at Jordan Plumbing, a commercial company, where I did my apprenticeship and then I went out on my own for a couple of years and was self-employed as a plumber. From there I went overseas and worked over in Canada for a year and a half, then continued travelling for another year after that and came home. I worked for a company called Western Plumbers on a nursing home in Mount Gambier for about two years. Then I worked at a global company called REHAU and I was manager for REHAU for eleven years.

How did you set up your own business?

I have been with Eco Building Supplies for three years. I was working at REHAU and always wanted to work on my own again, so what we do now is we sell different kinds of rainwater harvesting systems and underfloor heating systems to plumbers. We also run an eco-plumbing course for the Plumbing Industry Association of South Australia once a year. The most challenging aspect, besides the preparation of running a two day course, is public speaking, which is one of my greatest phobias.

What do you most enjoy about your job?

Probably working for myself—you can basically come up with your own ideas and follow your dreams. We've got a showroom warehouse now and we employ three other people at the moment.

What is one of your most memorable experiences on the job?

Working in Whistler, Canada. I was about 23 or 24 and I wanted to go overseas. I started off as a chef working in a kitchen and a plumbing job came up at a ski resort. It was fantastic. The thing with travelling at the end of the day is the different people you meet along the way—it's almost like a second education. The most challenging time there was working on a restaurant outside and having an air horn alongside when the bears came too close. Skiing home from work back to my chalet was great fun!

Can you describe your biggest challenge throughout your career?

I took a risk of taking out a loan of $80 000 to set up and start up. When we started the GFC was going on so there wasn't a lot of work then but now it's going well. The biggest challenge in starting a business is cash flow as we initially had to stock about $20 000 worth of rainwater tanks and pumps. In 2009 a company went broke and owed us $8000. This was a financial burden as our company was in its infancy. Then in 2010 a client owed $25 000, which almost sent us under, and I was forced to float $20 000 to pay our creditors. Now things are back on track.

What advice would you give to students?

The best advice I can give to any student is that you do not need to be the most academic or best plumber to make a career in the plumbing industry—just never give up. If you fail the first time at anything get back up and do it again learning from that first mistake, and making sure not to make the same mistake twice. Completing an apprenticeship is like a big safety net. You can travel, try a new occupation and it will still be there to fall back on if you need to. Not too many jobs offer that security.

Levelling

LEARNING OBJECTIVES

In this chapter you will learn about:

- **6.1 the levelling process**
- **6.2 levelling terms**
- **6.3 levelling tools**
- **6.4 the types of optical levels**
- **6.5 general levelling procedures**
- **6.6 the rise and fall method.**

INTRODUCTION

Accurate levelling is important as a design and installation requirement in plumbing and drainage systems. Transferring levels to enable trenches to be excavated and drains to be laid to previously determined values is a major application of the use of levelling tools and instruments. Design requirements determine the depth and grade of a trench, which are usually indicated on the drawings of the plumbing design. Some important levels related to the plumbing system are the elevation levels (EL) of the invert of the pipe, which are required to ascertain the minimum ground cover in the initial design. During the design stage, levelling techniques are often employed to obtain a contour plan to minimise excavation by taking advantage of the fall of the land (Figure 6.1).

Often, the only reference point for the plumber to work from is a temporary bench mark (TBM) as the building may not have been started or is in the early stages of construction. This TBM is the reference point for the total building design requirements and therefore all levels for the plumber and other trades are related to one reference point. Levelling results achieved by improvisation or guesswork will eventually lead to unnecessary and costly errors. When the correct method is applied, accurate results can be easily achieved. An example of a drainage design with appropriate levels can be seen in (Figure 6.2.)

FIG 6.1 (a) Contour lines, (b) drainage lines

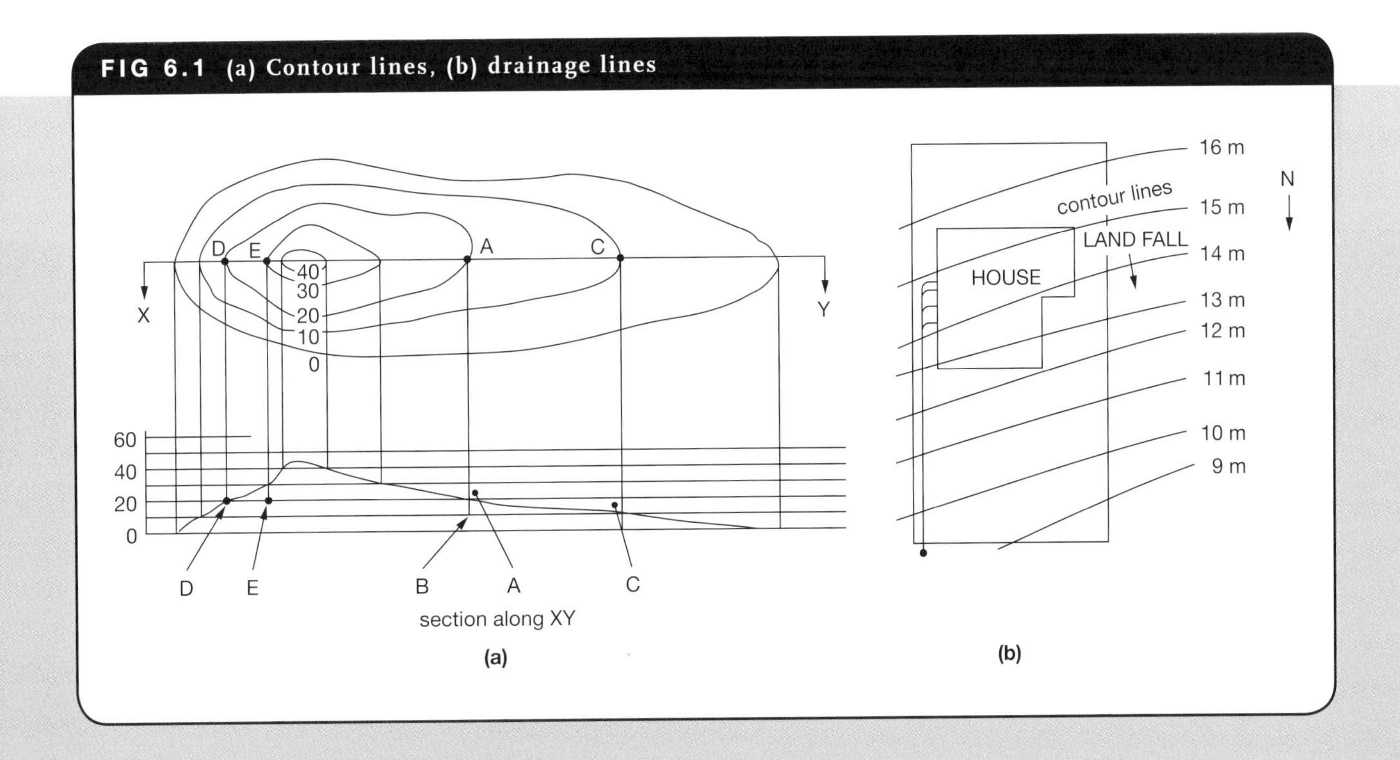

FIG 6.2 Drainage plan

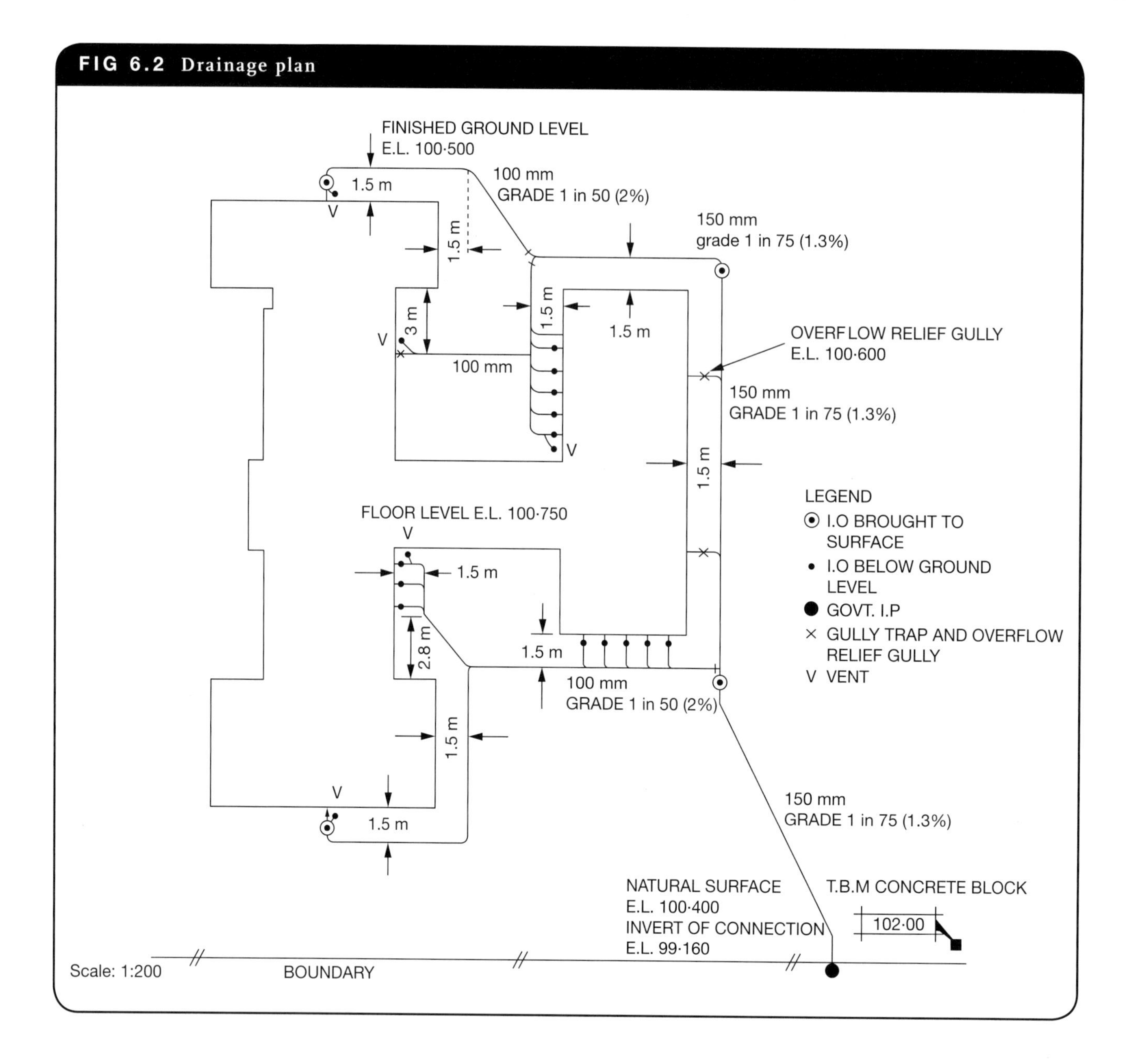

THE LEVELLING PROCESS

The process of levelling can be defined as a method for measuring variations in height, or particular points on the surface, in the setting out of construction works. This is achieved by means of an instrument called a 'level' that produces a truly horizontal line.

Levelling terms

- Datum is an arbitrary level surface to which heights of all points are referred to. This can be a local datum point established on a construction site.
- Reduced level (RL) is a distance recorded, in metres, as a height above or below the datum.
- Bench mark (BM) is a point of reference for a measurement. A bench mark is any permanent marker placed by a surveyor with a precisely known vertical elevation, to be used for projects.
- A temporary bench mark (TBM) is placed for a particular project. It is not designed to be a reference for other projects or for a long term use.
- Backsight (BS) is a sight taken to a BM or a TBM. It is the first sight taken after setting the levelling instrument up.
- Foresight (FS) is the last recorded staff reading, taken before the instrument is moved to a new position.
- Intermediate sight (IS) is the sights taken between a backsight and a foresight.
- Line of collimation is an imaginary line that passes through a levelling instrument, viewed at the cross-hairs.
- A contour line can be defined as a line drawn on a plan joining all points of all the same height above or below a datum.

FIG 6.3 Grading a trench bottom using a straightedge and level

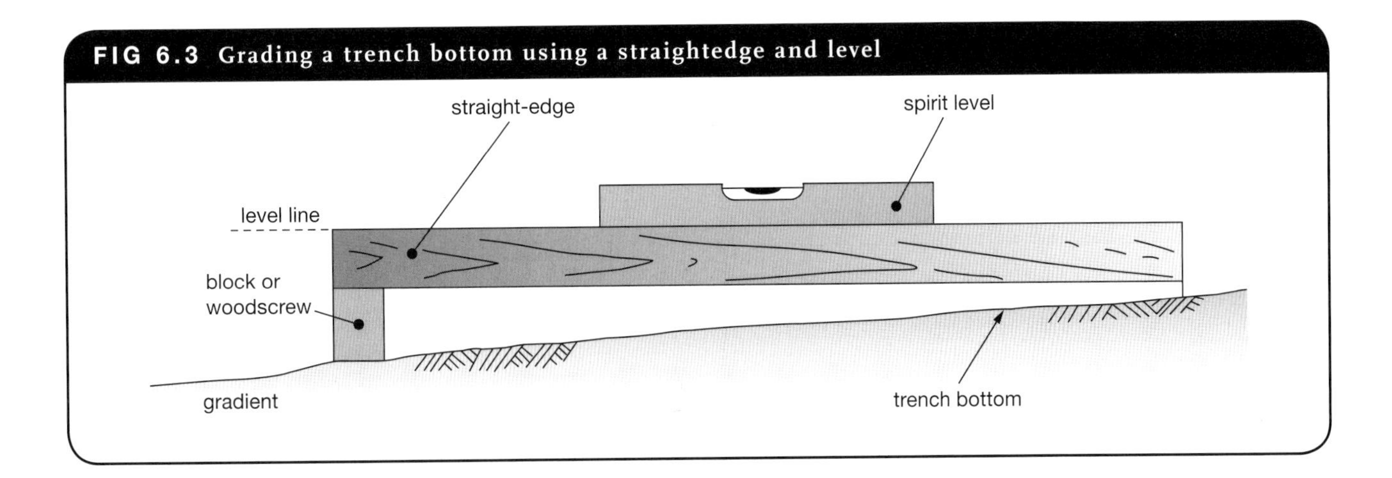

Levelling tools

The tools required for levelling are:

- straightedge
- boning rods
- water level
- plumb bob
- laser plumbline
- chalk line.

Straightedge and spirit level

A straightedge is usually made from timber or aluminium with the edges squared off and parallel. A spirit level is placed on top of the straightedge, midway between the pegs to be levelled, over relatively short distances (Figure 6.3). More accurate results are obtained by reversing the level and averaging the two results. Wooden straightedges should be checked regularly for bending or warping.

Boning rods

Boning rods are used for the purpose of setting intermediate pegs on the same grade between two fixed levels. In effect, the boning rods are three short travellers of equal length approximately 1 m long with a cross head 300 mm long. They are usually made of timber. (Figure 6.4) shows an application of boning rods.

FIG 6.4 Boning rods

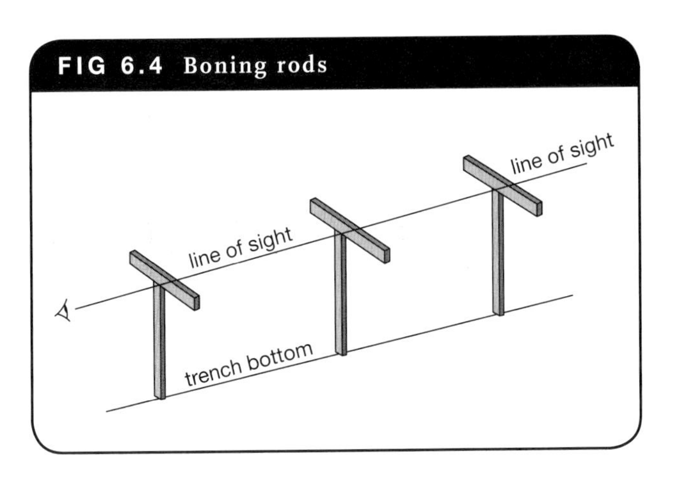

Water level

Water levels are simpler and less costly than automatic levels, or laser levels. They are inexpensive to build and consist of a length of clear plastic tube, 10 mm internal dimension, and filled with water (Figure 6.5). There is a saying that water finds its own level, and that is the behind this tool. One person holds the waterline on the datum mark, and the other principle, holding the other end of the tube, transfers the marks to the relevant points, It is important to ensure that there are no air bubbles in the tube as this will affect the accuracy.

FIG 6.5 Water level

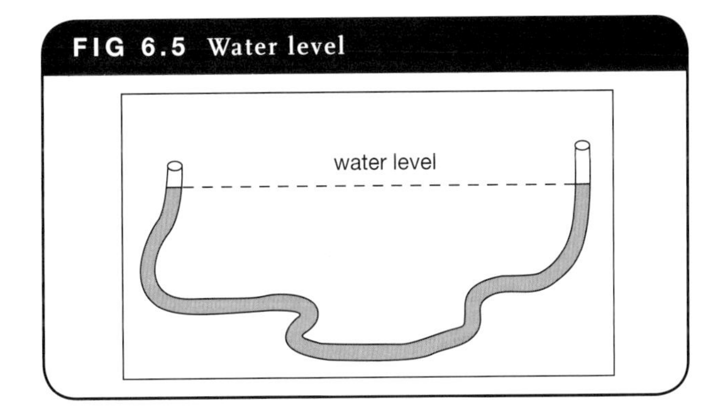

Plumb bob

The plumb bob is an old-fashioned tool that is still in wide use today because of its accuracy (Figure 6.6). A simple weighted string with a brass or metal bulb with a point on the end, using gravity to help builders and plumbers determine exact vertical alignment. When you need to straighten a doorway, wall, vertical pipework or any other tall vertical structure that is longer than a conventional level, a plumb bob can be used. When a plumb bob is placed on a nail on a wall by its string, it can indicate an accurate vertical line on the wall when the weight stops moving.

Laser plumb bob

The laser plumb bob is a plumb bob that uses a laser beam instead of the weighted string line (Figure 6.7). The laser is a popular alternative to the traditional plumb bob.

FIG 6.6 Plumb bob

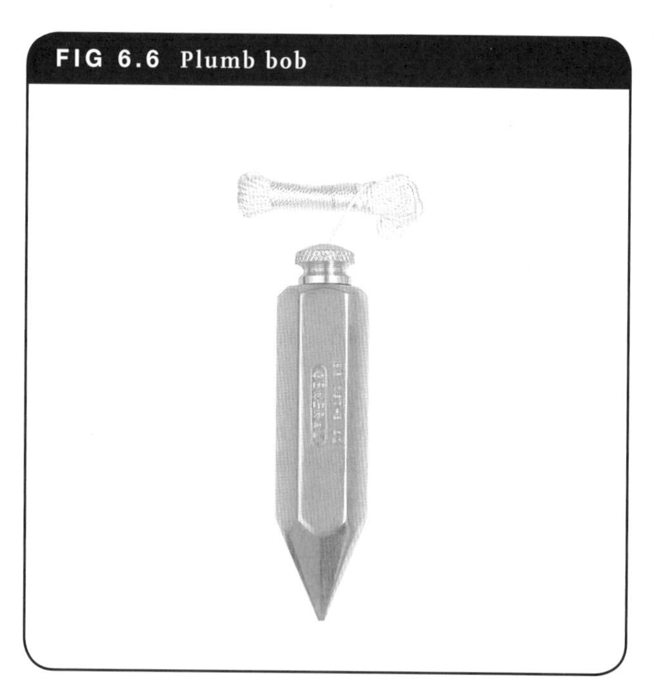

FIG 6.7 Laser plumb bob

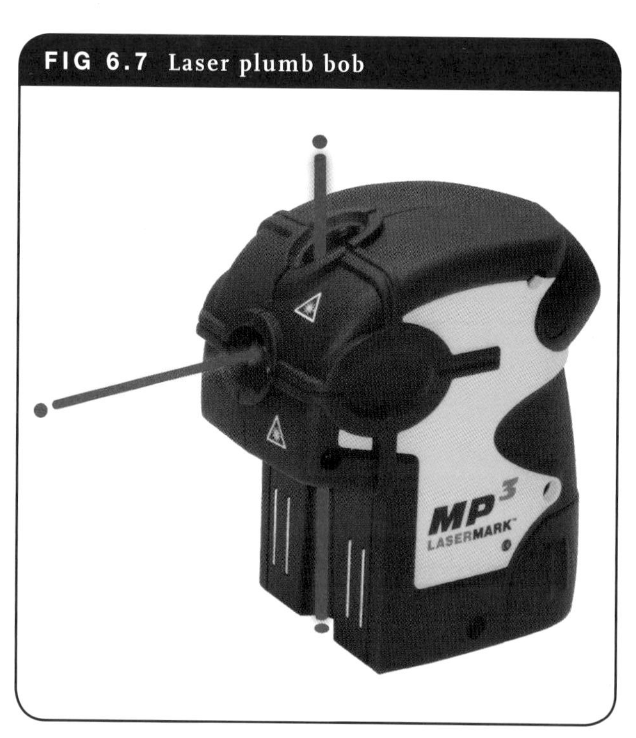

Chalk line

Plumbers, roofers and carpenters chalk several lines throughout their busy workday. Whether you have a basic chalk line or a geared box to rewind the string faster (Figure 6.8), they all basically work the same way to produce a quick and easy straight line for your project. Drawing a long, straight line on a wall or other large flat surface can be difficult. You need a straight edge that may be many metres in length. Fortunately, you can use a chalk line to do the job easily. The string is held taut on the surface between the two end points of the line to be made, the centre of the line is drawn away from the surface, and the string is released and snapped against the surface, leaving a line of chalk in the desired location.

FIG 6.8 Chalk line

Types of optical level

Three types of level are in common use:

- automatic level
- laser level
- theodolite.

Automatic level

This instrument has been in use for many years and is still a popular instrument for levelling (Figure 6.9). The levelling of the instrument is achieved by means of three foot screws and a bubble. Approximate levelling of the instrument is achieved by a circular bubble. There is no sensitive bubble

FIG 6.9 Automatic level

FIG 6.10 Compensator for auto level

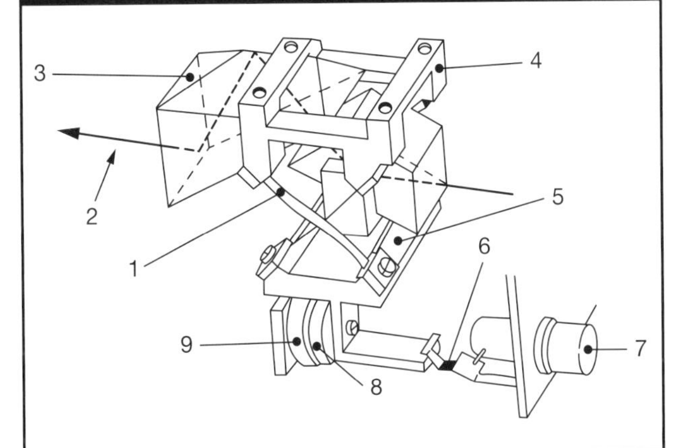

Parts of a compensator:

1 suspension tapes
2 line of sight
3 prism (fixed)
4 compensator housing, fastened to telescope body
5 pendulum with prism
6 spring that taps pendulum
7 press button
8 damping piston
9 damping cylinder

and the main advantage of this level is that it automatically gives a truly horizontal line of sight without the need to accurately set the bubble. The horizontal line of sight is achieved by suspended mirrors or prisms that are attached to a pendulum to control the light rays through to the eyepiece. A system of prisms (compensator) automatically stabilises the horizontal line of sight (Figure 6.10). Also, the automatic stabiliser within the telescope produces an upright image. Generally, automatic levels are instruments that save time in setting up and levelling.

Laser level

In the construction industry, the laser level is fixed to a standard tripod, levelled and then spun to illuminate a horizontal plane. The laser beam projector employs a rotating head with a mirror for sweeping the laser beam about a vertical axis. If the mirror is not self-levelling, it is provided with visually readable level vials and manually adjustable screws for orienting the projector. A staff carried by the operator is equipped with a movable sensor that detects the laser beam and gives a signal when the sensor is in line with the beam (usually an audible beep). The position of the sensor on the graduated staff allows comparison of elevations between different points on the terrain. They are very accurate, simple to use and an advantage is that one person can perform the levelling independently, whereas other types require one person at the instrument and one holding the staff.

For plumbing and drainage projects a laser that can do grades is needed. Whether for roofing, domestic drains or main sewage lines, lasers can have grade setting capability.

FIG 6.11 Laser level with the receiver on a staff

Laser safety and classification

The potential health hazard of a laser is indicated by its hazard classification as stated in the Australian/New Zealand Standard 2211 *Laser Safety* defines a hazard classification scheme for lasers:

- Class 1 lasers are considered safe under reasonable conditions of operation (used in CD players)
- Class 2 lasers are low power devices that are not completely safe, but eye protection is normally afforded by responses such as the blink reflex (Laser pointers and low powered Levelling lasers used in the construction industry)

FIG 6.12 Laser warning sign

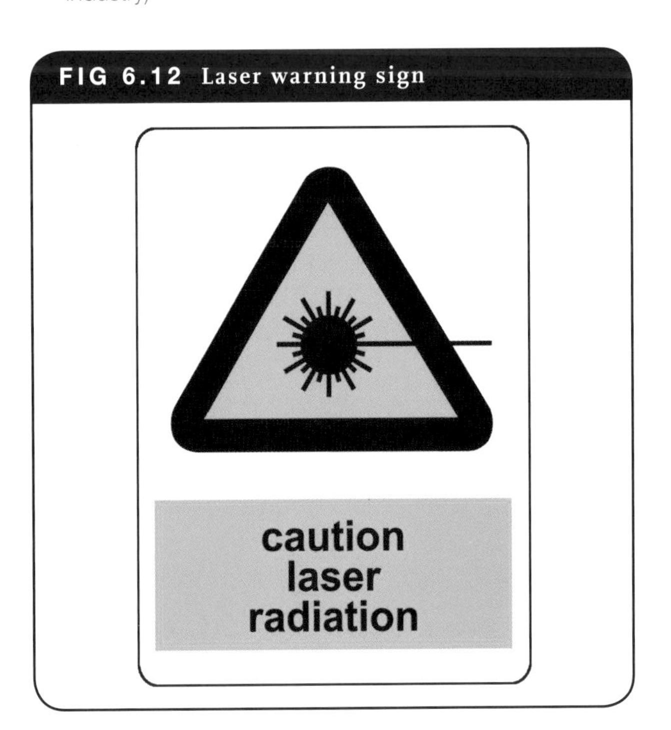

- Class 3A, 3B and 4 emit progressively higher levels of radiation and special precautions are required in their use.

Note that the Standard requires that operators of class 2, 3A, 3B (restricted) lasers must have appropriate training in the use of the laser and access to the relevant necessary information and manufacturers information for proper assembly, maintenance and safe operation to avoid possible exposure to hazardous laser radiation.

Protective and emergency equipment

- safety glasses
- protective clothing
- warning signs and barriers.

Pipe laying lasers

Pipe laying lasers can be used to establish grade, elevation and line reference for the installation of stormwater, sanitary and other gravity flow pipes. The laser produces a red, pencil thin beam of laser light, projected in the same direction and at the same grade as the pipe being laid. The beam is intercepted by a target mounted in the pipe, on the pipe, or on a target staff (Figure 6.13).

The pipelayer positions the pipe until the beam hits the centre of the target.

Theodolite

Although the theodolite can be used as a level, its primary function is to measure angles. A brief description only is therefore necessary, because of possible confusion with other levelling instruments. Basically a theodolite, as distinct from an auto level, is an instrument used for reading horizontal and vertical angles, setting out straight lines, and for special types of levelling. A theodolite is generally considered to be a precision surveying instrument (Figure 6.14).

FIG 6.14 Theodolite

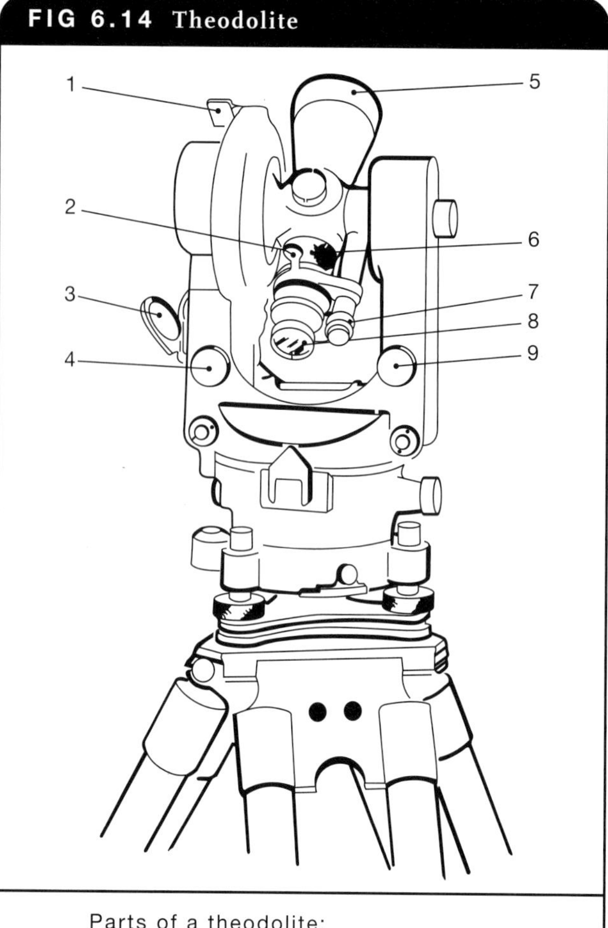

Parts of a theodolite:

1 level mirror
2 backsight
3 illumination mirror
4 level tangent screw
5 foresight
6 focusing ring
7 microscope eyepiece
8 telescope eyepiece
9 vertical tangent screw

FIG 6.13 Typical laser level set-ups

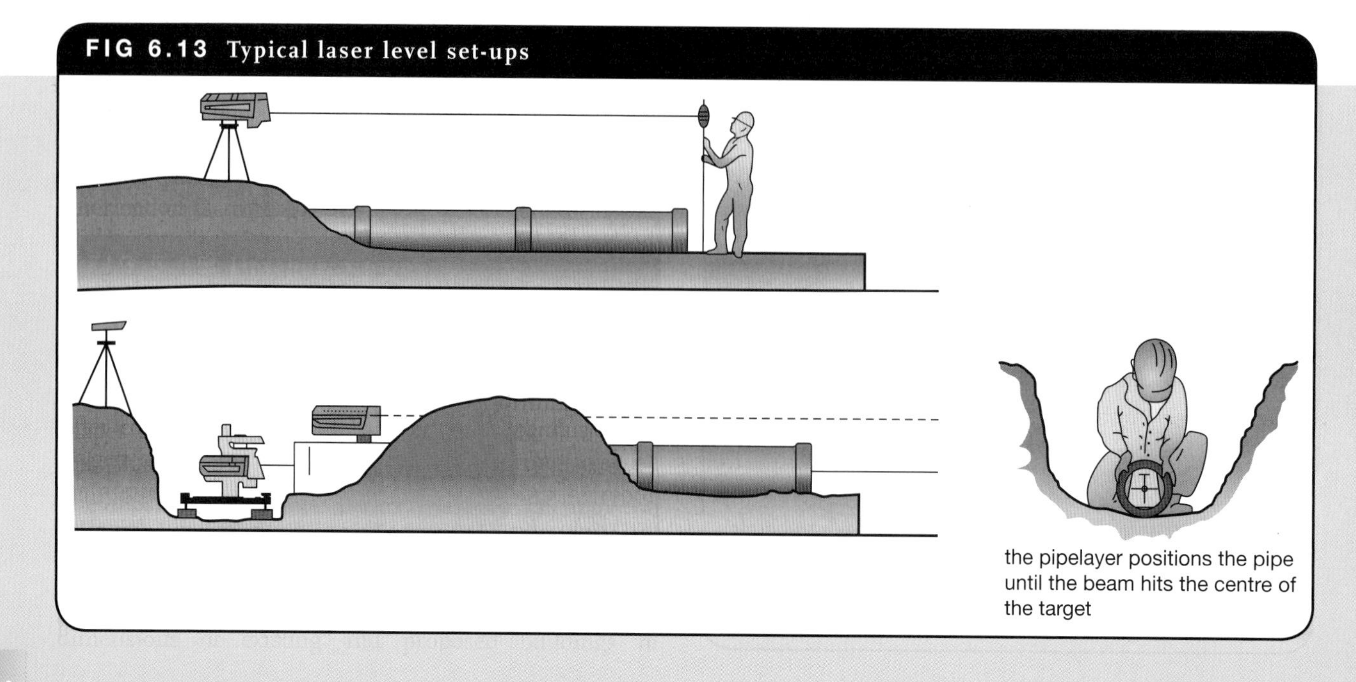

the pipelayer positions the pipe until the beam hits the centre of the target

FIG 6.15 Staff reading in metric 'E'

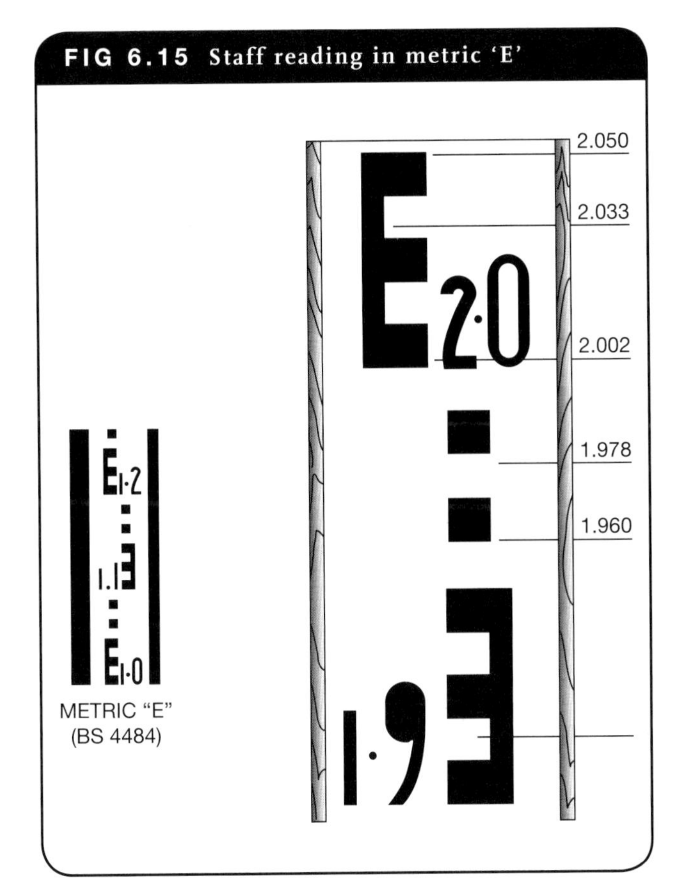

FIG 6.16 Level reading in Australian metric

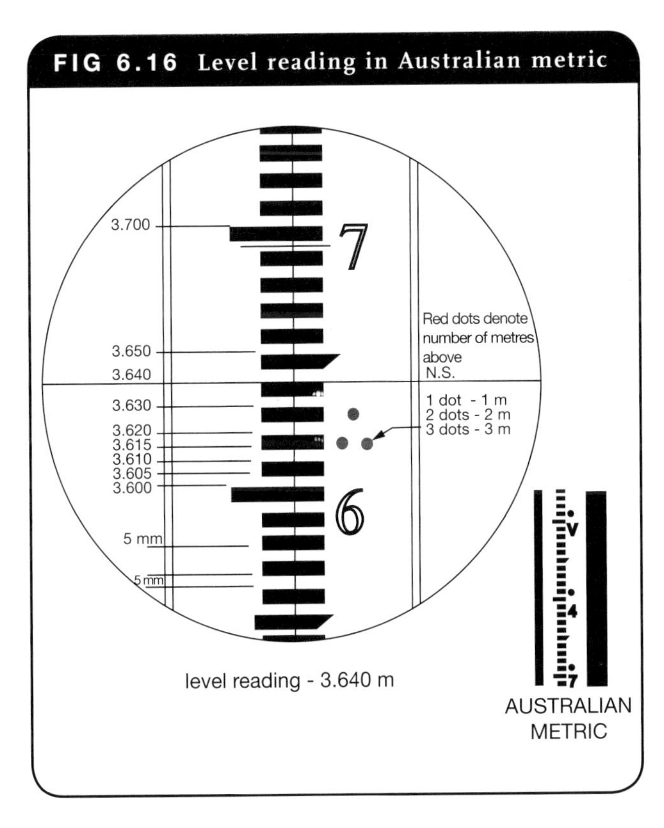

Levelling staffs

The vertical distance above or below a horizontal line of sight is read off a levelling staff. Most modern staffs are manufactured from aluminium and can be either folding or telescopic, extending to a length of four or five metres. Staff face markings normally available are metric 'E' (Figure 6.15) and Australian metric (w) (Figure 6.16).

FIG 6.17 (a) Clip-on and (b) adjustable staff level

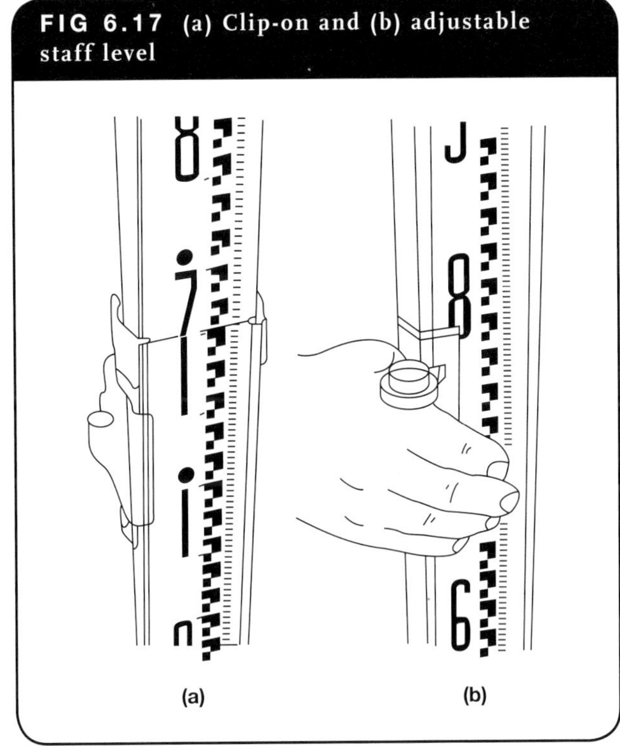

The staff face markings are in different colours, usually black and red, to show gradation marks in alternate metres on a white or yellow background. Major gradations occur at 100 mm intervals on the metric 'E' staff with the figures denoting the metres and decimal parts. Minor gradations are at 10 mm intervals with the lower gradation marks of each 10 mm division being connected by the letter E that covers 50 mm. The red E denotes the odd number of metres and the black E the even numbers. The Australian metric staff has black divisions of 5 mm and the white division in between is also 5 mm, while the dots indicate the number of metres. Staffs are available with clip-on levels or adjustable levels that consist of a circular bubble fitted on top of a metal angle plate to enable the staff to be held vertically, which is essential for accurate levelling see (Figure 6.17).

Errors in using equipment

The most common errors occur in reading the staff and recording the observations in the field book. Other common errors are:

- those caused by the instrument not being adjusted properly. These can be minimised by maintaining equal lengths between observations
- errors due to the bubble not being central when a reading is taken
- errors due to tilting of the staff. These can be eliminated by the use of a staff bubble level
- mathematical error in calculations.

FIG 6.18 Tripods

Tripods

The tripod is used to position, support, or stabilise instruments such as an auto level, theodolite, and laser. These tripods are normally hinged at the top, sturdily constructed of nonferrous materials (typically timber and/or aluminium), and are strong enough to support heavy measuring instruments. Most are fitted with metal ground spikes for use in soil or turf to stop the tripod legs spreading under the weight of the equipment attached to the top of the tripod. Equipment is attached via a Whitworth thread.

SETTING UP THE TRIPOD

1. Extend the legs of the tripod and, with the legs unopened, the height of the optical level base plate should be equal to the height of the user's chin.
2. Open the tripod legs and adjust the legs until the base plate is relatively level. Use the horizon as a visual guide to get the base plate level.
3. Attach the optical level to the base plate.

APPLICATIONS OF LEVELLING

Levelling, as applied to the trade of plumbing, is generally used in setting out prior to the excavation and during the laying of drains at specific grades. When very flat grades are encountered, a greater degree of accuracy becomes necessary. Often levels must be transferred from known TBMs to establish floor levels as the building progresses to enable floor waste gully risers (FWG) to be set at the correct level. Elevation levels are shown in Figure 6.19. Volume of the earth material can be calculated from spot levels or contours to enable an accurate cost estimate.

General levelling procedure

The method of establishing a difference in elevation between two points is illustrated in Figure 6.20. It can be seen that the distance down from the horizontal line of sight to A is 0.600 m and from the same line of sight to B is 3.250 m. The difference in elevation between the two points is therefore 3.250 − 0.600 = 2.650 m.

Series levelling

When the difference in between two points cannot be obtained in one set-up as described above, it is necessary to repeat the process. Figure 6.21 shows a plan and section of a typical series levelling exercise and the following calculation of levels will be taken from that exercise.

Reduced levels (RL)

Reduced levels can be calculated by using the rise and fall method and recording them in a field book.

Rise and fall method

This method is demonstrated using the readings obtained in Figure 6.21.

1. Referring to Figure 6.21, it can be seen that there is a rise of 0.550 from the TBM to A, obtained from

$$2.500 - 1.950 = 0.550$$

2. This reduction continues in the same way, always considering the rise or fall from one staff position to the next by subtracting the second reading from the first as though each were a simple backsight and foresight as follows:

BS FS

$$1.950 - 1.320 = +\ 0.630 \quad \text{(rise)}$$
$$1.320 - 1.602 = -\ 0.282 \quad \text{(fall)}$$

These readings are booked in their appropriate columns as shown in Table 6.1.

3. The staff is then read at change point C, and the rise or the fall is established between C and D, thus:

$$0.900 - 1.500 = 0.600 \quad \text{(fall)}$$

4. After the rises and falls have been reduced, the difference between the sum of each column will give

FIG 6.19 Plan and elevation of dwelling showing elevation levels

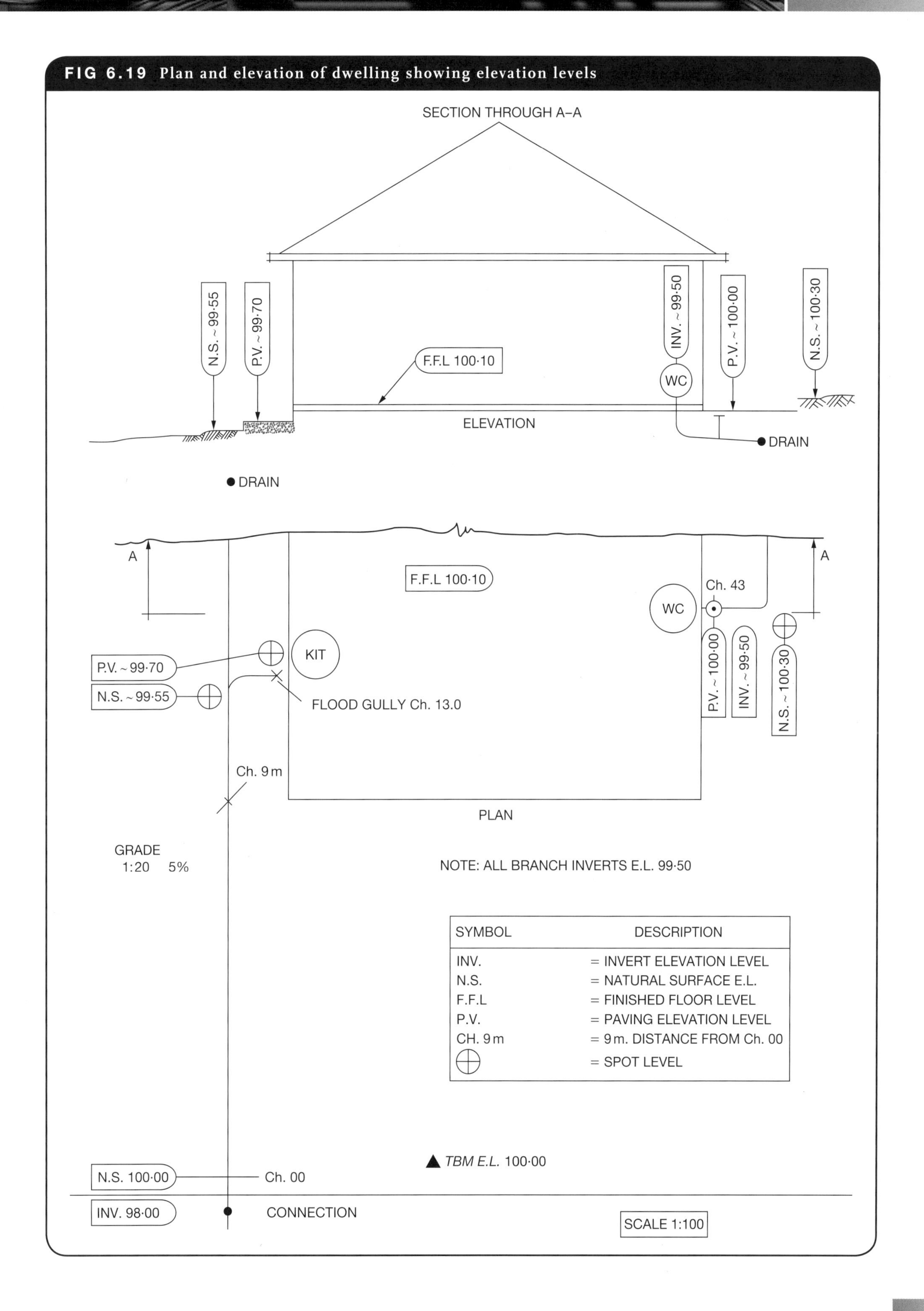

FIG 6.20 Difference in level between two points

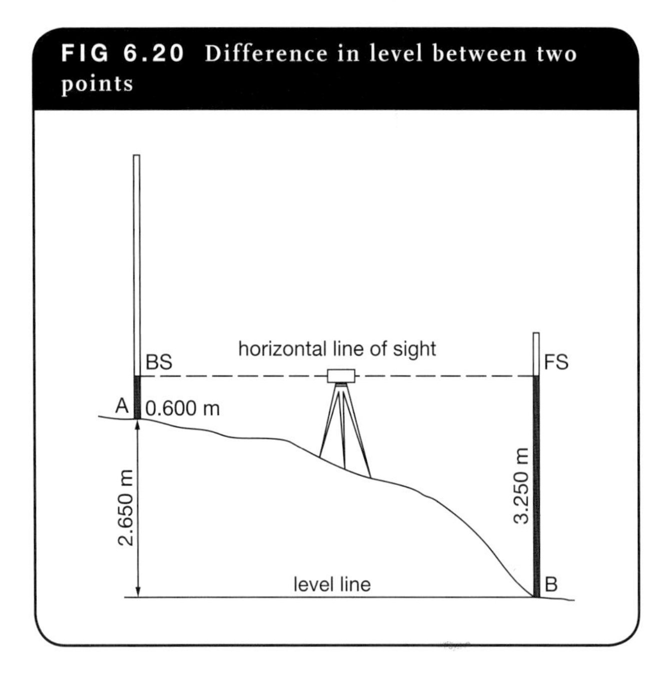

the total rise or fall. This must then be the difference between the sum of the BSs and FSs for the reduction to be correct. Therefore, the checked total rise in this example is 3.228.

5 After checking the rises and falls, the reduced level of each point is obtained by adding or subtracting the rise or fall to each preceding RL until the final RL is obtained.

Checking levels

To check the accuracy of the levelling observation, it is necessary to close the run of observations onto a TBM of known value or start point. An arithmetical check can be made by taking the difference between the sum of the BSs and FSs which should equal the known difference in height between the starting and finishing points. An arithmetical check has been made to the rise and fall method at the bottom of Table 6.1.

FIG 6.21 Series levelling

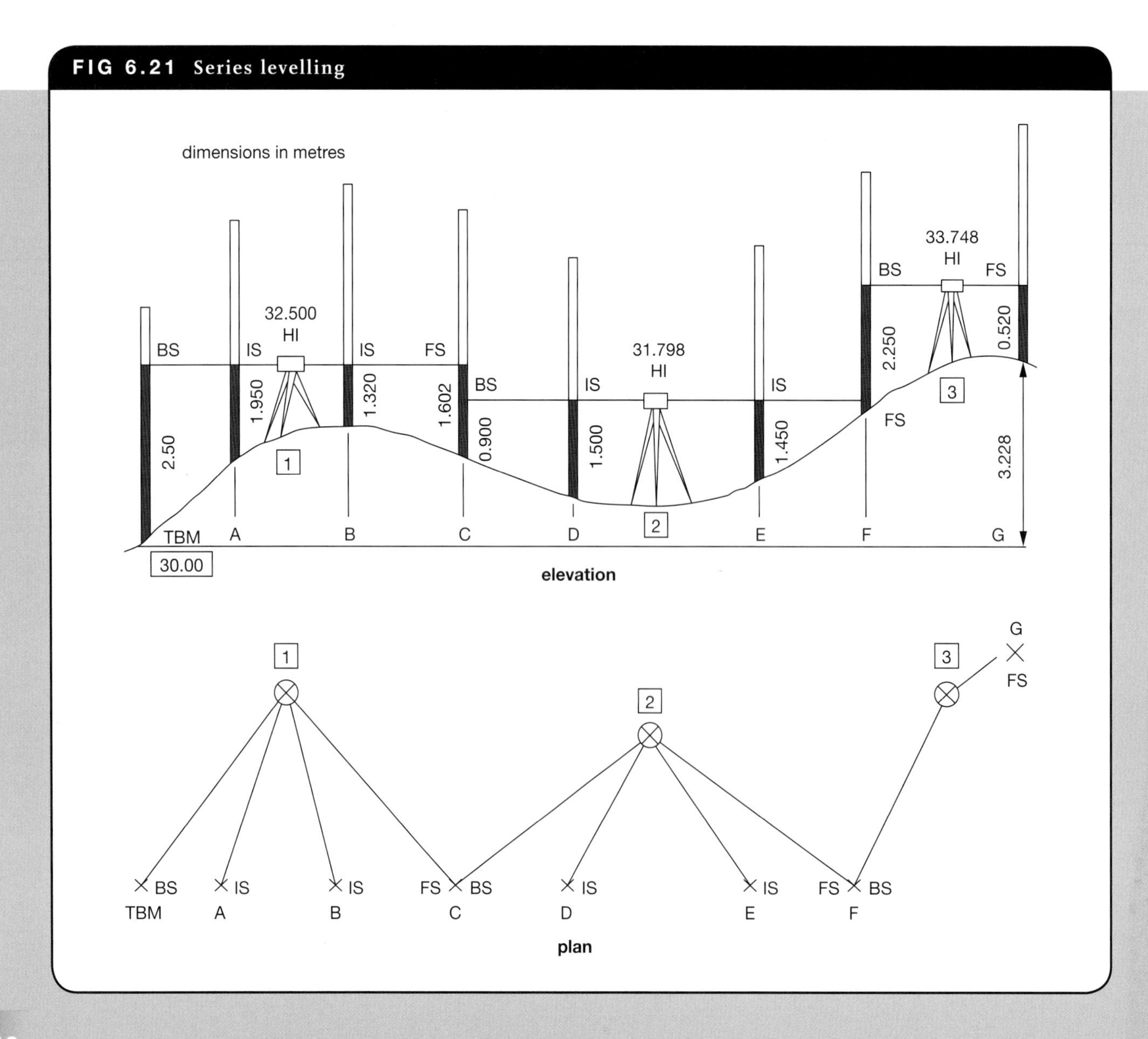

TABLE 6.1 Levels by rise and fall method

BS	IS	FS	Rise	Fall	RL	Remarks
2.50					30.000	TBM
	1.950		0.550		30.550	A
	1.320		0.630		31.180	B
0.900		1.602		0.282	30.898	C
	1.500			0.600	30.298	D
	1.450		0.050		30.348	E
2.250		0.300	1.150		31.498	F
		0.520	1.730		33.228	G
5.650		2.422	4.110	0.882	30.000	
2.422			0.882			
+ 3.228			+ 3.228		+ 3.228	

Note: Table 6.1 shows a method of booking levels and an arithmetical check using the rise and fall method to obtain the reduced levels.

FOR STUDENT RESEARCH

- Australian Standard AS 2397: 1993 Safe use of lasers in the building and construction industry
- Crawford, Wesley G (2002) Construction Surveying and Layout: A step-by-step field engineering methods manual (3rd edn), Creative Publishing,TX

PLUMBER PROFILE 6.1

TREVOR BATKIN

Job Title: Part-time teacher at North Sydney TAFE, Director of Trevor Batkin Plumbing Services, New South Wales

An old-school plumber who carries his ruler in one pocket and iPhone in the other, Trevor has been working in the plumbing industry since he started his apprenticeship in 1958. He has worked on an array of projects over the years and has witnessed substantial changes in industry standards since his earlier plumbing days.

How did you first start out as a plumber?

My apprenticeship started in January 1958 and at that time the training was at night. I do a tremendous variety of work now. I do maintenance work, and I have worked on heritage buildings, on MRI machines in hospitals. I have had a lot of apprentices, I now teach at North Sydney TAFE part-time. I employ people, I have done commercial projects—I have done basically everything!

What would you say the biggest difference is between working as a plumber a few decades ago compared with today?

Our trade in the pre-60s was a harder and lot more physical form of work. The drains were mainly done by hand and the drainage pipes and fittings were heavy earthenware pipes and fittings. Joints were of mortar mixed by hand. Threads of steel pipe were cut by hand—we all became very fit and strong back then. Plumbing today is easier, plastic and all that. It's like another world now compared to how we used to work.

What was a particularly challenging job you had to do in your earlier plumbing days?

We were called upon to run water and waste and other services under wharves and marinas. The environment was great, but there was difficulty with the rocking and raising and lowering of the tide, not to mention falling from a boat into the water while trying to use these difficult heavy tools.

How have the safety regulations changed over the years?

We often had to work on roofs high up on large heavy wooden ladders, around 50 feet tall, which are banned today. It wasn't scary because that's what we were told to do. Harnesses had not been thought of yet for plumbers. A friend of mine once fell into a palm tree! Also years ago the apprentice (this is unheard of today) often travelled in the back of the ute in winter or summer—the bosses in the front, and apprentice in the back, and the law allowed for this.

What health risks did you face as a result of the relaxed safety standards?

The 60s required working with lead which was a hazard due to lack of hand washing and many plumbers suffered the after effects, similar to asbestos. That period also involved working with asbestos sheeting and pipe lagging but we never knew. Many plumbers of my age have suffered because of that. I have worked with asbestos but luckily my lungs are OK. We were put in a lot of dangerous situations.

What is one of the most memorable jobs that you have worked on?

A lady rang us up in tears as her diamond ring was down the toilet, we found it and she said her husband had bought the ring 40 years ago and cried and hugged us. It was a nice feeling.

What advice do you usually give your students and apprentices?

To do the job correctly and be fair to your customers, give them good value.

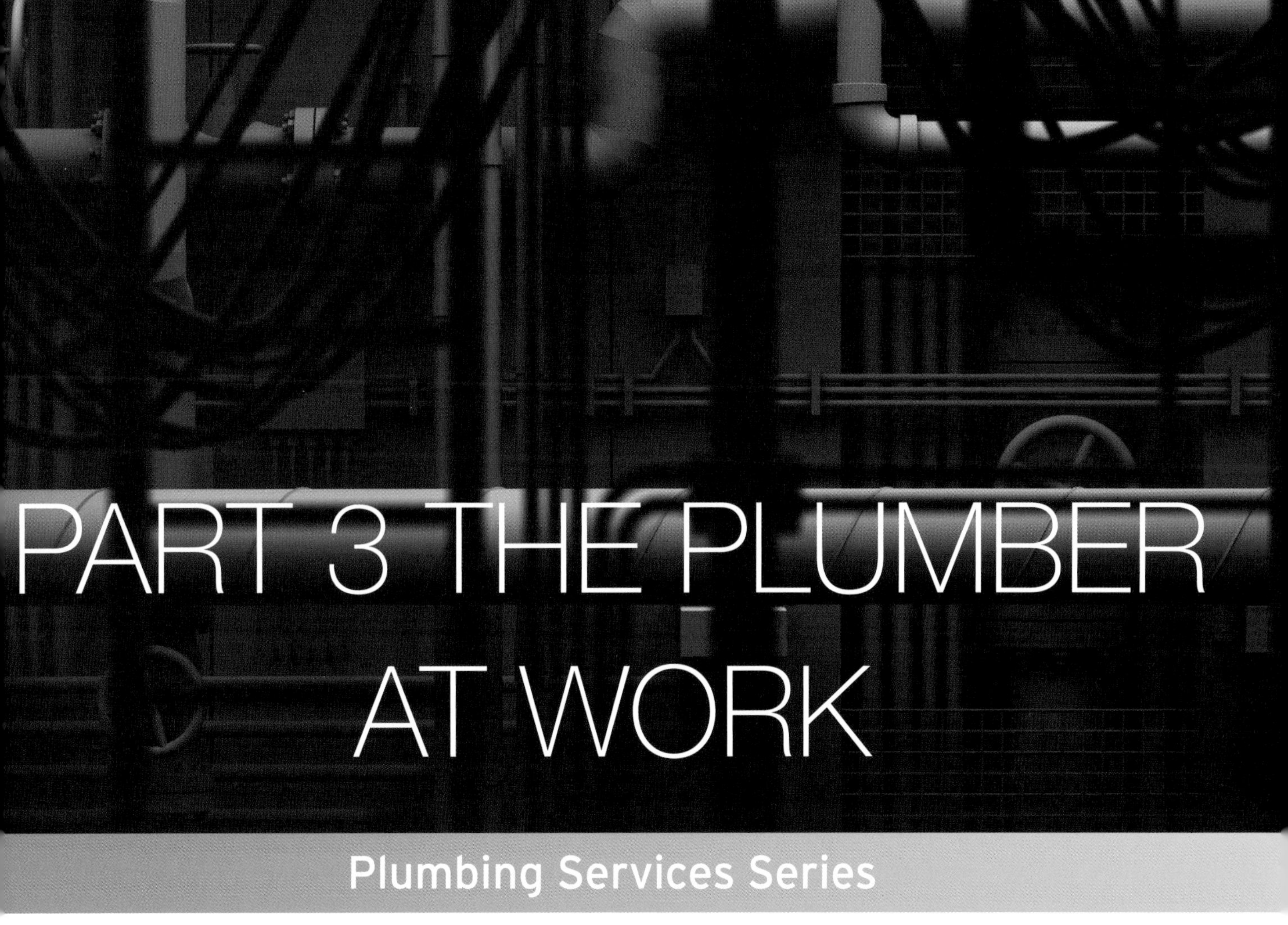

PART 3 THE PLUMBER AT WORK

Plumbing Services Series

7 The green plumber

8 Effective communication

9 A safe and healthy workplace

CHAPTER **SEVEN**

The green plumber

LEARNING OBJECTIVES

In this chapter you will learn about:

7.1 what sustainability is

7.2 how sustainability affects the plumbing industry

7.3 what plumbers can do to operate in a sustainable manner.

INTRODUCTION

It seems that the subject of sustainability has become a key focus of concern in society today. Wherever we turn, people are discussing the environmental impacts of how we live our lives and how we can do this without depriving future generations of a comfortable, secure and prosperous lifestyle. Why is there so much concern over sustainability? Is it that we have been forced to become aware of the enormous quantity of resources we consume every day and the very real impact this is having on our natural environment? Or is it that the cost of finding these resources is starting to hit us in the hip pocket? Whatever the reason may be, if we wish to preserve the natural beauty and diversity of our planet, we all need to embrace and promote sustainability principles in everything we do.

In this chapter we will take a look at how sustainability applies to you as a tradesperson working in the plumbing industry. You may think that right now as an apprentice or a trainee you don't have a lot of influence over what happens at your workplace. The fact is that you can share your sustainability learning with your workmates. You can even help your boss to understand and apply new ways of doing things; more efficient ways that may even help him become more profitable! The great thing about this is that by the time you are through your training and maybe even in business for yourself, you will find it easier to set up and operate in a sustainable way. The fact is our

FIG 7.1 Life out of death

communities want to deal with sustainable businesses, so if you operate yours in this way you will have a real head start.

The aim of this chapter is to help you understand what sustainability is about and how it applies to you as both a professional working in the plumbing industry and as an inhabitant of planet Earth. We will look at the ways in which sustainability affects a plumbing business and how you as a professional can operate in a way to ensure maximum conservation of resources and minimal impact on the environment.

ABOUT SUSTAINABILITY

Sustainability—what exactly does it mean? Depending upon the context, there are lots of different definitions of 'sustainability'. As we want to understand it from an everyday viewpoint we will take it to mean:

> '... the ability to maintain things or qualities that are valued in the physical environment'.

(from *Living well within our environment. A perspective on environmental sustainability*, P. Sutton)

Figure 7.2 gives further understanding of sustainability as a concept.

FIG 7.2 Balance and sustainability

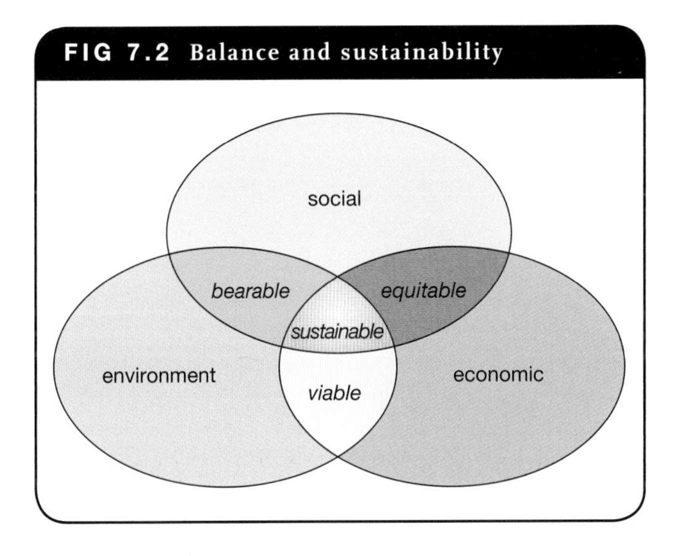

This diagram shows us the relationship between the three key influences affecting our lives; economic, environmental and social. Here we can see that when our activities are conducted in a way that meet environmental, social and economic requirements, then these activities are not only sustainable but also viable, bearable and equitable. When we conduct our business activities in a sustainable way and balance environmental issues with economic issues, we will have a 'viable' business. This is because we are not wasting resources, but rather using them in a responsible manner. If we run our businesses in a way that balances economic issues with social issues, then our communities are much more 'equitable'. This is due to the fact that all benefit from these economic activities and no-one is exploited. If we balance environmental concerns with social concerns, then our lives will be 'bearable'. We will enjoy life in our communities, as we are able to make use of our environment in a way that brings them benefit. When we hold all three of these areas in balance, then our business activities and our communities will be sustainable.

So what does sustainability have to do with plumbers? To answer this question we need to look at a number of key areas relating to the way plumbers operate in their businesses. These are:

- the way their business operations impact upon the environment
- the energy efficiency of their buildings and equipment
- the thermal efficiency of their operations (buildings, plant and equipment)
- the water efficiency of their business operations
- the amount of waste produced by the business in relation to business output
- the way businesses obtain goods, services and materials.

If we have an understanding of these six key areas, then we can start to formulate a plan of how to move toward more sustainable practices. We'll look at each one of these in a little more detail.

How do plumbing business operations impact upon the environment?

We've all heard stories in the media about businesses that have disregarded all the rules in pursuit of a good return for their shareholders. For example, the mining company that allowed leachate to get into underground watercourses and ruined the water supply for villages and towns for kilometres around; or the timber company whose sprays completely ruined the ecology of an entire river and marine ecosystem. What about the directional drilling company whose main claim was that their underground boring service made minimal impact on the environment? Their service didn't look so 'environmentally friendly' when they were caught illegally dumping their drilling slurry on a rural property just outside Sydney. This list is, unfortunately, endless. The choices we make in our day-to-day life will determine whether or not we deserve to be placed on the list of environmental vandals.

How do we work effectively and make a minimal impact when we are involved in activities that can cause considerable disturbance to the environment? Table 7.1, overleaf, presents a few examples of the environmental impact plumbing activities can have, and some possible solutions.

TABLE 7.1 Dealing with common environmental risks

Activity	Risk	Mitigation measures
Excavation work for installation of gas, drainage or water lines	Nuisance and entry of silt and dust into the environment	Installation of silt barriers to minimise run-off. Wetting down of site to prevent spread of dust. Change of work practices
Building and construction work	Disturbance to the environment through the production of excessive noise during construction activities	Change of work practices. e.g. use of diamond bits rather than percussion drills, or use of silenced air compressors, excavation plant and generators
Supply of tap-ware, fixtures and fittings	Water wastage due to the installation of inefficient equipment	Specification and supply, and promotion of 'WELS' and 'Green Star' rated tap-ware, fixtures and fittings
Installation of plumbing work	Environmental pollution due to the use of toxic plumbing materials, adhesives and sealants	Use of non-toxic plumbing materials, adhesives and sealants
Purchase and use of equipment and tools	Wastage of materials, energy and resources through use of poor quality, low cost, disposable type hand and power tools	Adopt a policy of purchasing higher quality tools designed to give a long service life. Wherever possible choose tools that can be repaired if they fail

We will discuss the use of energy and water efficient products in the following sections.

How efficiently is energy used in the plumbing business?

You might think that it is a bit irrelevant to ask how efficiently plumbers use energy in their businesses; however, when you stop and think, there is a lot of scope for plumbers to be energy efficient in many areas of their operations. Take for example some of the larger plumbing service companies you see around town. How many trucks do these companies run, how many men do they have working for them and what about their business premises? There are many substantial savings to be made in this area, for example:

- the use of well serviced, suitably selected modern vehicles running on alternative fuels such as liquefied petroleum gas, natural gas or diesel
- installation of photovoltaic back to grid energy systems in plumbing offices
- installation of alternative hot water units such as heat pump, solar or gas boosted solar in plumbing offices
- a move away from conventional heating, ventilation and cooling (HVAC) in business premises to more energy efficient alternatives such as evaporative coolers and natural ventilation systems
- promotion of energy efficient fixtures, equipment and practices to clients.

FIG 7.3 An example of a silent-running generator

Once again, plumbers can assist their clients to make informed and sustainable decisions during the design and implementation stages of their projects.

How thermally efficient is the plumbing business?

When we consider thermal efficiency we need to look at both our own business practices and how we advise our clients. The thermal efficiency of buildings is a very large subject so we will look at it only briefly here. In our case we are looking at our business premises and how we can improve them so they require less energy to heat or cool. Achieving greater thermal efficiency in our buildings may involve a little or a lot of work. We may not always be able to achieve our goals due to financial or time constraints, or because we may not own the building. These are a few of the measures we can take to improve the thermal efficiency of our buildings:

- Install appropriate insulation to roofs and walls.
- Install suitable glazing (double glazing or high efficiency glass) to reduce heat loss or gain into or out of buildings.
- Design the eaves and roof shapes of buildings for passive solar efficiency to heat during summer and increase warming during winter.
- Seal gaps around doors and windows and anywhere else heat is lost or gained.
- Use suitable window coverings to retain heat during winter and to block out unwanted heat during summer.
- Recommend the installation of high efficiency insulation to all hot water pipe-work and to air-conditioning reticulation lines. Inform clients of the availability of retrofit insulation for water heaters.

Figure 7.4 shows where heat escapes from the average house. This photo was taken using an infrared camera by a company that specialises in helping people improve the thermal efficiency of their homes. The yellow areas show the areas of greatest heat escape.

How efficiently does the plumbing business use water?

Water efficiency is certainly an area in which the plumber is an expert! More than ever before there is a great need to reduce the water used by our ever-increasing population, and so it is most important that we as plumbers have a thorough knowledge of water efficiency regulation and products. Although water efficiency regulations are set out in the Building Code of Australia, and Australian and New Zealand Standard 3500 (AS/NZS 3500), we see them in action through the WELS (Water Efficiency Labelling and Standards) scheme. The WELS scheme is a system designed to assist householders in the conservation of water through providing information on water saving fixtures and fittings. WELS is supported by AS/NZS 6400:2005, *Water-efficient products–rating and labelling*. The WELS labelling allows consumers to identify which items are the most efficient in terms of water consumption and which best suit their needs. The WELS rating system covers a large range of tested, rated and certified fixtures and fittings, including:

- showers
- tap equipment
- flow controllers
- toilet suites
- urinals
- clothes washers
- dishwashers.

FIG 7.4 Heat loss detection by infrared camera

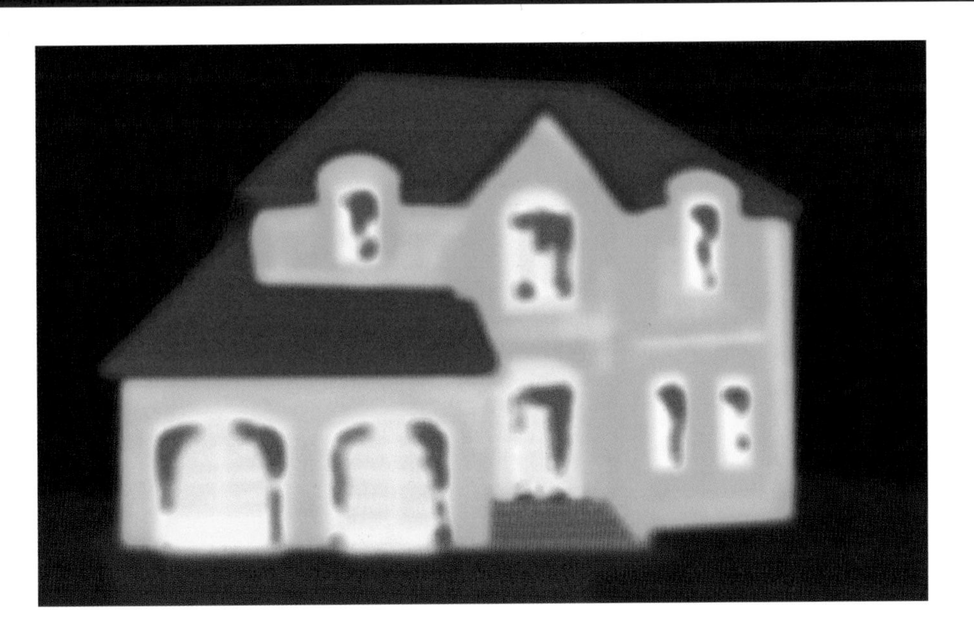

FIG 7.5 A typical WELS label

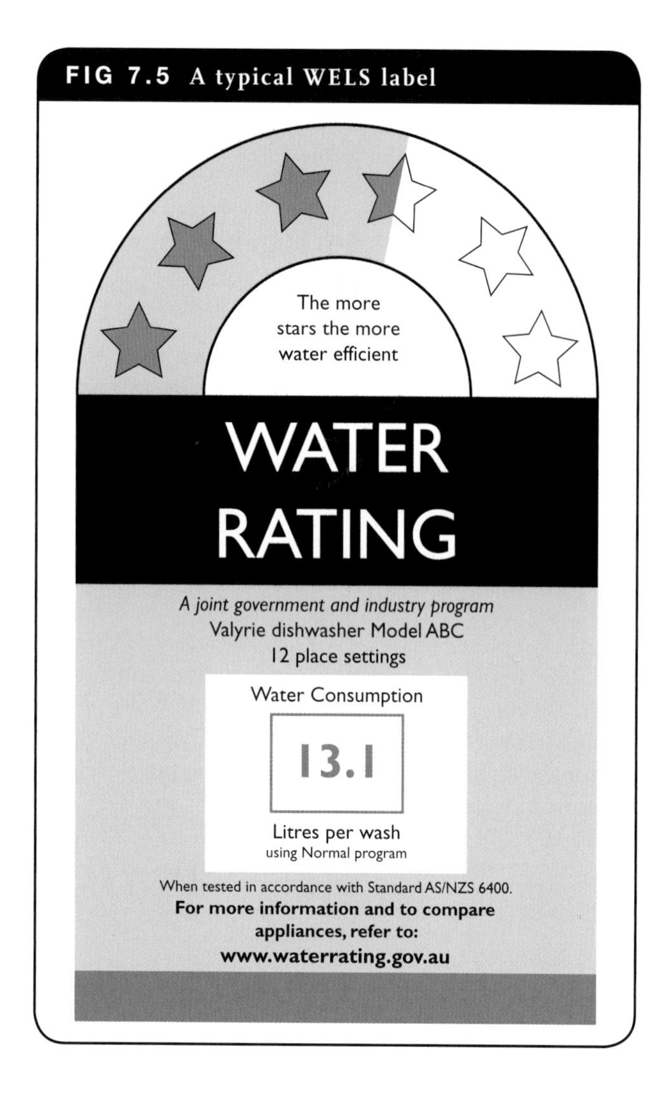

Efficient water usage for a plumbing business could include:

- installation of water-efficient (WELS rated) tap-ware
- installation of water-efficient reduced flush toilets and urinals
- use of drip or below ground irrigation systems
- installation of flow and return hot water circuits that remove the need to waste large quantities of hot water
- altering on-site work behaviour to eliminate water-wasteful practices and replace with more efficient techniques, for example, washing down with full flow hoses could be replaced with the use of high pressure water blasters
- actively promoting the awareness, installation and use of water-efficient appliances, fixtures and fittings by clients
- promotion of the installation and use of grey water recycling equipment in homes and businesses
- use of compressed air for the testing of sanitary drainage, sanitary plumbing and water lines to minimise water consumption.

You, as the plumber on-site, have an excellent opportunity to promote water efficiency awareness and practices. Figure 7.6 shows the break-up of water use in

FIG 7.6 Typical domestic water consumption

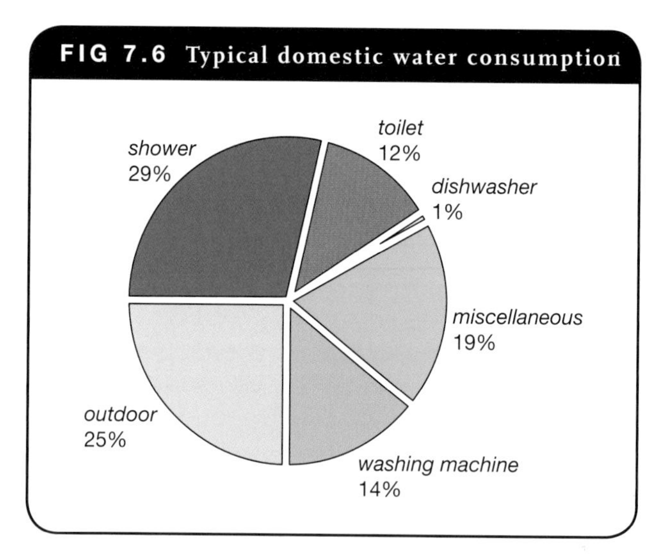

FIG 7.7 Commercial waste

a typical household. Can you see any opportunities for water savings?

How much waste does the plumbing business produce?

Probably the greatest difference between 'old school' hardware stores and plumbing suppliers is the way items are presented for sale. Many of the hardware stores sell their items from bulk bins and are placed in a cardboard box or bag when given to the customer. The rise of some of the larger hardware chains and plumbing suppliers has seen much of what we buy packed individually using shrink-wrapping or some other form of non-recyclable packaging. To see the high degree of packaging of building materials, we only need to go to the average building site and take a look at the site rubbish dump or skip.

Have you fitted out a new house or replaced taps or other fittings? How long did you have to fight with the packaging before you could do the job and when you were finished, how much refuse did you have to throw away? Remember, that's just *your* job. Now multiply that by all the plumbers working that day and then we'll have some idea of how much of a problem we have created by the way we do things. The encouraging part is that there are lots of bright people like you out there to help find a solution! What can be done?

As consumers, we have a lot more power to bring about change than we think. Simply by choosing to buy sustainably packaged products we can influence manufacturers and retailers to alter their practices. If we purchase plumbing supplies packaged in recyclable materials we can return our used packaging to the system for remanufacture and redistribution, and in this way operate our businesses in a more sustainable manner by conserving natural resources and reducing the quantity of material going to landfill.

Where does the plumbing business obtain its goods, services and materials?

The final area we will look at in our study of sustainable business operation is the question of how we choose suppliers for our business. Now that you have gone to the trouble of setting up your business in a sustainable way, it will probably make sense to try and source your goods, services and materials from companies that operate in a similar way. The process of sourcing goods and services is called 'procurement', and many companies have what is called a 'procurement policy' that governs the way they source their goods and services. When selecting a supplier for your business, you can request that your prospective supplier provides a copy of their procurement policy, and from this you can determine whether or not you would like to deal with them.

When selecting a supplier you could ask them some of the following questions:

- Do they source their products from companies that operate in a sustainable manner and what policies do they have in place to monitor this?
- Can they guarantee that all workers involved in the supply of their products are employed in an exploitation-free workplace?
- How do they deal with waste management issues in their business and do they have a recycling policy?
- Do they have provision for the return and recycling of used packaging?
- Do they have an ongoing plan to improve the sustainability of their business?

The great thing about taking the trouble to select suppliers who operate sustainably is that you will not only have a more efficient business, but you will also attract customers seeking to deal with plumbing contractors who operate in a sustainable way. With an ever-increasing

FIG 7.8 An advertisement promoting the sustainable use of paper

WHY REDUCE, REUSE, RECYCLE ?

PAPER is the largest percentage of material thrown away.

A ton of PAPER made from **recycled fibres** instead of **virgin fibres** CONSERVES:

- 7,000 gallons of water
- 17-31 trees
- 4,000 kWh of electricity
- 60 pounds of air pollutants

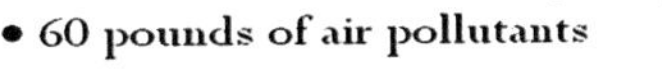

WASTE LESS... RECYCLE MORE...LIVE WELL.

emphasis upon sustainability in our communities, your business can only do well!

SUMMING UP

The purpose of this chapter has been to give you, the apprentice plumber, a clear understanding of what it means to operate in the plumbing industry in a sustainable way. We have had a look at the six basic areas a business needs to address if it is to operate in a sustainable manner, some issues involved and some possible solutions. Once again, these are the six areas:

1. the way their business operations impact upon the environment
2. the energy efficiency of their buildings and equipment
3. the thermal efficiency of their operations (buildings, plant and equipment)
4. the water efficiency of their business operations
5. the amount of waste produced by the business in relation to business output
6. the way businesses obtain goods, services and materials.

The most effective means of enhancing sustainability, whether in our own lives or in the business realm, is through behavioural change. If we all practice responsible stewardship of our natural resources, and if we are committed to applying the principles of sustainability to our everyday practice, then we will have taken a substantial step toward securing a standard of life for future generations that is 'equal or greater than our own'.

FOR STUDENT RESEARCH

- AS/NZS 3500: Part 1 Water services
- AS/NZS 3500: Part 4 Heated water services
- Building Code of Australia-section J
- AS/NZS 6400:2005 Water efficient products-Rating and labelling

Reference

Sutton, P *Living well within our environment. A perspective on environmental stability?* A paper for the Victorian Commissioner for Environmental Sustainability. www.ces.vic.gov.au/ accessed 23 September 2011.

PLUMBER PROFILE 7.1

BRENT PAPADOPOULOS

Online Learning Centre

Brent with the Victorian Premier John Brumby receiving the Premier's Sustainability Award for Small Business 2009

Job Title: Director of Sustainable Plumbing Solutions, Nunawading, Victoria

Three-time winner of the Master Plumbers and Mechanical Services Association of Australia's 'GreenPlumber of the Year' award, and with a further two nominations, Brent has been heavily involved in protecting the environment through sustainable plumbing since he finished his apprenticeship. His business, Sustainable Plumbing Solutions, won the Victorian Premier's Sustainability Award in both the Small Business and Recognition categories in 2009.

How did you first get started as a plumber?

I first went to uni to do arts commerce but it only lasted six months as I didn't enjoy it. I finished high school, which I think is really important, and I went to a private school and nobody ever talked about going into a trade—it wasn't even an option or consideration at the school I went to. And then I didn't find uni suitable, so I thought I had to do something and I decided a trade might be the way to go. So I gave it a shot and never looked back. I did the pre-apprenticeship which gives you a good understanding of what the trade is all about and a good insight if you want to continue your apprenticeship.

How did you set up your own business?

I started my business the day I finished my apprenticeship. I have never worked for another company and I employed my first employee after six months of being on my own.

What was the motivation behind going into sustainable plumbing?

We could see a market opening in that area. At the time there were no other plumbers even considering going into environmental plumbing and another important factor was my partner Ruth. She was a horticulturalist at the time and very driven by the environment and gave me a helping push in the right direction. I have always been passionate about it, but I started off on a pioneering track as nobody was doing it. It was new and exciting and different to normal plumbing. I enjoy the challenge and doing something for the environment.

What kind of projects have you been involved in?

Lots of things from large greywater treatment plants, recycling greywater for both domestic and commercial projects: so shower and bathwater—instead of going down the drain—it is getting captured and cleaned and then re-used back in the toilets. We have worked on large solar hot-water projects, heating the water with the sun so we are not relying on our precious resources like natural gas and electricity, and many, many rainwater harvesting projects. This involves capturing rain off the roof and utilising this throughout the house or building. There is a long-term effect involved as the things we are setting up will be present for the next 30 or 40 years and can make quite a difference.

What are the main challenges you face as a sustainable plumber?

At the start it would have been getting employees. I have had to train every employee, and not just train a little bit of knowledge here and there, for them it has been like doing another apprenticeship in the environmental sector. So, constant training for my employees, which is rewarding but also time consuming and costly. It takes a long time to train even the best of qualified plumbers.

Now my challenge is that there is a downturn. Environmental plumbing can add extreme costs to construction. If you want to add solar hot water, rainwater, it can add $60 000 to the construction of the house. They benefit householders in water consumption though; costs can come down if using solar and they are using their own water so there are cost savings in utilities of water and gas. It gives you the option, when there are restrictions and you can't water your garden. If you have your own water supply you can do what you want.

What is the most difficult job you have done?

There was one recently for the Victorian Government down in Anglesea. It was a very challenging job. They had 108 little rainwater tanks installed in the building and each tank had four connections to it. So we had to make sure every single connection was 100 per cent and not leaking, otherwise the whole building could have flooded. Another company had already attempted this project and failed, and we were brought in as the experts to fix the whole mess. We ripped out the whole project and started it again, but we succeeded and got them up and running. So it was a very challenging but extremely rewarding job.

Do you have a favourite project that you have worked on?

We were working for a large company called Solar Systems and did a test facility out in Bridgewater in Central Victoria, north of Bendigo. We spent around nine months out there, on and off, setting up the entire facility with rainwater and greywater treatment plants and solar hot-water treatments. The biggest and best project!

Do you have any tips for those wanting to move into sustainable plumbing?

Definitely training. Go to your local association, like in Victoria, it's Master Plumbers. They offer courses in sustainable practices. That is the best way to open a door to a career path in environmental plumbing. There is always more that needs to be learned and developed. We had a peak in 2009 and now it has dropped, but it will go up again. Plumbing always relates to the environment.

CHAPTER **EIGHT**

Effective communication

LEARNING OBJECTIVES

In this chapter you will learn about:

- **8.1 the definition of communication**
- **8.2 methods of communication**
- **8.3 non-verbal communication (body language)**
- **8.4 the communication cycle**
- **8.5 the barriers to communication**
- **8.6 how to use communication channels.**

INTRODUCTION

When was the last time you stopped and actually thought about communication? You probably haven't spent much time thinking about it at all. We have all been communicating in one way or another since before we were born but the question we need to ask, especially in the context of the workplace, is: 'How effectively do I communicate?'

We've all had those experiences where we've tried our hardest but have totally failed to get an important message across—when we've ended up in Newtown instead of Newport, wondering whether we speak a different language from everyone else on the planet!

COMMUNICATION IN THE WORKPLACE

Effective communication is important in all of areas of life and is often the factor that determines how much we enjoy our life experiences. This is particularly so in the workplace. How smoothly does your workplace run? What are the relationships like? Are there often difficulties in getting the job done just the way the client ordered and in a profitable manner? Do accidents occur frequently? The presence of these factors, and possibly many others, may indicate that the lines of communication in the workplace are not clear and this is hampering the way the workplace runs.

FIG 8.1 How effectively do we communicate?

"THE CLIENT'S NOT TOO HAPPY WITH THE WAY I INTERPRETED THE BRIEF..."

In this chapter we look at communication in the workplace. We will see that there are a number of different types of communication, and that our communication is usually a combination of these forms.

WHAT IS COMMUNICATION?

Communication has been described as the process of passing information or messages to another person or group. The problem with this description is that it ignores the fact that communication always involves interaction between those communicating, and so it would be better to describe communication as 'the creation and exchange of messages' (Goldhaber, 1993). Communication involves the exchange of many types of messages, whether they are written, spoken or those conveyed through facial expression, tone or body language. If we want to be effective workplace communicators we need a good understanding of how to make best use of its various forms.

Non-verbal communication (body language)

Non-verbal communication, or body language, forms a major and extremely important part of the way we communicate. When we interact with other people, we constantly send and receive a stream of messages that have a profound effect on our communication. Whether or not we are aware of them, the way we speak (how fast or loud), our facial expressions, what we do with our hands, our posture, our eye contact and how close we stand all send strong messages to the other party.

These signals can tell us whether or not the person we are talking to is listening, how interested they are in the conversation and how much they care. They can have a number of different influences on our communication, both positive and negative. Five influences that non-verbal communication signals have are:

- **repetition**—they can repeat the message the person is making verbally
- **contradiction**—they can contradict a message the individual is trying to convey
- **substitution**—they can substitute for a verbal message. For example, a person's eyes can convey a far more vivid message than words often do
- **complementing**—they may add to or complement a positive verbal message
- **accenting**—they may accent or underline a verbal message. Pounding the table, for example, can underline a message.

Non-verbal cues in conversation

Think about a recent conversation. What non-verbal cues were you sending and what influence may they have had? How do you think the other person saw you? What non-verbal cues can you remember receiving? If you think about it, it's likely those cues fell into the categories shown in Figure 8.2.

Effective communication is a key to success in both life and business. It is worth building our communication skills, and part of this is to be aware of the non-verbal cues we send. Use the guide in Table 8.1 overleaf to evaluate your non-verbal communication skills.

FIG 8.2 Elements of non-verbal communication and body language

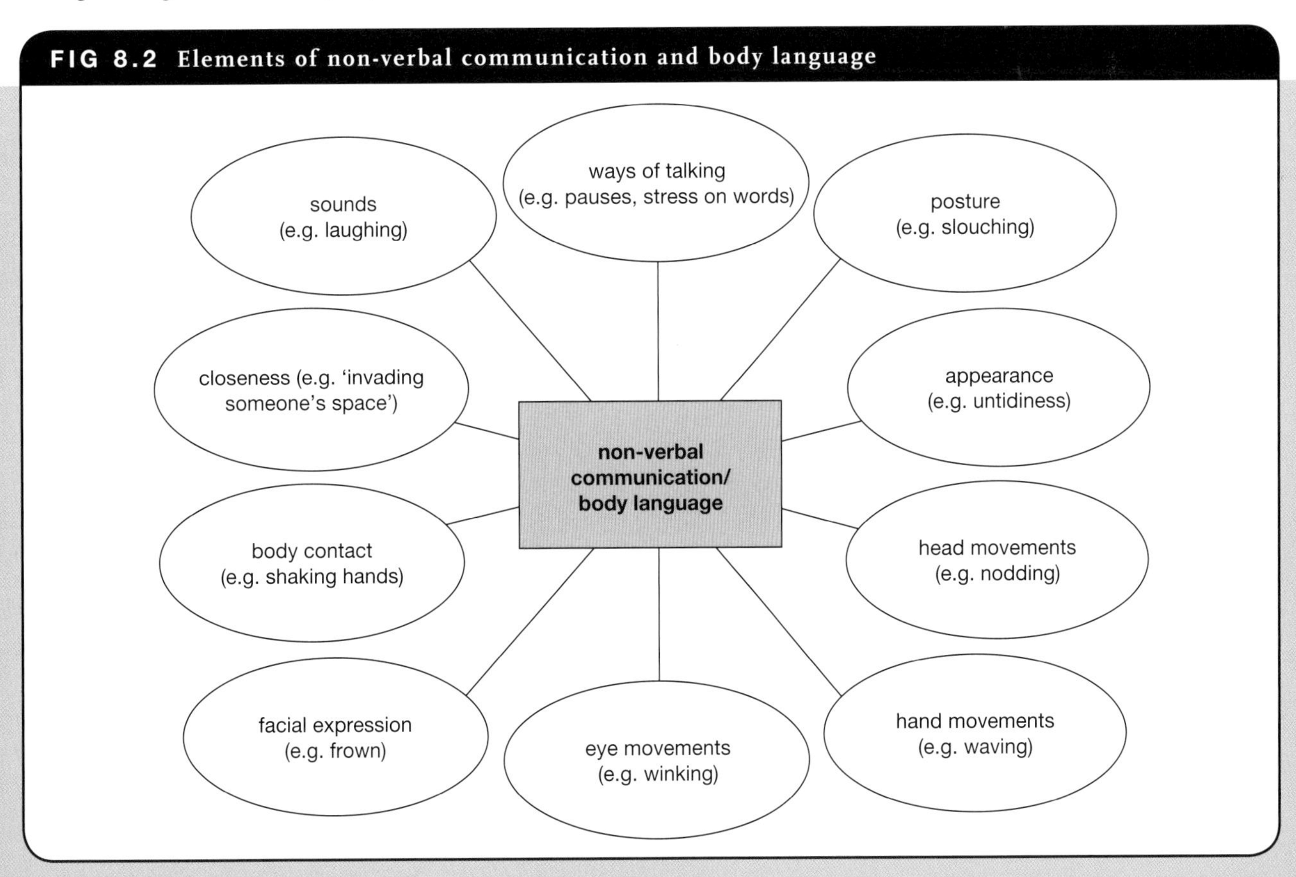

TABLE 8.1 Evaluating your own non-verbal communication skills

Non-verbal cue	Things to check
Eye contact	Is this source of connection missing, too intense, or just right in yourself or in the person you are looking at?
Facial expression	What is your face showing? Is it mask-like and unexpressive, or emotionally present and filled with interest? What do you see as you look into the faces of others?
Tone of voice	Does your voice project warmth, confidence and delight, or is it strained and blocked? What do you hear as you listen to other people?
Posture and gesture	Does your body look still and immobile, or relaxed? Sensing the degree of tension in your shoulders and jaw answers this question. What do you observe about the degree of tension or relaxation in the body of the person you are speaking with?
Touch	Remember that what feels good is relative. How do you like to be touched? Who do you like to have touching you? Is the difference between what you like and what the other person likes obvious to you?
Intensity	Do you or the other person you are communicating with seem flat, cool and disinterested, or over-the-top and melodramatic? Again, this has as much to do with what feels good to the other person as it does with what you personally prefer.
Timing and pace	What happens when you or someone you care about makes an important statement? Does a response—not necessarily verbal—come too quickly or too slowly? Is there an easy flow of information back and forth?
Sounds	Do you use sounds to indicate that you are attending to the other person? Do you pick up on sounds from others that indicate their caring and concern for you?

Source: Jeanne Segal (2008) *The Language of Emotional Intelligence, McGraw-Hill, NY*

FORMS OF COMMUNICATION

Human relationships are always complex, and can become more so when we take them into the workplace. Communication in the workplace occurs in many forms, including the following:

- letters, memos, notes, reports, plans, specifications, brochures
- telephone, SMS, MMS, radio communications
- email, internet-based media (Facebook, Twitter)
- conversation—face-to-face interaction
- meetings—small or large
- observing—watching and listening to a demonstration or instruction, awareness of surroundings and building site signs
- listening.

Some methods of communication have advantages over others. Some of these are listed in Table 8.2.

TABLE 8.2 Advantages and disadvantages of various communication styles

Communication channel	Advantages	Disadvantages
Face-to-face	• Gives immediate feedback • Offers opportunity to observe body language cues	• There may be personal issues between parties • Discussion may become emotive • May be costly in terms of time and money
Telephone, radios	• Rapid • Overcomes distance • Confidential • Can do conference calls for groups	• Messages can be misunderstood • Can be difficult to persuade people • Issues of tolerance with delays • Technology issues: drop outs, line issues • Unable to observe body language cues
Written messages	• Allows time to compare messages • Permanent record • Can be copied and distributed • Can reflect on message/content	• Feedback is delayed • Written messages can be lost • Can lack persuasive power • Once given, the written message is a permanent record
Graphic messages (signs, diagrams)	• Quickly and easily understood • Better than words with complex message • Overcomes language and cultural barriers	• Limited power as feedback cannot be given

Source: Adapted from Rossignol, 1999, p. 55–56

THE COMMUNICATION CYCLE

All of our communication is a process where there is neither a beginning nor an end. If you think about how we communicate, the cycle would look something like the diagram in Figure 8.3.

FIG 8.3 The basic cycle of communication

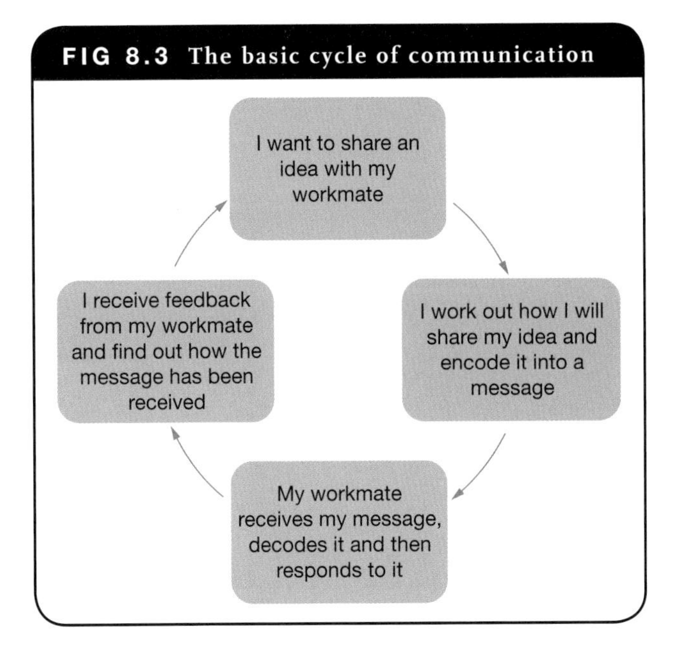

It would be wonderful if the process was as simple as this. The problem is that when we send and receive messages they pass through a series of filters and barriers made up of our life experiences, our mood at the time and a myriad of other influences. Our communication is also influenced by the means or channel we choose to convey our thoughts and ideas, and so the process looks more like that shown in Figure 8.4.

BARRIERS TO COMMUNICATION

So, depending upon the channel we choose and the amount of noise affecting our communication, our message passes through to the other party. The challenge we have is to choose a channel that is not easily affected by noise, so that the message gets to the receiver in the least corrupted form possible.

Some examples of 'noise' could be:

- difficulty in understanding the technical language used in a message
- difference in the understanding of language between sender and receiver
- cultural differences between sender and receiver
- body language sending a different message to the original one sent
- expectations and prejudices influencing the way the message is received
- time pressure or being too busy preventing full concentration
- environmental issues, for example hazardous, noisy or busy building sites
- our perception of our place in the 'power relationship' influencing the way we interpret the message for example a tradesperson speaking to an apprentice
- previous dealings with the other person.

An understanding of the way noise can affect our communication will help us to choose the most appropriate communication channel and to think about how we can express ourselves in a way that is easily understood by others.

COMMUNICATION CHANNELS

In this section we will look at some channels available to us as communicators. At the worksite we see many types of communication channels in use as we go about the business of maintaining a safe, efficient and comfortable workplace. Good communication is the key to achieving all of these things.

Channel 1: Face-to-face communication

Much of our day-to-day communication is done through this method. Face-to-face communication allows us to quickly and efficiently pass on our message and to receive and respond to feedback, thus resolving issues on the spot. When we communicate face to face, we have the advantage of being able to use all the available clues to clearly understand

FIG 8.4 The actual communication process

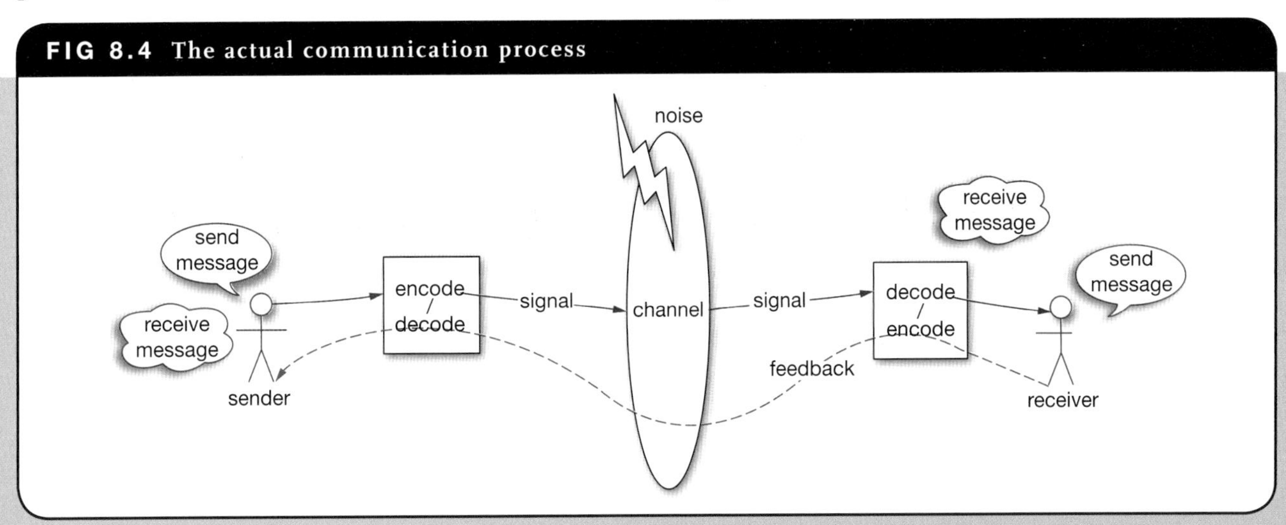

the message. We can watch for body language clues, and we can immediately give and receive feedback on what is being discussed, and we can listen and observe and seek clarification of items we don't understand. The downside of this immediacy is the strong possibility of 'noise' interfering with our message. Maybe you have had previous dealing with this person and don't quite trust them, or maybe they use lots of technical language that you find hard to understand. Maybe it's just been a bad day at work and everyone is a bit uptight! All of these things can hinder the way we hear and understand.

Perhaps the most effective communication tool we have is good listening. It's very frustrating having a conversation with someone who we know is only pretending to listen to us. You can see they are distracted and that their attention is somewhere else. On a worksite proper attention to, for example, your boss's directions may be the thing that prevents a serious accident or a disruption in the work program.

Here are some hints for better listening:

1 Show the other person that they have your full attention. Make eye contact and follow what they are saying by using appropriate body language (lean slightly forward, have arms uncrossed) and watch for signals from the other person. Allow for sufficient time for the conversation and try not to be interrupted; show the other person you are committed to the conversation. Avoid distraction and stay focused.

FIG 8.5 Are you listening?

2 Ask questions to make sure that you understand what is being said.

3 Participate in the conversation by repeating words and phrases back to the other person.

4 Demonstrate your understanding by summarising the other person's point of view.

5 Watch for the rise of emotions in the conversation—don't allow the discussion to become heated.

6 Demonstrate an interest in what the other person has to say. Show empathy for their opinion, even if you don't agree.

FIG 8.6 Let's negotiate!

Table 8.3 gives some examples of common work-related face-to-face communication scenarios in the workplace.

Channel 2: Telephone and radio

Ask any older tradesperson how it was in the days before mobile phones. They may speak of today's convenience of being able to simply pull the phone from their pocket and order their material or deal with client inquiries. They may also complain about having their work interrupted by endless incoming calls or the annoyance of drop-outs and disconnections. Whatever they may say, the speed with which we can communicate and source information has led to a new level of ease and responsiveness in business.

Like communicating face-to-face, communication by telephone and radio have the benefit of immediacy. Just imagine you are working on the same task but on different floors of a multi-storey worksite. Radios can be purchased inexpensively and allow work teams to stay in touch as they complete complex tasks. Photos can easily be sent by MMS to provide a greater level of information to

TABLE 8.3 Face-to-face communication scenarios

Form of face-to-face communication	Example
One-on-one	Discussing job details with your workmate
Toolbox talk	On-site occupational health and safety (OHS) briefing
Staff meeting	A meeting to brief staff on company policy or decisions
Site meeting	A meeting to discuss work progress, particular site conditions and issues. Could be all contractors or just your team
Union meeting	A meeting of union members to discuss award conditions, site conditions, OHS issues, pay claim, proposal to strike
Committee meetings	Meeting of a representative group appointed to manage issues such as OHS or social groups
Team meeting	A meeting of a small group to discuss special issues (e.g. the drainage team or the roof plumbing team)

workmates, suppliers and clients. This avoids confusion and consequent disruption of progress on the worksite or the delivery of incorrect goods.

Telephone conference calls can help overcome the potentially high cost of face-to-face meetings by providing links between people who could be anywhere on the planet. The cost of airfares, accommodation and lost time can easily be overcome simply by coming together over a pre-arranged multi-party telephone conference call. Video conferencing by web-cam takes this concept one step further and adds the possibility of gaining visual feedback from conference participants.

Getting our message across by phone or radio can present some difficulties:

- We may have trouble expressing exactly what we have to say and so find it hard to convince others.
- Verbal messages are easily misunderstood and therefore greater care is required in the way they are conveyed.
- Radio-based communication can be interrupted by network issues and radio interference. Reception is often weak within buildings and around structures.
- Unless we are doing a video call, we miss the opportunity to observe body language cues.

Table 8.4 gives some hints for better radio and telephone communication.

FIG 8.7 What's he doing wrong? See Table 8.4 to find the answer

Channel 3: Written messages

Written messages form an integral part of communication in the plumbing industry. Every day we encounter masses of technical information, instructions, directions and warnings that we need to be able to understand, interpret, apply and communicate. Some examples of written communication are:

- instruction manuals
- product information brochures
- plans and specifications
- hydraulic drawings

TABLE 8.4 Hints for more effective radio and telephone communication

Communication channel	Recommended technique
Radio	• Clearly identify yourself at the beginning of the call. • Clearly state the name of the person you are calling. • Speak slowly and clearly *across* the microphone, not into it. • Think about what you have to say before you press the send button • Request confirmation that the message has been received and understood. • Acknowledge that you have received and understood the message. • Be ready to write down any important information. • Don't block up the airwaves with unnecessary talk.
Telephone	At all times be aware of your tone! ***Making calls*** • Clearly identify yourself at the beginning of the call. • Clearly state the purpose of your call and your required outcomes. • Answer quickly and speak in a clear, friendly manner. • Be in charge of the call, ask questions—why, when, where, who. • Write down the details of your call and what has been agreed on. • Provide your name and number to the other party. • End the call politely and thank the other party for their time. ***Receiving calls*** • Answer incoming calls promptly using a warm, friendly manner. Identify yourself and your company. Remember that this conversation may be the other person's only contact with your company. First impressions are important! • Take notes as you go. This will help you to recall important points of the conversation. • Listen carefully and, if appropriate, repeat important points to the caller for clarification. • Give an indication of the action you will take in response to the call. • Finish the call on a pleasant note.

Source: Adapted from Rossignol 1999

- bills of quantities
- invoices
- delivery dockets
- contract documents
- diaries
- material safety data sheets (MSDS)
- safe work method statements (SWMS)
- standard operating procedures (SOPs)
- letters
- memorandums (memos)
- reports
- building site signs
- safety signs
- machinery operation manuals.

Written messages form an important part of workplace communication as they can provide a permanent and reproducible record of many types of information. They also provide an opportunity for users to carefully examine the information presented, and to duplicate and share it among the organisation. Due to the written form of the communication the writer needs to ensure that the information is presented in a manner suitable for the intended reader and that the message is clear and easily understood. Unless the reader is able to seek more information from the writer, the written message must be able to be understood without the need for further explanation. Written messages are often used as evidence in legal proceedings, so writers need to be conscious of the fact that whatever is written may be held and produced later as a record of events, communications or transactions.

Written messages and symbols also appear in the form of signs. Signs in the building industry are used to alert workers to a range of different messages, but most often are related to OHS.

FIG 8.8 The plumber's shopping list

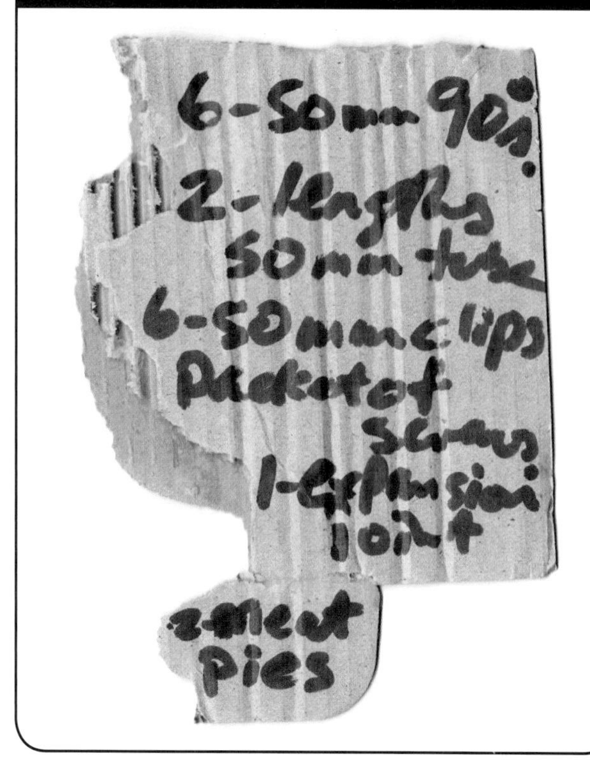

Channel 4: Graphic messages (signs and diagrams)

It is said that 'a picture is worth a thousand words'. If you think about how many times you have had something explained to you through a sketch or picture, you can see how this is particularly true in the plumbing industry.

Pictures provide a number of advantages: they can overcome cultural and language barriers; convey complex principles in one diagram; and can be quickly and easily understood. The only disadvantage of communication through pictures is that the opportunity for giving feedback is limited.

Safety signs

The picture most commonly seen in the plumbing industry is the safety sign. Table 8.5 shows the three main types of safety signs that are used in the workplace.

TABLE 8.5 Safety signs in the workplace

Type	Example
Picture signs. These use symbols and pictures.	
Text-based signs	
Picture and text signs. These sometimes include a short message. These signs reach the maximum number of people in the workplace. They are easily understood by people from non-English speaking backgrounds or with low reading ability.	

Colour and shape

Signage is also identified by colour and shape. These signs fall into seven categories, as shown in Table 8.6.

TABLE 8.6 Workplace signage

Type	Appearance	Message	Example
Prohibition signs	Black symbol, white background, red circle with diagonal line through it	Inform of activities you must not do	Authorised personnel only
Mandatory signs	Solid blue circle, white symbol, no border	Indicate what items of personal safety equipment must be worn	dust mask must be worn
Restriction signs	Circular red border, no cross bar, white background	Indicate a limitation on use or activity in an area	40
Hazard warning signs	Yellow triangle, black border, black symbol	Warn of danger or risk to personal health	CONTROLLED AREA AUTHORISED ENTRY ONLY
Danger hazard warning signs	White rectangular background with the word 'DANGER' in white on a red background, with a black border and black text	Warn of a hazard or condition that is potentially life threatening	CAUTION THIS EQUIPMENT STARTS AND STOPS AUTOMATICALLY
Emergency information signs	Solid green background with white symbol and/or text	Indicate the location of emergency equipment	
Fire signs	Solid red background with white symbols and text	Indicate the location of fire alarms and fire-fighting equipment	FIRE EXTINGUISHER

Source: Adapted from NSW Education and Training, 2008.

SUMMING UP

In this chapter we have looked briefly at the broad area of communication and learn that it is often the factor that determines how much we enjoy our life experiences. Whether we are looking for greater success in our personal lives or seeking to become a more effective trades or business person, working on our communication skills can only bring worthwhile results. This chapter will have provided you with some tools to help you along that path.

FOR STUDENT RESEARCH

Book

K Le Rossignol (1999) *Communication Skills in the Workplace*, Eastern House, Croydon, Vic

Websites

Non-verbal communication

http://helpguide.org/mental/eq6_non-verbal_communication.htm

Industrial signage

http://www.hsc.csu.edu.au/ind_tech/ind_study/3376/signage.htm

References

Goldhaber, GM (1993). *Organizational Communication*, Dubuque: Wm. C. Brown Communications, Inc.

NSW Education and Training. (2008). *Industrial Technology*, http://www.hsc.csu.edu.au/ind_tech/ind_study/3376/signage.htm, accessed 24 May 2011.

Rossignol, KLe (1999). *Communication Skills in the Workplace* Croydon: Eastern House Croydon, Vic

Segal, J, Smith, M & Jaffe, J (2011). *Nonverbal Communication*, http://helpguide.org/mental/eq6_nonverbal_communication.htm, accessed 24 May 2011.

PLUMBER PROFILE 8.1

TRAVIS MURPHY

Online LearningCentre

Job Title: Plumber with Cooke & Dowsett Pty Ltd

Originally from Victoria, Travis has been in the plumbing industry for seven years. He spent two years in Broome managing the non-for-profit plumbing organisation NUDJ, of which Cooke & Dowsett is a partner. NUDJ is involved in an Indigenous apprenticeship initiative working on community infrastructure. Travis was heavily involved in training and providing plumbing services to Aboriginal communities in the top end with this organisation.

How did you get involved in doing non-for-profit plumbing work in Broome?

I had travelled to Broome, through Cooke & Dowsett as part of NUDJ. I was there for 2 years. In Broome I worked with Indigenous communities. The culture difference was huge. It was totally different from what I was used to doing. I was doing high-rise buildings and hospitals in Victoria and then went to the dessert, which was really hard. We were doing jobs four or five hours away from the nearest town, and places with only 40 to 50 people.

What did your role entail?

I was managing the company so making sure everyone was happy, had the right tools, managing jobs, etc. There was plenty of travel as we were going to remote locations all the time.

What was one of your most challenging jobs there?

Bringing the sanitation conditions up to speed. There was a lot of rust throughout all the houses, mosquitoes, bacteria in the water. There is some work being done but a lot more needs to be improved. Funding is the challenging thing—there is a lot of work needed but most of these communities don't have enough money to pay for it. We are a non-for-profit company so we train up the kids there in plumbing as well.

What has been your most memorable experience as a plumber?

Definitely meeting all the people up in Broome! I really enjoyed the experience. Plumbing can take you anywhere—all over the world.

What are you working on now?

I was up there for two years. I'm back in Victoria working for Cooke & Dowsett on a desalination plant on Phillip Island, but I'm going to go back as the lifestyle is so good up there. I love the places plumbing can take you; I may want to travel overseas in the future too.

CHAPTER **NINE**

A safe and healthy workplace

LEARNING OBJECTIVES

In this chapter you will learn about:

- **9.1** **the *Occupational Health and Safety Act 2000***
- **9.2** **the Occupational Health and Safety Regulation 2001**
- **9.3** **how OHS operates in the workplace**
- **9.4** **duty of care responsibilities**
- **9.5** **hazards and the risk assessment process**
- **9.6** **how to deal with common hazards in the plumbing industry.**

INTRODUCTION

Stop for a minute and think back to the last time you were on a building site or maybe walking past one. What did you notice about it? What did it sound like? What was going on there? Were there many people working on-site? If so, what were they wearing? Were there many signs on the building site and, if so, what were they about?

If you have a think about the last two questions then you will probably start to get an idea of how much occupational health and safety (OHS) is a part of life in the building industry. What do we see on the building site? We see signs alerting us to various hazards or instructing us to don our personal protective equipment (PPE); we see tradespeople wearing bright fluorescent shirts and vests and walking around in helmets, safety glasses and steel capped boots. An entire industry sector is now founded on meeting the PPE needs of an increasingly safety conscious workforce here in Australia.

But let's take it back to a personal level and consider why we all go to work. Most of us go to work to meet our basic needs of food, shelter and clothing, but also to achieve the goals we have set ourselves. To be injured at work would prove to be at the least an inconvenience and may even involve the serious injury or death of yourself or others. Your plans to save money and go into business for yourself or to buy a new surfboard may now have changed because of a debilitating injury. But such situations can be avoided if we practise sound OHS principles.

The size of the problem

How big a problem is workplace injury? Australian Bureau of Statistics records show that in 2006 10.8 million people worked and of these 689 500 or 6.4 per cent of these people suffered workplace injuries. Of this number, the largest injury group was men between the ages of 20 and 24. That large number may be due to the fact that men are often engaged in more hazardous employment than women. As Figure 9.1 shows, plumbing apprentices are included in the group with the highest incidence of injury.

FIG 9.1 Demographic split of Australian workplace injuries 2006

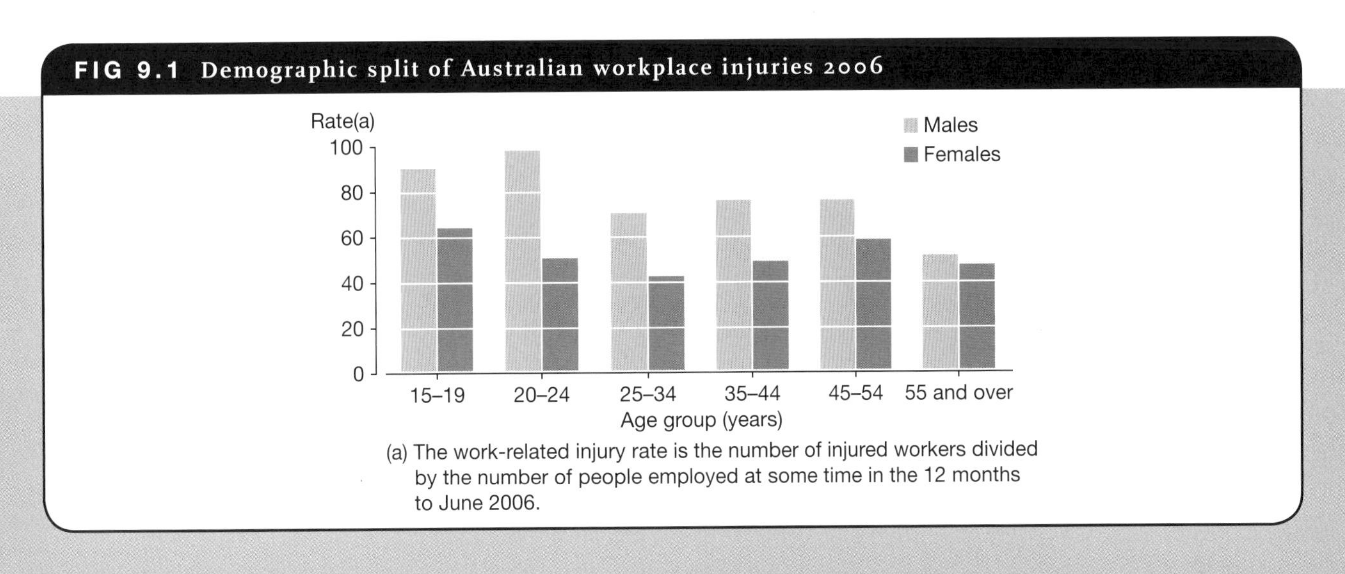

(a) The work-related injury rate is the number of injured workers divided by the number of people employed at some time in the 12 months to June 2006.

In this chapter you will gain an overview of the critically important area of Occupational Health and Safety. Together we will examine how OHS applies to you, your employer and all those you work with, and how your workplace can be managed in a safe and efficient manner. We will also identify some hazards commonly found in the plumber's workplace and how these can be managed in a way that ensures a safe, comfortable and profitable work environment for all.

FIG 9.2 Keep it safe!

OHS ACT 2000 AND OHS REGULATION 2001

What is OHS all about and who makes the regulations? In 2008 the decision was made by the Australian Government to form a body that would have the responsibility of overseeing OHS on a national level. SafeWork Australia came into being in 2009 and is now working in partnership with the government, employers and unions in many areas including the following:

- improvement of OHS outcomes and workers' compensation arrangements in Australia
- formulation of model OHS legislation for adoption as a law of the Commonwealth, the States, the Territories, workers and employers.

Following the formation of SafeWork Australia, the decision was made to formulate a federal OHS Regulation. At the time of writing, each state or territory in Australia has its own OHS regulations which can prove difficult for businesses that have operations in more than one state and/or territory. The aim of the new federal legislation, due for introduction in 2012, is to implement a uniform set of regulations across Australia, thereby making it easier for business to be OHS compliant.

The NSW context

For the purposes of this chapter we will look at OHS regulation in NSW. WorkCover NSW is responsible for the implementation of OHS regulations across the state and administers all areas relating to workplace safety. The following is an extract from the NSW *Occupational Health and Safety Act 2000* and outlines the purpose of the regulation and of the principles of OHS in general: The objects of this Act are:

- to secure and promote the health, safety and welfare of people at work
- to protect people at a place of work against risks to health or safety arising out from the activities of persons at work
- to promote a safe and healthy work environment for people at work that protects them from injury and illness and that is adapted to their physiological and psychological needs
- to provide for consultation and co-operation between employers and employees in achieving the objects of this Act
- to ensure that risks to health and safety at a place of work are identified, assessed and eliminated or controlled
- to develop and promote community awareness of OHS issues
- to provide a legislative framework that allows for progressively higher standards of occupational health and safety to take account of changes in technology and work practices
- to deal with the impact of particular classes or types of dangerous goods and plant at, and beyond, places of work.

WorkCover NSW provides numerous resources for businesses, employers and employees. You can access this information through their website, www.workcover.nsw.gov.au.

Codes of practice

Codes of Practice are produced by WorkCover as a guide to achieving the standard set out by the OHS Act and the OHS Regulation. They have been formulated in consultation with industry, employers, employees, government representatives and special interest groups to produce industry-specific safe work solutions.

Although the Codes of Practice are not legal documents and cannot be used as evidence in a court of law, they may be used to demonstrate failure to observe industry standard practice. Persons cannot be prosecuted for not observing Code of Practice recommendations.

Plumbers will find a wealth of useful information in the Codes of Practice. Some examples of relevant codes are:

- the storage and handling of dangerous goods
- risk assessment
- cutting and drilling concrete and masonry products
- excavation work
- safe work on roofs, part 1 and 2
- electrical practices for construction work.

A full list of codes of practice can be accessed from the Workcover NSW website, www.workcover.nsw.gov.au/formspublications.

Rights and responsibilities under the OHS Act

Workplace safety is the responsibility of all those involved in workplace activities. These parties include employers, employees and those organising activities on-site, such as building site foremen or construction managers.

DUTY OF CARE RESPONSIBILITIES: YOURS AND YOUR EMPLOYER'S

The OHS Act is based on the principle of duty of care and covers all workplaces in NSW.

What is duty of care?

The principle of duty of care relates to the fact that workplace safety is everyone's responsibility. If I see my workmate doing something potentially dangerous to himself or someone else, it becomes my responsibility to either alert him to the fact, or to inform someone in authority of the possible danger. It is not acceptable for me to simply walk away and say that it's someone else's problem. The fact is that if there is an accident, I may be held liable for not taking positive action to remedy the situation.

The OHS Act sets out the duty of care responsibilities of employers and employees.

Employers' duty of care

An employer must ensure the health, safety and welfare at work of all the employees of the employer. That duty extends (without limitation) to:

- ensuring that any premises controlled by the employer where the employees work (and the means of access to or exit from the premises) are safe and without risks to health
- ensuring that any plant or substance provided for use by the employees at work is safe and without risks to health when properly used
- ensuring that systems of work and the working environment of the employees are safe and without risks to health
- providing such information, instruction, training and supervision as may be necessary to ensure the employees' health and safety at work
- providing adequate facilities for the welfare of the employees at work.

In addition to the preceeding points, an employer must ensure that people on site (other than the employees of the employer) are not exposed to risks to their health or safety arising from the conduct of the employer while they are at the employer's place of work.

Employees' duty of care

An employee, while at work, must:

- take reasonable care for the health and safety of people who are at the employee's place of work and who may be affected by the employee's acts or omissions at work
- cooperate with his or her employer or other person so far as is necessary to enable compliance with any requirement under this Act or the regulations that are imposed in the interest of health, safety and welfare on the employer or any other person
- not intentionally or recklessly, interfere with or misuse anything provided in the interests of health, safety and welfare under occupational health and safety legislation.

Offences and penalties

Non-compliance with OHS regulations can lead to the application of penalties by WorkCover. Penalties and offences under the Act can apply to employers and employees when breaches of the OHS regulations occur.

The Act applies penalties for offences according to 'penalty units'. Each penalty unit is valued at $110 (as at 2006) and can be applied at the maximum rates shown in Table 9.1.

TABLE 9.1 Penalties under the OHS Act 2000

Party involved	Maximum penalty units	Fine
Corporation	5000	$550 000
Individual	500	$55 000
Repeat offender (corporation)	An additional 2500	$275 000
Repeat offender (individual)	An additional 25	$27 500

Source: http://sydney.edu.au/ohs/ohs_manual/legislation.shtml

The OHS Act also allows for the application of penalties in addition to the above.

General induction and OHS training

Under the OHS Act and the OHS Regulation, employers must provide training for all workers intending to work on a building site. This training comes in three forms:

- general induction
- site induction
- task-specific induction.

FIG 9.3 Sample OHS safety induction card (White card)

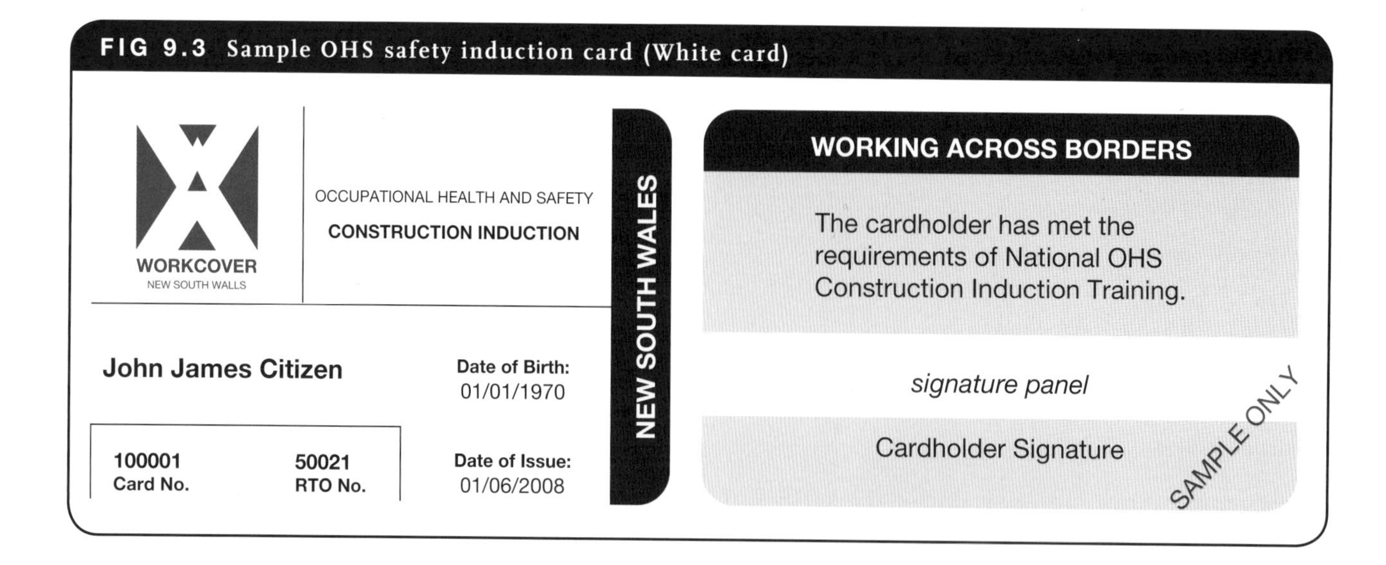

General induction

The purpose of this training is to provide workers with a thorough understanding of all elements relating to workplace safety, encompassed by:

- the rights and responsibilities under OHS law
- common hazards and risks in the construction industry
- basic risk management principles
- the standard of behaviour expected of workers on construction sites.

All of these elements are found in the unit 'Work safely in the construction industry' that is part of the BCG03 General Construction Training Package.

On completion of general induction training, the trainee will be presented with a statement of attainment that details the units completed. The trainee will also be presented with a general induction card similar to the one shown in Figure 9.3 can be used for proof of training:

Site induction

This training provides site-specific information and instruction to anyone engaged on a particular construction site with knowledge of the contractor's rules and procedures for site safety, emergency management, the supervisory and reporting arrangements, and other site-specific issues.

Task-specific induction

This training provides information and instruction to anyone undertaking a particular construction activity of the risk factors and control measures relating to that task.

HAZARDS AND THE RISK ASSESSMENT PROCESS

What is a hazard? The OHS Regulation defines a hazard as:

> Anything (including work practices or procedures) that has the potential to harm the health or safety of a person.

Hazards can be described as either an acute hazard or a chronic hazard:

An acute hazard is one that will cause sickness or injury after a short exposure, for example, burns from a newly-welded joint on copper pipe.

A chronic hazard is one that will cause sickness or injury after exposure for a longer period of time, for example industrial deafness or lead poisoning.

Classifications of common building site hazards

The wide range of workplace hazards encountered by plumbers can be classified into six groups as shown in Table 9.2.

TABLE 9.2 Hazard categories

Hazard type	Examples
Biological	Bacteria, viruses, insects, plants, birds, animals and humans, e.g. contact with effluent
Chemical	Dusts, strong acids and alkalis in solid, liquid or gaseous form, e.g. drain cleaners, soldering fluxes
Ergonomic	Repetitive movements, incorrect stance when working, e.g. digging or lifting
Physical	Radiation, magnetic fields, pressure extremes, noise, heat or cold, e.g. excessive sun exposure, working with noisy equipment, cold or windy conditions
Psychosocial	Stress due to excessive workload, violence, workplace bullying
Safety	Slips/trips/falls, injury due to ineffective or missing machine guards, failure of tools or equipment, incorrect operation of machinery, negligent or careless work practices

Source: Canadian Centre for Occupational Health and Safety *Hazard and Risk*, November 2009

FIG 9.4 Homer at it again!

THE HAZARD IDENTIFICATION AND CONTROL PROCESS

To ensure the safety of workers in the workplace we need to be able to identify all hazards and formulate a strategy to either eliminate hazards or reduce them to an acceptable level. The first step in this process is to carry out a risk assessment.

Through this process a ranking is given to each hazard on the basis of:

1 the consequences to worker safety if an accident involving this hazard occurred
2 the probability of this occurring.

Risk assessment form

A risk assessment form is used to standardise the process. By using a risk assessment matrix we can assign a risk level to each hazard we identify. Figure 9.5 presents an example of a risk assessment matrix. Table 9.3 provides the key to this matrix.

FIG 9.5 Sample risk assessment matrix

Likelihood	Consequences				
	Insignificant	Minor	Moderate	Major	Severe
Almost certain	M	H	H	E	E
Likely	M	M	H	H	E
Possible	L	M	M	H	E
Unlikely	L	M	M	M	H
Rare	L	L	M	M	H

Safe work method statement

The safe work method statement form is used to formally record all the details of a proposed workplace activity and should at the very least include the following items:

- a description of the work to be undertaken
- foreseeable hazards associated with the work
- the step-by-step sequence in doing the work
- the safety controls that will be used to minimise these hazards
- all precautions to be taken to protect health and safety
- identification of all health and safety law, standards or codes applicable to the work.

TABLE 9.3 Sample risk assessment matrix key

Rating risk level	E	Extreme risk—death or permanent disability
	H	High risk—long term or serious illness
	M	Moderate risk—lost time injury
	L	Low risk—first-aid treatment
Likelihood	A	Almost certain—expected in most circumstances
	B	Likely—will probably occur in most circumstances
	C	Possible—could occur at some time
	D	Unlikely—not expected to occur
	E	Rare—exceptional circumstances only
Consequences:	5	Severe—would stop achievement of functional goals/objectives
	4	Major—would threaten functional goals/objectives
	3	Moderate—necessitating significant adjustment to overall function
	2	Minor—would threaten an element of the function
	1	Negligible—lower consequence

Source: Department of the Prime Minister and Cabinet, *Guidelines for Cabinet Submissions and New Policy Proposals*, November 2010

For major works, the following details should be added:

- the names and qualifications of those who will supervise the work and inspect and approve it
- work area, work methods, protective measures, plant, equipment and power tools
- description of what training is given to people doing the work
- identification of plant and equipment needed on-site to do the job (e.g. ladders, scaffolds electrical leads, welding machines)
- details of the inspection and maintenance checks that will be, or have been, carried out on the equipment listed.

The safe work method statement form allows work tasks to be listed along with their associated hazards, their hazard levels and proposed risk mitigation levels. Through the use of the safe work method statement, contractors and workers are able to demonstrate that they have thoroughly thought through all aspects of their proposed project and have planned for all likely eventualities. Table 9.4 presents a safe work method statement giving an example of how the hazards on a roadway excavation job could be dealt with.

As an apprentice on a work site you probably won't be responsible for writing a safe work method statement but if you do it should be with the help of your supervisor. You should, however, be aware of what safe work method statements mean, how they are put together and what impact they have on your work and the people you work with.

The people who need to complete a work method statement are those who:

- build or construct anything on-site
- use plant or machinery (e.g. forklift, boomlift, scissor lift, crane or tilt tray)
- manually handles weights over 25 kg
- perform work requiring welding, grinding or other 'hot work'
- use equipment that makes excessive noise (e.g. jackhammers or compressors)
- carry out work in a confined space
- any job that requires dangerous substances or chemicals, in which case a MSDS (materials safety data sheet) must also be supplied
- carry out a task that could possibly constitute a risk.

TABLE 9.4 Safe work method statement sample

Safe work method statement			
Company name & details:	**Statement prepared by:** **Position:** **Date:**		
Description of job:	**Signature:**		
Work site:	**Commencement date:**		
Critical steps in this job	**Potential hazard**	**Hazard ranking (from matrix)**	**Safety controls**
Excavation of roadway for pipe-laying	Collapse of road surface	E (extreme)	Use alternative methods i.e. bore under roadway to reduce excavation, (preferable) **OR** Install shoring system to support roadway during excavation work
	Injury to pedestrians due to fall into trenches	M (moderate)	Install and maintain traffic control and barricades at all times. Redirect pedestrian traffic around work area
	Injury to workers from noisy machinery	M (moderate)	Wherever possible, use silenced or low noise emission equipment (preferred) **OR** Provide hearing protection to workers and instruct on appropriate use. Supervise use through period of job
Contractor staff/personnel details—all staff to sign after reading the statement			
Name	**Position**	**Signature**	
OHS Officer use only			
Date received		**Event name**	
Approved by		**Event date**	
Signature		**MSDS required/ supplied**	

DEALING WITH COMMON HAZARDS IN THE PLUMBING INDUSTRY

According to the OHS Regulation there is a hierarchy of control measures that can be applied to hazards. This means that there is a preferred order in which we apply the various means to reduce or eliminate a hazard, and is clearly set out in the Regulation:

> For the purposes of this Regulation, an obligation to *control* a risk to health or safety (in any case in which the elimination of the risk is not reasonably practicable) is an obligation to take the following measures (in the order specified) to minimise the risk to the lowest level reasonably practicable:
>
> (a) Firstly, substituting the hazard giving rise to the risk with a hazard that gives rise to a lesser risk
> (b) Secondly, isolating the hazard from the person put at risk
> (c) Thirdly, minimising the risk by engineering means
> (d) Fourthly, minimising the risk by administrative means (for example, by adopting safe working practices or providing appropriate training, instruction or information)
> (e) Fifthly, using personal protective equipment

An example of how this hierarchy works is contained in the excavation example contained in Table 9.4. You can see that two solutions have been given for the risk of 'Collapse of road surface'. The first (preferred) solution involves doing the job using a technique that greatly reduces the amount of excavation required and so essentially eliminates the risk. The second solution of using shoring reduces the hazard by preventing road collapse but does not eliminate the possibility of this occurring at some stage during the job. If this technique was to be used, personnel should be kept away from the excavation whenever the excavation was unsupported, and/or during installation and removal of the shoring.

The least desirable hazard reduction technique of all is the use of personal protective equipment or PPE. PPE should only be used when the hazard cannot be reduced or eliminated by other means, but only when workers' safety is not compromised. Table 9.4 uses the example of personnel working near noisy machinery. This hazard has been given a rating of 'moderate' from the risk assessment matrix due to the fact that hearing damage usually occurs over an extended period of time. Two solutions have been given, the first, and most desirable, is the use of low noise emission plant and the second is the use of ear muffs or ear plugs. Once again, it is preferable to eliminate the hazard rather than to find a method of dealing with it.

Some common hazards

This chapter will conclude with a brief look at some of the hazards encountered by plumbers in their day-to-day work and how these hazards can be dealt with. In many cases, a number of control measures may need to be applied to ensure worker safety on a job site. On your job site it is up to you to keep a watch out and to identify hazards that may jeopardise the safety of you and your workmates. Table 9.5 provides some examples of hazards you should look out for.

TABLE 9.5 Hazards in the workplace

Job task	Associated hazard	Control measure
Clearing a choked sewer	Infections	Change your work practice to limit exposure to risk. Use gloves, face mask and overalls to reduce exposure to contagions. Keep disinfectant hand wash in your truck and use it at the completion of a job and before eating or smoking. Wear barrier cream to prevent exposure of skin to contagions. Vaccinate against hepatitis
	Cuts from broken earthenware pipe.	Change your work practice to find alternate ways to do the job. Wear appropriate PPE (gloves, overalls, boots) to eliminate risk of cuts
	Injury from drain cleaning equipment	Use the equipment in strict accordance with manufacturers' instructions. Avoid working when tired or when concentration is low
	Injury from heavy lifting	Avoid situations where this could occur by changing work practices. If heavy lifting is required (e.g. man-hole covers or drain cleaners), do this using correct techniques e.g. correct posture, obtaining help with the lift, using lifting equipment
	Burns from drain cleaning chemicals	Carefully follow manufacturers' instructions for product use. Consult Material Safety Data Sheet (MSDS) for handling instructions and emergency procedures. Wear appropriate PPE (rubber gloves, safety glasses, respirator) whenever handling these materials
Roughing in for new pipe-work in brickwork	Injury to lungs from dust inhalation	Use dust-free processes wherever possible (e.g. wet saws, dust extractors). Use extraction fans to remove dust from area. Wear close-fitting, good quality, standards-approved respirators (not the paper type)

(continued)

TABLE 9.5 Hazards in the workplace (*continued*)

Job task	Associated hazard	Control measure
	Damage to eyes from flying particles	If dust-free methods cannot be used ensure all equipment is fitted with appropriate guards. Wear good quality, standards-approved eye protection
	Damage to hearing from angle grinders or chasers	If there is no quiet alternative, use good quality standards-approved hearing protection, preferably well fitting ear-muffs.
Laying a new sewer line	Death or injury from cave-in	If possible, use alternative methods that eliminate excavation (e.g. directional boring). Observe requirements of WorkCover Code of Practice: Excavation (e.g. maximum unsupported trench excavation depth of 1.5 m). Use appropriate shoring over this depth and if possible carry out work from outside the trench. Load all spoil at least a metre away from edge of trench to avoid slippage due to surcharge loads
	Back injury from manual digging	Where possible use machinery for all excavation. Build 'core strength' to increase resistance to injury. Use correct digging techniques to avoid undue stain on the back and other parts of the body. Use a long handled shovel to increase leverage and to reduce the need to bend the back.
	Sunburn	Wear a wide brimmed hat (maybe incorporated with a safety helmet) and appropriate level sunblock. Minimize exposure time, in particular from 11 am to 2 pm
	Head injury from falling objects	Remove all loose objects from the trench area and ensure there are no unnecessary protrusions from the trench wall. Wear an approved safety helmet
	Injuries to feet	Use appropriate work practices to avoid the risk of injury. Wear approved reinforced-toe safety boots
Disconnecting a water meter or altering pipework	Electrocution	After notifying building occupants, turn off power and affix danger label to switch. Use bonding straps across the section you are working on to ensure continuous earthing. If in doubt, call a licensed electrician

FIG 9.6 A common mandatory PPE sign

WHERE TO NOW?

In this chapter we have had a brief look at the nature of occupational health and safety and how it has an impact upon you as a worker in the Australian plumbing industry. Hopefully what you have learned here will reinforce the training you have already received in the workplace and provide a platform for greater understanding and awareness of OHS issues and their application. Your greatest tool as a tradesperson working in an ever-changing industry is your ability to locate and apply new information as it becomes available. This is one of the keys to maintaining a safe, comfortable and productive work site, where you are free to achieve your goals without the constant fear of injury.

FOR STUDENT RESEARCH

- NSW *Occupational Health and Safety Act 2000*
- OHS Regulation 2001
- WorkCover Codes of Practice:
  - The storage and handling of dangerous goods
  - Risk assessment
  - Cutting and drilling concrete and masonry products
  - Excavation work
  - Safe work on roofs, Parts 1 and 2
  - Electrical practices for construction work.

References

Canadian Centre for Occupational Health and Safety (2009). *Hazard and Risk*, http://www.ccohs.ca/oshanswers/hsprograms/hazard_risk.html, accessed 24 May 2011.

Department of the Prime Minister and Cabinet (2010). *Guidelines for Cabinet Submissions and New Policy Proposals*, http://www.dpmc.gov.au/implementation/policy.cfm, accessed 24 May 2011.

PLUMBER PROFILE 9.1

ALAN GRASSET

Job Title: Retired State Manager, Plumbing Services, NSW TAFE.

Alan has been associated with the plumbing industry for 50 years—27 of which were served working in TAFE. He has taught in seven different colleges both in the city and country and has served in every position in the plumbing educational sector of TAFE—at the time of his retirement he was responsible for the training programs in that area. Alan was heavily involved in WorldSkills and still maintains an active interest in the organisation. With a strong belief in the role of education, Alan has completed an array of certificates and diplomas throughout his career, including the Plumbing Trade and Post Trade Certificates, Health and Building Inspectors Certificate, LP Gas, Advanced Gas, Thermostatic mixing valves, Diploma of teaching, a degree in Education and also studies in Fine Arts.

As a result of his knowledge, experience and dedication Alan is now a Life Fellow of the Institute of Plumbing after serving as President for eight years, Patron of the Guild of Health Science Teachers, and an Honorary Life member of the Hydraulic Services Consultants Association. He also was a Teacher Member of the Master Plumber Association for 22 years and was presented by that organisation with an award for dedication to the plumbing industry. Alan was awarded an Order of Australia medal in 2009 for his services to vocational education and training, particularly in the plumbing industry.

How did you first start out in the plumbing industry?

I started with a pre-apprenticeship in 1960. I was going to continue at school but my father suggested I start a trade. I was then given a job in '61 as an apprentice. The training now is very different from the 1960s. I worked with some wonderful people. The boss and the TAFE teachers always encouraged me, pushed me. It was a great experience and a chance to build a career.

How did your career progress from your apprenticeship?

In my last year—the fifth year in those days—the teacher I had was really inspiring and he suggested I carry on with my studies, which I did, and I started the health and building inspection course which took me seven long years, as I was a young man in my twenties and typically I had too many things going on just to concentrate a lot on studying!

What kind of work were you doing alongside your studies?

I was still working as a plumber throughout those years. I was doing a mixture of on-the-job construction work and I was in the office to relieve senior people, which involved things like estimating and design.

What direction did your career take upon completion of your health inspector course?

Towards the end of that period my boss had lined up a job to work as a junior health inspector. But I had met a fellow during my health inspection course who was a plumbing teacher. He said to me, 'You should apply', and another friend of mine who was there said to me, 'You would never make a teacher'. And to say that to me, you get the opposite reaction! I was fortunate enough to be given the job in 1970. I started at North Sydney TAFE and I worked with some marvellous men, one in particular who was my head teacher in those days. He pushed me and I suppose he saw a bit of a spark there hidden in my foolishness. I stayed at that college for five years.

What was happening with your education during this period?

I completed health inspection, did some other training as well in my own time, technical training to upgrade my own skills and knowledge, which is an ongoing thing. Most people can't accept that it's ongoing, you can't stay at the same level, you must improve, and the only way to do that is through experience and education. Each year of my working life I did further studies, it was necessary.

How did your teaching career progress after starting at TAFE North Sydney?

I was transferred to work at both Bathurst and Orange. Typical of the department in those days I was told Friday afternoon to start work at Bathurst on Monday morning. I stayed there for three years. At that time I used to work with Bob Puffett, and of course he is the author of the books, and I got to know him fairly well at that stage and over the years we have been close friends. He has pushed me into things I wouldn't have been involved in otherwise. I was then promoted to head teacher at Gymea and I stayed there for a few years, then transferred to Granville, and I was head teacher there. But the person in charge of the section was working as the assistant to the head of school. So I ran the section, which was in those days the largest plumbing section in Australia. It was a pretty hectic job. I was promoted to a senior head teacher's position at Miller TAFE and I was appealed against by a senior man in the department, he won the appeal. I was then sent me back to Gymea. The following year I was promoted to senior head teacher at Meadowback. Sent there, appealed against, and was beaten again by a more senior man. Anyhow, I went back to Granville.

How did you feel about all the appeals?

Everyone has the right in the department to appeal against the decision. It was just part of the system. When I went back to Granville I was given the opportunity to continue my teacher education. I had been teaching for ten years at that stage. I finished the Diploma of Teaching and was then offered a position to do a degree in Education.

How did you first get involved in WorldSkills?

When I was sent to the Head of Schools Unit in 1984, I was approached to undertake a task with the WorldSkills Australia Foundation. That was to design a project for regional competitions for plumbing. All of the skill areas in TAFE were having these competitions. So it was my job to design the project for plumbing. I became the first national and international project designer and judge. Japan was my first international experience representing Australia. Over my working life I have travelled to every state in Australia and many other countries with my involvement in these activities.

What's your most memorable international experience?

The Japanese one was extremely memorable. This is a classic story: I had been going to the competition each day and working with young people who are using their hands and you get dirty so I wore a pair of jeans and a t-shirt. I had been out with the Korean and English judge. I didn't get to find out we were meeting some VIP's at the competition the next day. So I turned up in the morning in jeans and a t-shirt and everyone else is in a suit with a tie. I asked, 'what's going on?' and they said, 'it's the Emperor and Empress of Japan being introduced to us today'. I was in the line and the Emperor shakes my hand and welcomes me to the country. Then I was introduced to the Empress and she looks me up and down and she said, 'You would have to be the Australian'!

Have you been in any difficult situations in your travels?

I was travelling with a colleague to Taiwan, we were coming from England and Germany where I had been working and we were asked to pay a $30 airport tax to get into the country. We didn't have the cash, only a few Deutschmarks and English pounds, back then there were no credit cards. So, I was held at machine-gunpoint by this security guard with his finger on the trigger while my colleague went to change travellers cheques. He had to go through customs and come back in to pay the fee. I said to him, 'Where have you been?' as it had been around an hour, and he said, 'It was hot! I had to go grab a beer!'

Which job did you find particularly challenging?

I was assistant to head of school for two years … at the end of the two years I was offered a job by WorldSkills to take on the technical coordination of the competitions for Australia and I thought this would be a different experience so I did this for two years. It was the most intense working situation I have ever undertaken. I would come home on a Wednesday evening with a bag of dirty clothes and pack new clothes and leave on Thursday morning. I worked all over Australia. The trip to Taiwan was a part of that. I met thousands of people. WorldSkills does amazing things to people, you just grow. It's hard work though.

What happened when you finishing working with WorldSkills?

I have never seen my career as a challenge, more a set of opportunities. I was in the right place at the right time all the time! When I was finished at WorldSkills I came back to TAFE and was appointed as a head of division. I took over as head of school in the interim period when TAFE decided to have a major change to the structure. I became the plumbing industry specialist with less of a role as a result. after a few years the structure was changed again and I was appointed as the state manager for plumbing services. Some time after that I was told that I had cancer and I retired. That was 13 years ago. I remained with WorldSkills though and have been involved in many different and interesting projects. There are so many wonderful experiences to look back on. I was very lucky.

With all your work, what did you do in your free time?

I enjoy cartooning, which stems from my involvement in painting, which started as a kid. I was fortunate to have some great art teachers who also encouraged me to 'have a go'. I've have had a lot of fun with the cartoons.

When you look back at your career, what would you say you are particularly proud of?

I was president of the Institute of Plumbing and when I retired they made me a life member which was great. Teachers back then gave Bob Puffett and me a hell of an honour by making us dual patrons of the Plumbing Teachers Guild. Life membership from the Hydraulic Consultants Association and the award from the Master Plumbers Association were also unexpected honours. A couple of years ago I was given the Order of Australia for work in the plumbing industry which I didn't expect and was out of the blue. It's not just me who was given the Order, it's all the people I have worked with. You're not one person, you are part of a team.

Do you have any advice for students?

I can't emphasise it too much; you should always have a go. What's the worst thing that can happen? By putting my hand up I filled my life with opportunities and experiences. If you don't put your hand up you just don't get those opportunities. They should continue their studies and keep learning. Our society can't survive without the plumbing industry so it is critical that plumbers have the best possible education and to understand their importance to the health and welfare of our society.

Glossary

acetone an inflammable and volatile liquid used as a solvent in acetylene cylinders to dissolve and stabilise acetylene under pressure

acetylene a highly combustible gas composed of carbon and hydrogen (C_2H_2) used as a fuel gas in oxyacetylene welding and cutting. When burned with oxygen in the correct proportions it produces a flame temperature of approximately 3000°C

act legislation passed by parliament to control a specific area of undertaking, e.g. Sewerage Act

alignment the setting in a straight line of a number of points, e.g. pipe lengths in a pipeline

alloy a mixture of two or more metals united by melting together, e.g. bronze as used in filler rod

annealing the process of gradually cooling a metal part after welding, or reheating it to make it soft enough for mechanical working. Annealing will relieve stresses in an existing metal or stresses that may be set up by welding operations

approved approved as directed by the relevant water, sewerage or drainage authority

arc voltage the voltage across a welding arc

aspect the direction something faces, e.g. a northerly aspect

atmospheric pressure the pressure exerted by the atmosphere at a given point, e.g. sea level

Australian Standards approved standard for materials, equipment, techniques or procedures as set down by the Standards Association of Australia

authority the appropriate body authorised by statute to exercise jurisdiction over the installation of plumbing, gasfitting, sewerage and drainage works

backfill material used for filling trenches and excavations after a pipeline has been laid

backfire a loud snapping or popping noise which occurs when a blowpipe flame suddenly or momentarily goes out during oxyacetylene welding

backflow a flow of water in the reverse direction to that intended

backhand welding (rightward, backward) welding with the blowpipe flame pointing in the direction opposite to that in which the weld progresses, that is, towards the finished portion of the weld. The opposite of "forehand welding"

backing strip material used for backing up a joint during the welding process to ensure a sound weld

backsight (BS) the first reading taken on a levelling staff and logged in the appropriate column in the field book after the initial setting up of the levelling instrument

base metal metal on which another metal is to be welded

bead front edge of a gutter, usually circular to provide stiffening to the gutter edge. *See also* weld bead

bedding material beneath and "cradling" a pipe or drain

bench mark (BM) stable reference points, the elevations of which have been accurately determined. Bench marks are taken as permanent reference points during levelling operations. Temporary bench marks (TBM) are used as interim marks

bevel to prepare metal so that the edge is at an angle to the surface; also the sloping edge so produced

blowpipe an instrument for bringing together and properly mixing a fuel gas with oxygen or air in such a way that the mixture, when ignited, will produce a controlled flame

branch the intersection of two pipes

branch drain any branch off a main property drain

branch pipe a discharge pipe to which two or more fixture traps at any one floor level are connected

branch vent a graded vent at any one floor level interconnecting two or more individual trap or group vents

braze welding a joining process that unlike brazing does not depend on capillary attraction. The parent metal is not melted but the joint design is similar to that which would be used if a fusion weld were made. The filler metal is generally a non-ferrous metal or alloy with a melting point above 500°C

brazing a joining process in which the molten filler metal is drawn by capillary action between two closely adjacent surfaces. The filler metal is a non-ferrous metal or alloy with a melting point above 500°C but lower man that of the metal being joined

BSP thread British Standard Pipe thread

building any building used as a workplace, residence, place of business, entertainment institution, place of human habitation or as a factory, which contains plumbing fixtures

butt weld a weld in which the two edges of metal to be united are butted together

capillary action the force of adhesion existing between a solid and a liquid in capillarity—occurs between two close-fitting smooth surfaces

capillary joint a joint in which the parts are united by the flow of filler metal by capillarity along the annular space between the outside of the tube and the inside of the fitting

carbon a non-metallic element

carburising flame a mixture of oxygen and acetylene gas in which mere is an excess of acetylene

catchment area the area of land from which run-off is taken to fill a dam or reservoir

caulking rendering a joint watertight by compressing the sealing material

ceiling joist a timber member fixed to wall plates to enable ceiling linings to be fixed

chalk line string or twine containing chalk dust, The line is drawn tight then plucked to produce a chalk line on the work surface

chase a recess cut into brickwork to allow for downpipes, flashings etc.

cladding wall covering, especially of moulded metal sheets

clout flat-headed galvanised nail

cold crack a crack occurring in weld metal, or in the heat-affected zone of a base metal, after cooling
colloid an apparently soluble substance which can be strained from a liquid
combined soil and waste pipe any pipe which receives the discharge from both the soil and waste fixtures and conveys them to the drain
common rafter the rafter which runs from the fascia to the ridge
compaction the process of consolidating backfill by mechanical or other means
compression joint a joint made by fittings in which the end of the pipe is held under compression:
1. *manipulative* a fitting in which the joint is made by compression of a ring or sleeve or part of the fitting on the outside wall of the tube
2. *non-manipulative* a fitting in which the direction of compression is through the axis of symmetry of the ring

concave fillet a fillet weld having a concave face
conduit 1. a pipe or channel usually of large size used for the conveyance of liquids
2. a pipe of large diameter through which a smaller pipe passes

cone that part of an oxyacetylene flame that is conical in shape and located at the end of the welding tip, heating tip or cutting tip. *See also* inner cone and outer envelope
connection that part of the drain which connects a property drain with the main sewer
contour line a line that indicates points of equal height, generally spaced at one metre intervals on construction drawings but may be greater distances apart on maps
convex fillet fillet weld having a convex face
cover glass a clear glass used to protect the filter in goggles from damage by spattering material during welding operations
crater a depression at the termination of a weld bead
cross-connection faulty plumbing design which may permit the entry of contaminated water into a drinking supply
cross-vent a vent which interconnects a stack with its relief vent
cylinder a portable steel or aluminium container for storing and transporting industrial gases such as compressed oxygen or dissolved acetylene

datum plane (or datum) a horizontal plane of known height to which elevations of different points can be referred. The mean sea level is the level surface generally adopted as a datum. The Australian Height Datum (AHD) is based on the mean sea level determined by the tide gauge readings around the Australian coast
depth of fusion a welding term expressing the distance that the fusion extends into the base metal
dewpoint the temperature at which water begins to form from vapour
dezincification the selective corrosion of copper alloys (brasses) in which the alloy loses its zinc component and is converted into a porous shell of copper which has poor mechanical strength
diameter straight line passing through the centre and from one side to the other of a circle or solid circular figure
dimensions measurements showing the size and extent of an object
discharge pipe any pipe for the conveyance of sewage or trade waste
domestic fixture a fixture or appliance which is designed for use in residential situations only. A fixture or appliance of this type may be installed in a non-residential building, but the waste which it discharges must be similar to that of its domestic counterpart
downpipe a pipe conveying rainwater from the gutter to the water table or stormwater drain
downpipe clip a special clip used to secure downpipes to a wall
downstream towards a lower level
drain the line of pipes, normally laid underground including all fittings for the conveyance of sewage and/or trade wastes to the sewer
duct an enclosed area to accommodate pipework

earth (or ground) lead *see* work lead
eaves overhanging edges of roofs
eaves lining a lining used to close in the bottom of an overhanging edge of a roof. *See* soffit
effective cover the width of roof sheet less the lap
electrode filler metal in the form of a wire or rod, either bare or covered, through which current is conducted between the electrode holder and the arc
electrode holder the metal arc electrode holder that secures the electrode during the welding process
electrode lead conductor between the source of current and the electrode holder
electrolysis corrosion corrosion produced by the contact of two dissimilar metals in the presence of an electrolyte
elevation level *see* reduced level (RL)
evaporation conversion of a liquid to the vapour state by the addition of latent heat
expansion joint a joint used in long runs of gutters, designed to allow for expansion and permit relative axial movement of the jointed parts

fall the difference in elevation between two given points (e.g. on a gutter, which allows water to flow to the downpipe)
filler rod a metal rod or wire used in welding operations which is melted and deposited in the weld and used to supply additional metal
fillet weld a weld made in a corner, as in a lap or tee joint
filter (lens) a filter, usually of glass, designed to protect the eyes from glare and harmful radiation during welding and cutting operations
fitting any component of a sanitary plumbing installation other than a fixture or pipe
fixture a device, the operation of which results in a discharge into the sanitary plumbing installation
fixture discharge pipe the pipe to which the single fixture trap is connected
flame (excess acetylene) *see* carburising flame
flame (excess oxygen) *see* oxidising flame
flanges raised flat-faced fittings attached to the ends of pipes and connected together by bolts
flashback the burning back of the flame into the blowpipe, or the ignition of an explosive mixture in one of the gas lines
flashing metal configuration used to prevent the ingress of water between two surfaces
floor waste grated inlet within a graded floor, intended to drain the floor
flux a chemical powder or paste used to dissolve oxides, clean weld metal of undesirable inclusions and prevent oxidation of a metal during welding or brazing operations
forehand welding (leftward, forward) welding operation in which the blowpipe flame points in the direction in which

the weld progresses, that is, towards the unfinished seam. The opposite of "backhand welding"

foresight (FS) the last recorded staff reading for a levelling position. It is logged in the appropriate foresight column in the book before moving the levelling instrument to a new position

fusible plug *see* safety device

fusion the melting together of a filler metal and a base metal resulting in coalescence (fusion) of the metals

fusion welding welding in which the metals to be joined are melted and completely fused together without pressure, and in which the filler rod, if used, is of similar composition to the parent metal

gable triangular upper section of a wall at the end of a ridged roof

galvanising a zinc coating used on steel to prevent rusting

girt a steel purlin or structural member

grade the angle of inclination expressed as the ratio of unit rise to horizontal distance or as a percentage, i.e. 1:50 or 2 per cent

graded pipe a pipe or drain installed at a grade

graphics the art of making drawings, especially in mechanics, in accordance with mathematical rules

graphite a type of carbon in mineral form, used as a lubricant

grating a framework of metal strips fitted over the inlet of waste outlets and gully traps, to prevent the ingress of large solids

ground the surface of earth, soil or rock which conforms to the established finished grade at a specific location after all excavations have been backfilled and all surface treatment completed

ground lead *see* work lead

ground water water occurring in the subsoil

gutter an open conduit for the collection and dissipation of rainwater from roofs

hard facing the process of covering areas of metal subject to heavy wear with wear-resistant metal facing by welding

hatching method of shading with multiple lines; often used to show a cross-section through a solid object

head vent the vertical pipe which is the continuation of a drain at its upper end

header vent the vent interconnecting the tops of two or more relief vents or stack vents

heat a form of energy capable of performing work

heat-affected zone the region beneath a weld bead which has not melted, but whose mechanical properties have been altered by the heat of the welding process

heel the outside of a bend

height of collimation (HC) sometimes referred to as "height of instrument", is the imaginary line passing through the intersection of cross hairs and the optical centre of the object lens. This is the horizontal line to which all staff readings are taken. It is important to note that the collimation line is only horizontal when an instrument is in perfect adjustment and is set up and levelled

hose *see* tubing

hot crack a crack occurring in weld metal soon after the metal starts to solidify

hot-dip galvanising a process by which iron or steel is immersed in molten zinc to provide protection against corrosion

hydraulics the branch of science and technology concerned with the mechanics of fluids, especially liquids

hydrology the science that treats the occurrence, circulation, distribution and properties of water and its reaction with the environment

hydrostatic test a test that subjects a pipeline to a static head of water pressure

hydrostatics the study of liquids at rest and the forces exerted on them or by them

impact a force exerted when one body collides with another

inner cone the brilliant, short part of an oxy-fuel gas flame immediately adjacent to the orifice of a welding tip, or the preheat orifices of a cutting nozzle. The same as *cone*.

inspection opening an access opening in a pipe or drain sealed with a removable plug or cover used as an access for the purposes of inspection, maintenance and hydraulic testing

insulation a material which reduces the transmission of heat, sound, electricity or moisture

intermediate sight (IS) all readings recorded between a backsight and a foresight

intermittent welding welding in which the continuity is broken by leaving recurring unwelded spaces

invert the lowest point of the internal surface of a pipe at any cross-section

ion an electrically charged atom or molecule. "Ion exchange" is a chemical reaction in which mobile hydrated ions of a solid are exchanged, equivalent for equivalent, for ions of like charge in solution

isometric drawing a method of non-perspective pictorial drawing in which the object being drawn is turned so that three mutually perpendicular areas are equally foreshortened. All dimension and projection lines are at 30° to the horizontal

jointing the lapping of two pieces of material so that they can be joined by soldering, riveting or other approved methods

kerf the space left during the process of flame cutting by the removal of metal in the form of oxide

lack of fusion a weld fault in which there is inadequate fusion of the weld and base metals

lagging *see* insulation

lap weld a type of joint formed by two overlapping plates where the edge of each plate is welded to the face of the other

laser a fine beam of light

latent heat (also known as hidden heat) the amount of heat required to change the state of a substance without changing the temperature, e.g. ice at 0°C to water at 0°C

layer a quantity of filler metal deposited in a joint to completely cover the filler metal previously deposited A welded joint on heavy plate may require several layers of weld metal for completion

leeward the side away from the direction from which the wind is blowing

leg of fillet weld the distance from the root of a joint to the toe of a fillet weld

long bend a pipe bend greater than 45°C having a centre-line radius equal to or greater than 1.5 times the diameter

main the principal pipeline in a water or drainage reticulation system

main drain that drain which determines the depth of the connection

mandatory required by law

MIG metal inert gas welding; a process that uses inert gas to enshroud the weld to prevent oxidisation

mitre bend a pipe bend made with the use of a mitre cut (angle cut)

mitre fillet fillet weld in which the face of the weld is approximately flat

mixing chamber that part of a gas welding or cutting blowpipe in which the gases are mixed for combustion

neutral flame an oxy-fuel gas flame in which the inner cone, or that portion of the flame used, is neither oxidising nor carburising. It is characterised by an almost colourless outer envelope and a sharply defined inner cone without feather or secondary flame

nominal size standard sizes of pipes and fittings in accordance with the relevant Australian Standards

offset the pipes and fittings used to provide continuity between pipes of parallel axis but which are not on line

orthogonal drawing method of producing views of an object by projecting straight perpendicular lines from that object to a viewing plane

orthographic drawing *see* orthogonal drawing

osmosis diffusion of liquids through a porous layer of skin

outer envelope the secondary phase of combustion in an oxy-fuel gas flame which surrounds the inner cone

outlet (nozzle, pop, spout) an opening in a sanitary fixture, appliance or vessel serving to discharge the contents

oxide a compound of oxygen and another element or substance. Rust and mill scale are examples of iron oxides

oxidising flame an oxy-fuel gas flame in which the inner cone, or that portion of the flame used, has an excess of oxygen. It is characterised by the length of the inner cone when compared with the cone of a neutral flame

oxygen a colourless and odourless gas which supports combustion and is present in the atmosphere to the extent of approximately 21 per cent by volume. When the correct mixture of oxygen and acetylene is burned, a flame temperature of approximately 3000°C is obtained

pan that section of a metal decking between the ribs of a profiled sheet

parent metal (base metal) the metal of a part to be welded, as opposed to the metal that is added from the filler rod

pass one general progression along the line of a weld in which metal is deposited, also the metal so deposited. The result of a pass is a "weld bead"

penetration the depth of fusion obtained in a welded joint

plumb bob a pointed weight suspended at the end of a string line to test for perpendicularity

polyvinyl chloride (PVC) polymer of vinyl chloride; tasteless, odourless and insoluble in most organic solvents. A member of the family of vinyl resins and used for a wide range of pipes and fittings

porosity gas pockets or voids in a metal surface

potable water water suitable for drinking, culinary and domestic purposes

preheating the application of heat to a base metal before welding commences

prevailing wind direction the direction from which the wind blows most frequently

profile the shape of something when viewed from the side

prohibited discharge a waste considered by the water authorities to be dangerous to the disposal system or to the employees at a treatment works

projection the system of producing lines on a flat surface from features that may be flat or curved.

radiant energy heat transmitted from one body to another without heating the intervening medium

radius a straight line from the centre to the circumference of a circle

reduced level (RL) the height or elevation above a given point, adopted as a datum

regulator a device for controlling the delivery of gases at a chosen constant pressure regardless of variation in cylinder or pipeline pressure

ridge the highest point of a roof; occurs at the apex where the common rafters meet

rise the amount by which a given point is higher than the previous point

riser a straight length of pipe fitted to the inlet of a trap and extending to floor level, fitted with approved grate or seal, providing access to the trap

root that portion of a joint to be welded where the members approach closest to each other. In cross-sections, the root of a joint may be either a point, a line or an area

root face the unbevelled or ungrooved portion of a fusion face at the root

root weld the zone on the side of the first run furthest from the welder

safety device (welding) devices fitted to acetylene cylinders, either bursting discs or fusible plugs. These are designed to vent the cylinder contents in the event of an unsafe condition arising in the cyclinder

sanitary plumbing installation an assembly of pipes, fittings, fixtures and appliances used to convey wastes to the sewerage system

scribe to mark by scratching or drawing with a sharp pointed tool on metal workpieces

self-tapping screw a specially hardened screw that taps its own thread in sheet metal and steel

sensible heat that heat which may be felt and measured

sewer the conduit or piping which transports sewage or trade waste, including all accessories

sewerage system the whole system of sewage disposal, excluding the individual property installations

shoe a mitre joint at the foot of a downpipe to connect the downpipe to the stormwater drain

silicon sealant a fluid, resin or elastomer, most frequently used in a jointing compound used in jointing zincalume since it cannot be soldered

silver-brazed joint a welded joint in which the parts are joined with a filler metal which has silver as one of its components

silver brazing (silver soldering) a low temperature brazing process in which a silver alloy is used as a filler metal

single vee weld a weld in which the two edges to be joined are each bevelled at one edge only

siphon a pipe system comprising a rising leg and a falling leg, typically in the shape of an inverted U

slag dross, scum or non-metallic foreign matter that rises to the surface and hardens upon cooling

slag inclusion non-metallic material entrapped in a weld

slip joint a joint utilised in guttering to allow fixing of long lengths in situ

socket the 'female' end of a pipe having a larger internal diameter for the reception of the plain or spigot end of another pipe or pipe fitting

soffit
1. (drainage) the highest point of the internal surface of a pipe at any cross-section
2. (roof plumbing) the underside of the overhanging eaves

soil fixture a WC pan, urinal, slop-hopper, autopsy table, bed pan steriliser or sanitary napkin disposal unit

soil pipe the pipe which conveys the discharge from soil fixtures

solvent-welded jointing a method of joining uPVC materials in which the surfaces to be joined are coated with a solvent

which softens or melts the surfaces enabling them to fuse together when brought into contact with each other
spatter metal particles expelled during the welding process which do not form part of the weld
specific heat the ratio of the amount of heat required to raise a given mass of a substance through a given temperature range to that required to raise the same mass of water through the same temperature range
spigot the 'male' untreated plain end of a pipe or fitting
spill level the maximum level to which water will rise while overflowing a fixture or storage tank when all outlets are closed, and when the water supply system is discharging at the maximum rate at the nominated pressure
spouting another name for guttering
stack a vertical soil or waste pipe extending more than one storey (e.g. 2.5 m) in height
stack vent the extension of a discharge stack above the highest connected discharge pipe. *See also* head vent
stop end the turned-up end of a gutter which prevents water from flowing out
structural slab a self-supporting, concrete floor
sullage a general term for domestic waste liquid other than from soil fixtures, i.e. from bathrooms, laundries and kitchens
swarf the material removed during cutting and drilling

tack weld a shallow, temporary weld used to hold two edges of metal to be joined in proper alignment while the proper weld is being made
temporary bench mark *see* bench mark
test the approved test of pipelines for soundness, as determined by the water authority
test opening an adequate size of opening in a pipe to enable entry of a plug for the purposes of testing
thermoplastic a plastic material that will repeatedly soften when heated and harden when cooled
thermoset a plastic material which is hardened irreversibly by the action of heat and/or a catalyst
thermosiphon the circulation that occurs when gases and liquids are heated; this circulation is modified by varying densities within the substance
throat the inside of a bend
thrust blocks blocks, usually of concrete, placed at intervals along a pipeline and in other positions, adjacent to valves and changes of direction or grade to anchor the pipeline
TIG tungsten inert gas welding; uses a tungsten electrode enshrouded in an inert gas to prevent oxidisation
tinning (tin coating) the process of placing a protective coating over a brass or gunmetal fitting. The coating used is usually an alloy of lead and tin (solder). The term is also applied in brazing and braze welding where the spreading out of a thin layer of fluxed brazing metal ahead of the main deposit to form a "tinning coat" provides a strong bond between parent metal and deposit
tip the removable forward end of a welding or heating blowpipe; it contains the orifice from which the mixed gases issue
toe boundary between weld face and parent metal or between weld faces
top plate in a timber-framed cottage, the member which takes the main weight of the roof
trade waste waterborne wastes from business, trade or manufacturing premises other than domestic sewage and as defined by the appropriate authority
triangulation a method of determining true dimensions by using triangles extending from known fixed points
truss a prefabricated roofing frame. Used as a quick method of pitching roofs
tubing (welding hose) the means by which the gases are supplied from the source (the cylinder) to a welding or cutting blowpipe. Made of reinforced rubber hose, it is strongly built to resist the pressure of the gases and to withstand constant bending and twisting

under bead or **hard zone crack** a crack in the heat-affected zone during welding operations which may or may not extend to the surface of the base metal
undercut a groove melted in the base metal adjacent to the toe of a weld and left unfilled by weld metal
UPVC unplasticised polyvinyl chloride. *See* polyvinyl chloride
upstream towards a higher level

valley the internal intersection of two roof slopes
valve (general) a device used to control the flow of water in a piping system
variable speed a range of speed from one control
vent a pipe provided to limit the pressure fluctuations within a discharge pipe system
vertical any pipe which is at an angle equal to or more than 45° to the horizontal. A pipe with a grade of not less than 1:1
vertical welding welding in a position in which the axis of the weld is approximately vertical

wail plate a timber strip placed on top of a load-bearing brick wall on which rest the common rafters
waste fixture any waste fixture other than a soil fixture
waste pipe a pipe which conveys only the discharge from waste fixtures
water-seal the water retained in a trap which acts as a barrier to the passage of air and gas through the trap
WC the abbreviation for 'water closet'(toilet)
weathering any process carried out to improve weather proofing
weave bead a weld bead made with a slow, oscillating movement of the electrode
weld a union of two pieces of metal joined at faces rendered plastic or liquid by heat or pressure or both. Filler metal may also be used
weld bead a deposit of filler metal from a single welding pass
weld metal metal in a welded joint which has been melted in making the weld, including both the filler and parent metals and as distinct from the unmelted parent metal
welding blowpipe an apparatus designed for gas welding and brazing, wherein oxidising and fuel gases are controlled and mixed to produce the required flame
work lead conductor electrical lead between the source of current and the work or work table

zincalume an alloy of zinc and aluminium bonded to steel by means of a continuous hot-dip process

Symbols and abbreviations for sewerage and sanitary plumbing

Element	Symbol
Sewer line	S Green
Industrial sewerage	I S Green
Soil pipeline	S P Blue
Waste water pipeline	W Yellow
Vent pipeline	V&VP Red
Acid or chemical waste	A W Green
Vent (soil vent) pipe	V (S V)
Agricultural pipe drain	APD
Manhole (all types)	M H
Manhole (all types)	M H
Inspection pit (nature of pit, e.g. dilution, neutralising, designated below symbol)	
Boundary trap	
Inspection shaft	
Grease Interceptor	G
Yard gully with tap	Y G X
Dry pit	
P trap	P
Reflux valve	R
Cleaning eye	
Vertical pipe	Vert
Waste stack	WS
Septic tank	ST
Lamphole	L H
Pumping station	PS
Floor waste	FW
Ejector or pump unit	E

Element	Abbreviation
Down cast cowl	D C
Induct pipe	I P
Mica flap	M F
Junction for future use	J N
Tubs	T
Kitchen sink	K
Water closet	W
Educt vent	E V
Bidet	B I D
Bath waste	B
Handbasin	H
Shower	S
Washing machine	M
Vitreous clay pipe	V C P
Wrought iron pipe	W I P
Cast iron pipe	C I P
Floor waste	F
Urinal	UR
Drinking fountain	D F
Wash trough	W T
Dishwasher	D W
Glasswasher	G W
Bar sink	B S
Cleaners sink	C S
Laboratory sink	L S
Slop sink or slop hopper	S S
Potato peeler	P P
Disposal unit	D U
Cuspidor	CUS
Steriliser	St
Autopsy table	A T
Water meter	W M R
Stainless steel (Corrosion resistant steel)	s s CRS
Hot water unit	H W

Symbols and abbreviations for water supply and stormwater drainage

Element	*Symbol*	*Element*	*Symbol*
Water supply		**Stormwater drainage**	
Watermain (pipeline)		Stormwater line	
Domestic water service Cold water	CW	Down pipe	
Domestic water service Hot water	HWS	Inspection pit	IP
Fire water	FWS Red	Inlet gully	IG
Fire hydrant standpipe with cradle and direction of millcock	FH	Inlet sump	IS
Spring bail hydrant	SH(SBH)	Double grated gully pit	DGGP
Fire hose reel	HR(FHR)	Single grated gully pit	SGGP
Stop valve	SV	Grated drain	GD
Stop tap	ST	Open lined drain	OLD
Reflux valve	RV	Open unlined drain	OUD
Hose tap standpipe	HT	Reinforced concrete pipe	RCP
Water meter	WM	Vitrified clay pipe	VCP
Crossover		Glass-reinforced plastic pipe	GRP
Flanged joint			
Socket and spigot joint			
End capped off			
End blank flanged			
End plugged off			
Flow meter orifice			
Taper			
Cast iron cement-lined pipe	CICLP		
Galvanised mild steel cement-lined pipe	GMSCLP		
Asbestos pipe	ACP		
Copper pipe	CP		

Symbolic representation for pipeline drawings

Element	Symbol	Element	Symbol
Globe valve		Pipe size (nominated)	20mm
Globe valve with maximum flow adjustment		Reduction in size with branch	50mm 25mm 25mm
Ball valve (spherical plug cock)		Reduction in size without branch	50mm 25mm
Hand operated valve		Pipe bend	
Solenoid operated valve		Pipe elbow	
Diaphragm operated valve		Pipe tee	
Lock shield valve		Flanged connection	
Plug cock (tapered)		Blank flange	
Thermostat	T	Union joint	
Humidistat	H	Anchor	A
Thermometer dial	T	Hanger	H
Pressure gauge with pet cock	P	Orifice plate	OP
Sight glass		Open vent	
Centrifugal pump solid casing		Anti-convection loop	
Centrifugal pump split casing		Pipe guide	
Dirt leg with spherical plug cock		Air cock	AC
Injector or ejector		Automatic air cock	AAC
Coil (heating or cooling)		Air vessel	
Bib cock		Drain (plug cock with hose tail)	
Bib cock with hose tail		Thermometer pocket	T
Pipe crossing		Tundish	
Vertical pipe	S	Expansion bellows	
Rising vertical pipe	SR	Expansion bend	
Dropping vertical pipe	SD	Expansion bend (lyre bend)	
Direction of flow	FLOW	Strainer (Y type)	
Rise in direction of flow	RISE	Strainer (line type)	
Fall in direction of flow	FALL		

Symbolic representation for pipeline drawings

Element	Arrangement drawings: Normal	Arrangement drawings: Skeleton	Diagrams A key to the symbols to be provided
Pipe	Large Small		
Flanged joint			Not required
Spigot and socket joint			Not required
Pipe hanger			Not required
Pipe anchor	Outline of anchor to be drawn	Anchor	Not required
Stop valve			
Stop valve (motor operated)			M
Stop valve (right-angle type)			
Stop valve (three-way)			
Spring loaded safety or relief valve			
Counter-weight lever safety or relief valve			
Counter-weight lever safety or relief valve (right-angle type)			
Float operated valve			
Float operated valve (right-angle type)			
Non-return or check valve			
Non-return or check valve with lock			
Reducing valve			
Butterfly valve			
Fire hydrant	Outline of hydrant to be drawn	Simplified outline of hydrant to be drawn	
Strainer	Outline of strainer to be drawn	Simplified outline of strainer to be drawn	
Suction pipe strainer			
Suction pipe strainer with foot valve			
Steam trap	Outline of trap to be drawn	Simplified outline of trap to be drawn	
Flow meter (non-recording)			
Flow meter (recording)	R	R	R
Flow meter orifice	Flow meter orifice		
Exhaust head on ventilator			
Pressure or vacuum gauge	Outline of gauge to be drawn	Simplified outline of gauge to be drawn	
Thermometer	Outline of thermometer to be drawn	Simplified outline of thermometer to be drawn	
Steam separator	Outline of separator to be drawn	Simplified outline of separator to be drawn	Alternatives

Conduit diameters (coded)	
20 mm	+
25 mm	+ +
32 mm	— — — — —
40 mm	–
50 mm	/
100 mm	O

Symbolic representation and abbreviations for pipeline drawings

Element	Abbrev.
Air	A
Brine	B
Boiler blowdown	BD
Condensate	C
Cold water	CW
Condenser cooling water	CCW
Oil cooling water	OCW
Engine cooling water	ECW
Chilled water	ChW
Chilled drinking water	Ch DW
Distilled water	DW
Drain or overflow	D
Fuel oil	FO
Feed water	FW
Town gas	TG
Liquefied petroleum gas	LPG
Hot water supply	HWS
High temperature hot water	HTHW
Medium temperature hot water	MTHW
Low temperature hot water	LTHW
Lubricating oil	LO
Oil	O
Nitrogen	N
Oxygen	OX
Nitrous oxide	NO
Refrigerant (Show number if any)	R
Steam	S
Superheated steam	SHS
Vent	V
Vacuum	Va
Demineralised water	DMW
Heat transfer oil	HTO
Pneumatic tube	PT
Fire water service	FWS
Fire foam	FF
Fire sprinkler	FS
Fire CO_2	F CO_2

Element	Symbol
Main piping visible	———— S ————
Main piping concealed	– – – – – S – – – – –
Piping by others visible	—— - —— S —— - ——
Piping by others concealed	- - — — - S - — — - -
Future piping visible	— - - —— S —— - - —
Future piping concealed	— - - — — S — — - - —
Existing piping	———— - S - ————
Piping to be removed	////////////S////////////

Index

Exercises

chapter one

- **Read the following questions and answer them in the spaces provided.**
- **You will need to show all your workings where required. Do not use a calculator.**

Exercise 1

1 Addition. Add the following columns, down and across:

	(a)	(b)	(c)	(d)	**total**
	63 360	9870	5430	1230	....
	17 290	7650	3210	2380	....
	5 430	1015	3450	7860	....
	100 005	3450	7890	1270	....
	3 260	5680	9870	5405	....
totals	______	______	______	______	

2 Subtraction. Subtract and check the following:

(a) $624 - 317$

(b) $6556 - 4876$

(c) $86\,107 - 57\,218$

3 Multiplication

(a) 59×36

(b) 654×47

(c) 925×589

(d) 3178×12

(e) 50301×201

(f) 1.25×0.8

(g) 7.38×0.05

4 Division

(a) $829 \div 4 =$ ______

(b) $1313 \div 16 =$ ______

(c) $1429 \div 47 =$ ______

(d) $466 \div 0.15 =$ ______

(e) $5387 \div 353.7 =$ ______

(f) $13.64 \div 200 =$ ______

Exercise 2

1 Percentages

(a) Convert the following percentages to decimals:

80% = ______ 48% = ______ 38.4% = ______ 16.8% = ______

(b) Convert the following percentages to fractions:

50% = ______ 25% = ______ 45% = ______ 16% = ______

(c) Convert the following fractions to percentages:

7/9 = ______ 3/20 = ______

2 The square root
Find the square root of the following numbers:

(a) 576

(c) 1936

(b) 1156

(d) 2116

Exercise 3

1 Calculate the perimeter of a rectangular block of land with sides 28 m and 42 m.

Perimeter = ______

2 Calculate the circumference of a mains water pipe with a diameter of 280 mm.

Circumference = ______

3 Calculate the circumference of a drainage pipe with a diameter of 360 mm.

Circumference = ______

4 Calculate the areas of the geometrical shapes shown in Figure 1.46.

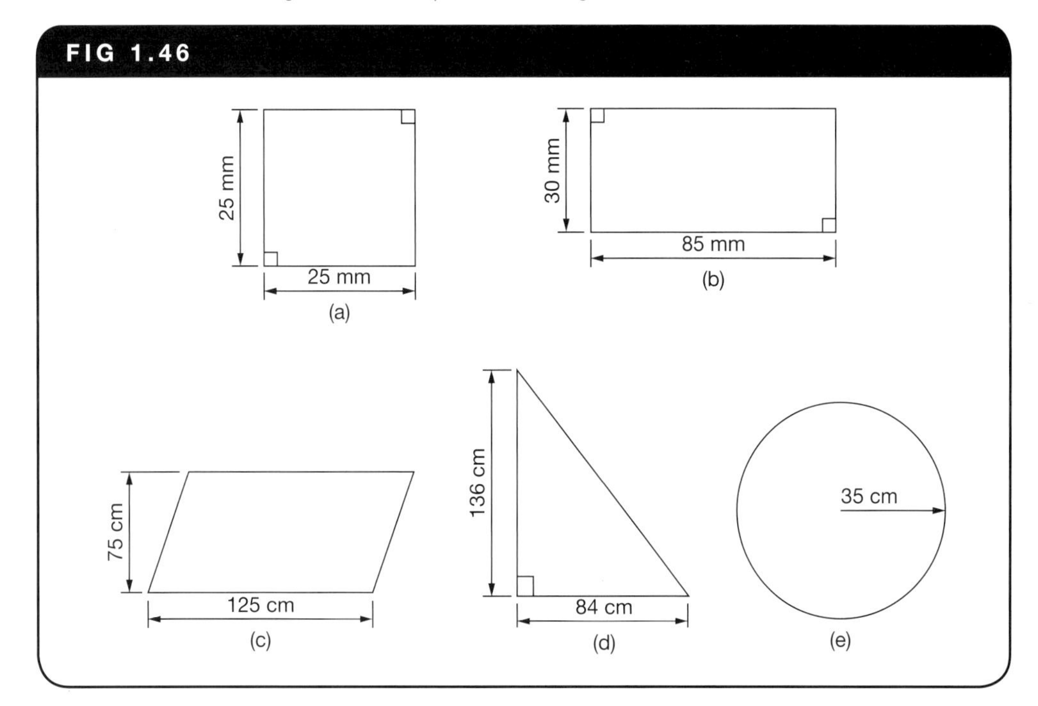

(a) ....

....

(b) ....

....

(c) ....

....

(d) ....

....

(e) ....

....

5 Calculate the volumes of the solids (a) to (d) shown in Figure 1.47.

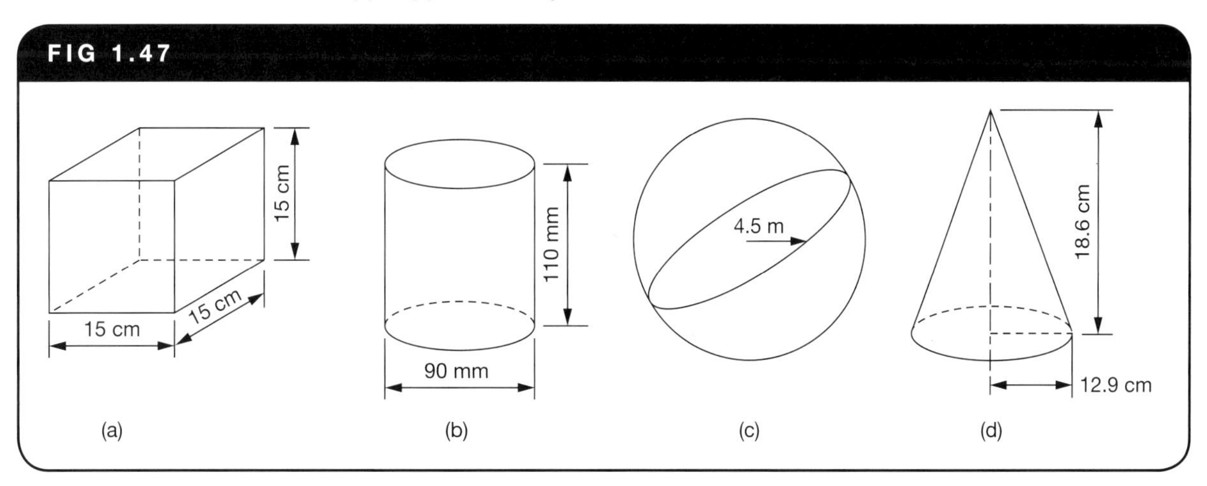

(a) Volume =

(b) Volume =

(c) Volume =

(d) Volume =

6 What length must the rafter measure for this gable roof?

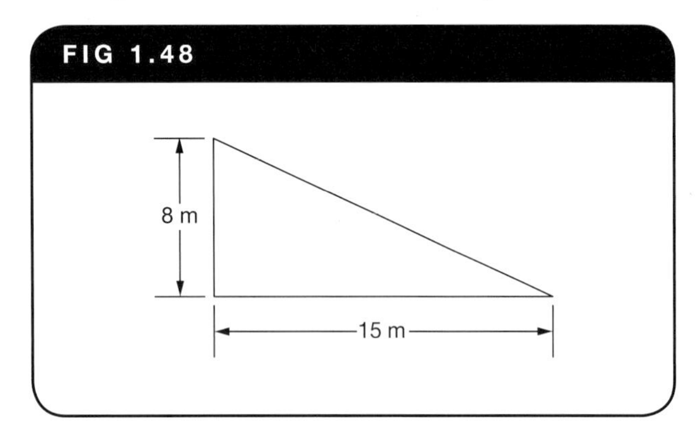

Rafter =

7 What length must the rafter measure for this gable roof?

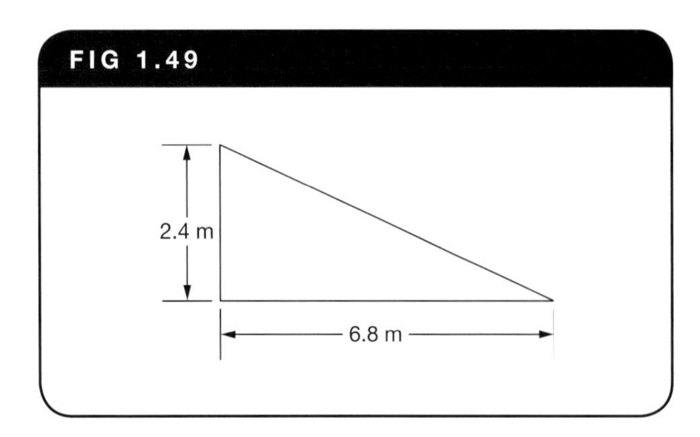

Rafter =

Exercise 4

1 Calculation of fall
Calculate the depth of the connection of the drain at the property boundary using the grade as a ratio, given the following:

Distance = 50 m

Grade = 1:100

Head of drain = 300 mm

Depth at the connection = ______________________

2 Calculation of fall
Calculate the depth at the head of the drain using the grade as a ratio, given the following.

Distance = 38 m

Grade = 1:100

Depth of the connection at the boundary = 1.4 m

Depth at the head of drain = ______________________

3 Pipe bending
Calculate the range of bends given the following:

(a) 30 mm pipe 45° bend 2 pipe diameters

(b) 50 mm pipe 90° bend 1.5 pipe diameters

4 Mechanics of water supply

(a) Convert the following heads to pressure:

20 m = ______ 7.5 m = ______ 16 m = ______

(b) Convert the following to metres of head:

500 kPa = ______ 350 kPa = ______ 50 kPa = ______

(c) Calculate the velocity at which water will flow from a supply tank 40 m above a flushometer.

Velocity = ______

chapter two

- **Read the following questions and answer them in the spaces provided. You will need to show all your workings.**

Exercise 1

1 Name the six types of quadrilateral shapes.

1 ______

2 ______

3 ______

4 ______

5 ______

6 ______

2 With the aid of a drawing list, explain the six types of triangles.

1 ______

______

2 ______

______

3 ______

______

4 ______

______

5 ______

______

6 ______

______

3 Why are isometric drawings used in the plumbing industry?

______

______

______

______

4 Using the notes for orthogonal projection as a guide, develop a front view and view from the right-hand side of the tee piece below. The tube is 40 mm × 40 mm for both sections.

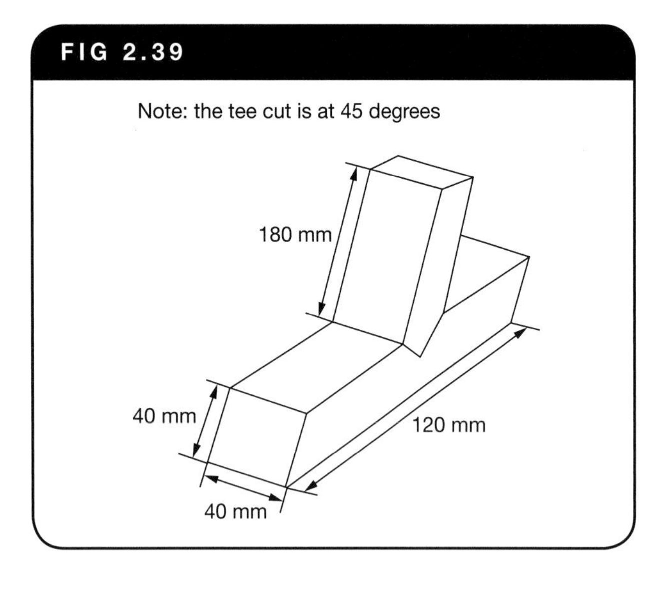

Exercise 2

1 Name two types of drawings used in the construction industry and explain their functions.

1

2

2 List three items of information a plumber could obtain from an elevation view.

1

2

3

3 Explain briefly why section views are used.

4 List the information found on a floor plan.

1

2

3

4

5

5 List the information included in a plan title block.

6 What are the appropriate scales used to create working drawings for the following?

Site plans

Floor plans

Elevations

Sections

Details

7 What are the actual lengths of each of the lines below?
Using an ordinary ruler (1:1), measure each of the following lines, and multiply them with the appropriate scale to obtain a full size measurement.

FIG 2.40

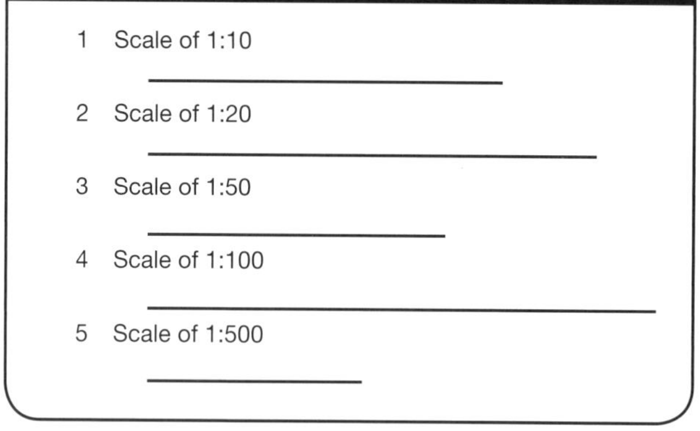

1 Scale of 1:10

2 Scale of 1:20

3 Scale of 1:50

4 Scale of 1:100

5 Scale of 1:500

1

2

3

4

5

8 Referring to Figures 2.26 and 2.28, answer the following:

(a) What direction is the house frontage facing?

(b) How far back is the house set from the street?

(c) What are the dimensions of the block of land?

(d) What are the overall dimensions of the house?

(e) What are the dimensions of bedroom 3?

(f) What is the scale of the site plan?

9 Explain briefly why a specification is necessary.

10 Give the full meaning of the abbreviations below (using the Symbols and Abbreviations on page 155).

WC

BID

H

S

WT

DW

HW

CP

K

Exercise 3

1 Using the notes for parallel line development as a guide, develop a pattern for a truncated cylinder pattern.
Details:

- The radius of the cylinder is 38 mm.
- The perpendicular height of the truncated cylinder is 80 mm.

The developed pattern may be cut and folded and the seams joined to see if the pattern has been correctly developed.

chapter three

- **Read the following questions and answer them in the spaces provided.**
- **Remember you can use drawings to help explain the questions.**

Exercise 1

1 List the main factors that influence the life of hacksaw blades.

2 What type of hacksaw blade would you recommend to cut downpipe made from galvanised sheet iron?

3 How would you obtain maximum leverage when using tin snips?

4 Give the main reason why the full blade of a pair of snips should not engage the metal being cut.

5 Explain why a multi-wheeled chain cutter is used to cut UPVC pipe, in preference to other pipe cutting tools, inside trenches.

Exercise 2

1 What is the most common drill point angle?

2 Excessive heat generated during drilling can dull the drill's cutting tips. How can this be overcome during drilling operations?

3 When sharpening drills, constant cooling is necessary. Explain why this is so.

4 List three different types of drill. Give a brief description of how each is used.

5 When tapping and threading, once the tap or die has commenced to cut downward pressure is no longer required. Why is this so?

6 Taps are generally used in sets of three. Explain why this is so and explain the difference between the taps.

chapter four

- **Read the following questions and answer them in the spaces provided.**
- **Remember you can use drawings to help explain the questions.**

Exercise 1

1 When making a grooved seam the grooving tool should be approximately 2 mm wider than the fold. Why is this necessary?

2 Both the 'cramp' and 'pan break' folder can be found in workshops. Explain the basic difference between these two machines and their different applications.

3 What type of heating flame should be used when hand-bending copper tube?

4 What stresses tend to occur during pipe and tube bending?

5 What is likely to happen if an oxyacetylene flame is not moved continuously during the annealing process?

6 State three different methods of bending copper tube and discuss the advantages and disadvantages of each.

7 Why is the bending of plastic pipe not recommended?

8 Describe a suitable method of manually forming a 50 mm × 80 mm × 45° junction in copper tube.

chapter five

- **Read the following questions and answer them in the spaces provided.**
- **Remember you can use drawings to help explain the questions.**

Exercise 1

1 When using plastic fixing anchors, over-tightening of the screws should be avoided. Why?

2 Fixing anchors into brick joints should be avoided. Why is this so?

3 What is the benefit of using powder-powered fasteners?

4 Does the operator of a powder-powered fastening tool have to be trained?

5 What materials can the basic powder-powered fasteners be used with?

Exercise 2

1 What is the purpose of the soft soldering process?

2 Explain the meaning of the term 'intermolecular penetration'.

3 What is the purpose of using a flux?

4 Explain in detail the procedure for reconditioning and tinning a soldering iron head.

Exercise 3

1 Explain the difference between 'braze welding' and 'fusion welding'.

2 Acetylene in a free state becomes unstable above a certain pressure. What is this pressure?

3 Why do fuel gas cylinders and connections have left-hand threads?

4 List the three types of flames required for oxyacetylene welding and explain (with the aid of diagrams) the difference between the flames.

5 Explain the terms 'backfire' and 'flashback' and how could you rectify these problems.

6 List the general safety requirements when using oxyacetylene welding equipment.

Exercise 4

1 How should joints be prepared for silver brazing?

2 What is the purpose of fluxing in silver soldering operations?

3 Explain the procedure for silver brazing copper tube.

4 What is the theory of the braze-welding process?

Exercise 5

1 List the suggested amperages for the following electrodes:

(a) 1.75 mm ____ (b) 2.5 mm ____

(c) 3.25 mm ____ (d) 4.00 mm ____

2 Non-metallic particles included in a weld are called 'slag inclusion'. For each of the following faults explain the correct remedy:

(a) lack of penetration, with slag being trapped beneath the weld

(b) rust preventing full fusion

(c) slag trapped in undercut from previous run.

3 List the general safety requirements when using arc welding equipment.

Exercise 6

1 List three approved methods of achieving a positive seal for threaded pipe joints.

2 What is meant by a 'capillary fitting'?

3 What types of joint fittings are used for joining copper tube?

4 What method would you use to join UPVC pressure pipe?

5 List three methods of joining polyethylene pipe.

6 (a) In what circumstances are flanged joints used?

(b) What advantages do flanged joints have over screwed joints?

7 Explain why metal sections need to be riveted in conjunction with a suitable sealant.

8 List four important properties that sealants must possess.

chapter six

- **Read the following questions and answer them in the spaces provided.**
- **Remember you can use drawings to help explain the questions.**

Exercise 1

1 Explain the following levelling terms:

(a) Contour

(b) Levelling

(c) Reduced level

2 What is meant by a 'temporary bench mark'?

3 Describe the basic differences between:

(a) a theodolite

(b) an automatic level

4 Explain how you would hold and position a staff to ensure accurate readings in a levelling operation.

5 Describe the following in relation to levelling:

(a) Boning rods

(b) Plumb bob

(c) Water level

6 (a) Draw an elevation view, as illustrated in Figure 6.21, in the space provided below. Use the given readings from the table in (b) to complete your drawing.

(b) Complete the following table and an arithmetical check, using the rise and fall method to obtain the reduced levels.

BS	IS	FS	Rise	Fall	RL	Remarks
2.600					30.000	TBM
	1.850					A
	1.220					B
0.800		1.500				C
	1.480					D
	1.400					E
2.100		0.350				F
		0.450				G

chapter seven

- **Read the following questions and answer them in the spaces provided.**
- **Remember you can use drawings to help explain the questions.**

Exercise 1

1 What is meant by the term 'sustainability'?

2 For a society to be sustainable what three key factors need to be in balance?

3 What six key sustainability areas relate to the plumbing industry?

4 In the 'mitigation measures' column, write some solutions to the following plumbing-related environmental risks.

Activity	Risk	Mitigation measures
Excavation work for installation of gas, drainage or water lines	Nuisance and entry of silt and dust into the environment	
Supply of tap-ware, fixtures and fittings	Water wastage due to the installation of inefficient equipment	
Installation of plumbing work	Environmental pollution due to the use of toxic plumbing materials, adhesives and sealants	
Purchase and use of equipment and tools	Wastages of materials, energy and resources through use of poor quality, low-cost, disposable hand and power tools	

5 List three ways businesses can reduce their energy consumption.

6 List three ways businesses can make their premises more thermally efficient.

7 List three ways businesses can reduce their water consumption.

8 What is the name of the system used to rate water fittings and what is the Australian Standard that governs it?

9 How can businesses reduce the amount of waste they produce?

10 Write three suggestions for how a business can source its goods and services from sustainable sources.

chapter eight

- **Read the following questions and answer them in the spaces provided.**
- **Remember you can use drawings to help explain the questions.**

Exercise 1

1 How would you define 'communication'?

2 What are some of the unspoken messages that make up non-verbal communication (body language)?

3 You are discussing some important job details with a workmate. How can you tell if he/she is listening and interested in what you are saying?

4 List five points you could check if you wanted to improve your own non-verbal communication skills.

5 You need to place a material order with your local plumbing supplier. What communication channel(s) would you choose to ensure you get the right goods as quickly as possible?

6 Give some advantages and disadvantages for the communication channel you have chosen in question 5.

7 Why is it important to give and receive feedback during the communication process?

8 How can 'noise' affect our communication? Give four examples of noise.

9 In the table below, write down two communication channels and list some advantages and disadvantages for each channel.

Communication channel	Advantage	Disadvantage
	1	1
	2	2
	1	1
	2	2

10 List the seven categories of signs that are defined by colour and shape.

chapter nine

- **Read the following questions and answer them in the spaces provided.**
- **Remember you can use drawings to help explain the questions.**

Exercise 1

1 According to the statistics, what age group and gender has the highest rate of workplace injuries in Australia?

2 What is the aim of SafeWork Australia and when was it formed?

3 What is the name of the organisation responsible for workplace safety in NSW and what is the name of the Act that it enforces?

4 List four aims of the above Act.

5 What is a 'code of practice'? List three codes of practice relevant to plumbers.

6 What does the term 'duty of care' mean to you?

7 Does an employer have a duty of care to employees? If so, list what they are.

8 What is the maximum fine that can be applied to an individual for a breach of OHS regulations?

9 What is a 'White Card' and what does it allow you to do?

10 According to the Occupational Health and Safety Regulation 2001, what is a 'hazard'?

11 What is the name of the document used to assess risk?

12 What is the name of the document used to describe how workplace hazards will be managed?

13 Where would we find details of how to deal with dangerous chemicals?

14 Name some hazards you have found at your workplace. Write down how you dealt with them.

15 You see a workmate at your site about to do something potentially hazardous. What would you do?

16 Whose responsibility is workplace health and safety?

17 Where can you find more information on workplace health and safety?